Holography Market Place

third edition

HMP

*The Reference Text and Directory
of the Holography Industry*

The original version of this book contained holograms from the vendors and many of them are no longer in business. Therefore this version of the book contains everything that was in the original version but it has no holograms.

Edited by Brian Kluepfel and Franz Ross

CIP Information

Holography Market Place - third edition.

Bibliography:
Includes index.
1. Holography.
2. Holography industry
Directories.

ISBN 978-0-89496-096-3
Copyright ©1991 by Ross Books

For information about purchasing mailing labels of the database, please write to the Holography Editor at the address below.

PREFACE

The Holography Market Place is an annual publication which provides an overview of the holography industry, detailed information on holography, and a database of bus inesses in the industry.

In compiling our database we accommodated all questionnaires returned by our deadline and made extensive phone calls to ensure the accuracy of our addresses. Eighty percent of all USA, UK and German holography businesses listed in this edition were telephoned to verify their listings.

If your business is new or did not receive our questionnaire, please mail us information about your company on your letterhead to the address below.

In this third edition, all ($) signs represent US dollars, and (£) represent British pounds sterling. A comma is used to separate thousands (i.e. 2,000 means two thousand as opposed to 2.000).

We realize we are not infallible in our knowledge of the industry. If you feel we have omitted anything or made any factual errors, please write us. We do read the mail and will make every effort to improve and expand in subsequent editions. Address all correspondence to:

Editor, HMP
Ross Books
P.o. Box 4340
Berkeley, California 94704
USA
Telephone: (IX415) 8412474
FAX: (IX415) 841 2695

About the cover:

The hologram on the cover is an example of a photopolymer created by Polaroid's "Mirage" process. The following description has been provided by Polaroid:

Working from your original artwork, model or object, we will supervise the step-by-step production of your hologram. Virtually anything can be used as reference, including sculptures, toys, watches, small objects, scale models and of course two-dimensional artwork, drawings or photographs. The important thing to remember is that the holographic image is never enlarged or reduced. It must always appear the same size as the original. From there, the process is simple. Our artists, designers and engineers will control the entire production process, showing you proofs, and gaining your approval, throughout the origination and manufacturing cycle which takes 8 to 10 weeks.

1. The process begins with your supplying of the camera-ready artwork or other 1:1 source material. We will consult with you closely to determine its effectiveness as final holographic art.

2. A three-dimensional representation of your artwork is produced by our team of modelers. (Two original sculptures were prepared by Polaroid to produce this fascinating two-channel Mirage Hologram of the human skull.)

3. Holographic mastering then takes place and a proof sample is submitted for your approval. While representative of the final hologram, this sample will vary somewhat from the final in terms of color and brightness.

4. Final production (5-6 weeks). Once we have your approval on the hand-made sample, we prepare your hologram from final production. Mirage photopolymer film is ganged and mastered. The images are then laminated with pressure-sensitive adhesive and die-cut to specifications.

Specifications: Original source: 1:1 models, sculptures, flat art, photography multiplex, computer-generated or pulsed images.

Image size: Mirage holograms can be as large as 10" x 14" or as small as 0.5 square inches.

Applications: Machine - or Hand-applied

Please note: Mirage holograms require no hotstamping.

Our custom and stock holograms have produced phenomenal marketing successes for such companies as AT&T, IBM, G.D. Searle, and 20th Century Fox. Whatever your communication needs--from advertising, to direct mail, packaging to point-of-purchase--Polaroid's Mirage holograms hold the perfect image.

Interested parties please contact:
Polaroid Corporation at (800) 237-5519 or (617) 577-4307

Table of Contents

Chapter 1

Introduction 7

Making A Simple Hologram 8
Reflection Holograms 9
Through Thick and Thin 10
The H-I And Master Hologram 12
Sunlight-viewable And Laser-viewable Holograms 13
Rainbow Holograms 13
Lighting 14
Recording materials 16
Lasers 16
Choosing A Subject 17

Chapter 2

Varieties of Holograms 19

Embossed Holograms 19
Holographic Stereograms 19
Making A Holographic Stereogram 20
Other Holographic Stereograms 20
Flat Holographic Stereograms 21
Embossed Holographic Stereogram 21
Alcove Holographic Stereograms 21
Holodisk® 22
Color Variations in Holograms 22
Bird's Eye View Of Holograms 24

Chapter 3

Recording Materials 25

Silver Halide Emulsion 26
Dichromated Gelatin 29
Photopolymer 30
Photoresist 32

Chapter 4

Lasers 35

A Brief Historical Background 35
How a Laser Works 36
Laser Applications In Holography 38

Chapter 5

Holographic Optical Elements 41

Diffraction Gratings 43
Holographic Lenses 43
Holographic Mirrors 44
Holographic Collimators 44
Holographic Mirror Beamsplitter 44
Other Hoe Applications 46

Chapter 6

NDT: Holographic Non-Destructive Testing 49

Real Time Holographic Interferometry 50
Double Exposure Holographic Interferometry 50
Sandwich Holographic Interferometry 51
Holographic Interferometry Through Dense Media 52
Time-Averaged Holographic Interferometry 52
Strobed Holographic Interferometry 52

Chapter 7

Computer-Generated Holograms (CGH) 55

Techniques for Outputting Computer-Generated Holograms 56
Binary Detour Phase Holograms 56
The Kinoform 57
ROACH 57
Computer-Generated Interferograms 57
Three Dimensional Computer-Generated Holograms 57
Uses of Computer-Generated Holograms 58

Contents

Chapter 8   *Holography Education 61*

Chapter 9   *Embossed Holography 63*

Successful Applications 63
Buying An Embossed Hologram 64
Preparing Artwork For Embossed Holograms. 67
Step-by-step Making Of An Embossed Hologram: 68

Chapter 10   *Holography Market Place & The Year In Review 75*

Copyright Holder 76
The Manufacturer 76
The Chain Of Distribution 76
1be Distributor 78
The Wholesaler 78
The Retailer 78
Holography 1990-- The Year in Review 80

Chapter 11   *Holography Businesses 87*

Chapter 12   *Businesses By Category 125*

Chapter 13   *Individuals 137*

Chapter 14   *Bibliography 157*

Chapter 15   *Glossary 167*

Index To Advertisers

Advanced Holographies 59

AH Prismatic 11

American Banknote ....insert page 32

Archeozoic ....inside back cover

Bridgestone Graphics. ....insert page 64

Cherry Optical 45

Coherent Laser Group 37

Control Optics 45

Dazzle Enterprises 86

Diffraction 65

DZ Company 12

Envision Enterprises 81

Foreign Dimension 69

Holocrafts 24

Holographic Products 15

Holography Dev. Group 77

Holography News 85

Ian Ginn Holography 23

Ibou 86

Keystone Scientific. 34

Laza 81

Lenox Laser .... inside back cover

Leonardo Journal/I.S.A.S.T 47

Light Impressions ... insert page 64

Polymer Holographies 27

Rainbow Symphony, Inc 59

Reconnaisance Ltd 49

Richard Bruck 18

Starlight Holographic 48

Textile Graphics 85

Towne Laboratories 27

UK Gold ....insert page 96

UK Optical Supplies 47

White Light Work 62

Wonders Of Holography .. insert page 96

1

Introduction

Most of us see holograms every day, though we may be unaware of their presence. They are on our credit cards. They are used in supermarket check-out lines to scan the labels on our food. They are on postage stamps, the covers of magazines, and in some countries, even appear on "paper" money - a $10.00 Australian bill and $50.00 Singapore bill are already in circulation. During the year 1991, bills of most denominations in Australia will begin to be introduced to replace the existing currency. These notes will be made of thin plastic and have embossed holograms of historical people on them to curtail counterfeiting.

We are getting ahead of ourselves, however, by citing holographic applications before we explain what a hologram actually is. We can start by contrasting holography with its neighboring field, photography.

Although often compared with photography, holography is really a completely different medium with different applications. They are the same only because they both are ways of capturing an image, and, at times, similar chemical processing is used in making both items. There are, however, major differences between a photograph and a hologram.

3d, Or Not 3d?

A photograph can be made under any ordinary lighting condition but the resulting image is only twodimensional. That is, if you move a photograph from side to side you do not see around the image - you will always see nothing but a flat image, even though many gimmicks (e.g. 3-D glasses) have been marketed over the years to trick the eye into thinking it is viewing a three-dimensional image.

A hologram, however, actually is a 3-D image. With some holograms, the image is actually formed in air in front of the holographic plate. Also, you can look around the object just as you would in real life. The distance the image forms in front of or behind the holographic plate and the degree to which you can look around the object depends on how the hologram was made; there are a large variety of holograms, each with its own good and bad points.

Although some holograms may be viewed in ordinary daylight, they all require a very narrow, almost single, wavelength of light to be made. Because of this, lasers, which can put out a single wavelength of light, are almost universally used to make holograms.

To borrow a phrase, a picture (or in this case, a hologram) is worth a thousand words. Therefore, to fully grasp what holograms look like, one should go to a local shop or gallery which displays them or order some from one ofthe many mail-order houses found in this publication.

Armed with mere words and diagrams, our task in this volume is to communicate, to both professionals and beginners, the most comprehensible facts about both the science of holography and its wide-ranging applications.

Perhaps the best way to begin this task is by explaining, step-by-step, the making of a simple hologram. After discussing some of the fundamentals of holography, the more exotic varieties of holograms will be discussed in Chapter 2.

Making A Simple Hologram

Suppose one enters a studio where a simple hologram is about to be made. There is a table in the room. On the table is a laser, some mirrors and a photosensitive plate in a plate holder. Everything on the table is arranged, or set-up, in a carefully measured manner. In the center of this holographic set-up is the object to be holographed.

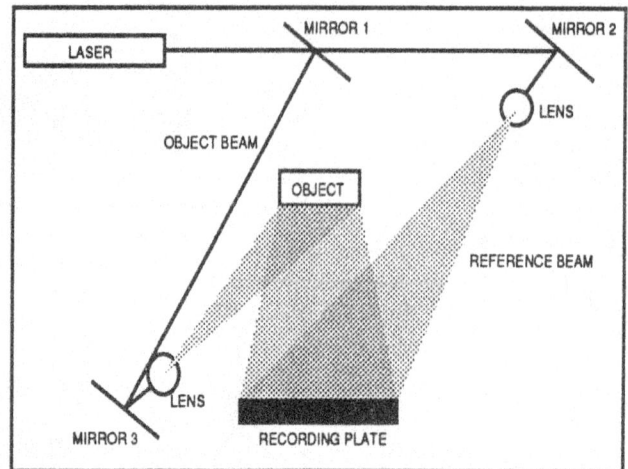

Figure 1.1. Transmission Hologram

To make the hologram, we turn on the laser. The laser beam strikes mirror 1. Due to the fact that mirror 1 is only partially mirrorized, part of the beam is reflected toward mirror 3, and the other part passes through mirror 1 to mirror 2. For this reason, mirror 1 is referred to as a beamsplitter.

The beam which passes through mirror 1 is called the reference beam because it never actually strikes the object being holographed. After the reference beam strikes mirror 2, it is reflected through a lens toward the photographic recording plate. The lens' function is to spread the beam so that it will cover the entire plate. (In some cases, the lens is placed in front of mirror 2; in either case its function is the same - to spread the beam).

At the same time, the other beam (which we call the object beam) reflects off mirror 3 and passes through a lens. This lens spreads the beam out so that it illuminates the entire object. The beam then reflects off the object (hence the name object beam) and strikes the photographic recording plate.

The two beams must travel exactly the same distance so the light waves in the beams will be synchronized with each other. After exposure, the photographic plate is developed, and the resulting developed plate is a hologram. Holding the developed plate up to light, we see that the plate is semi-transparent. On closer inspection we see that the darkness of the plate is caused by developed emulsion. The plate seems to have countless swirls of thread-like developed emulsion which are called fringes. The patterns the fringes make look like the swirls that make up your fingerprints or the boundaries of topology maps. There appears to be no order to the swirls.

Viewing The Finished Hologram

To see the completed hologram, the photographic recording plate is developed and positioned on the table in exactly the same place it was for the exposure. The object and mirrors 1 and 3 are then removed from the table.

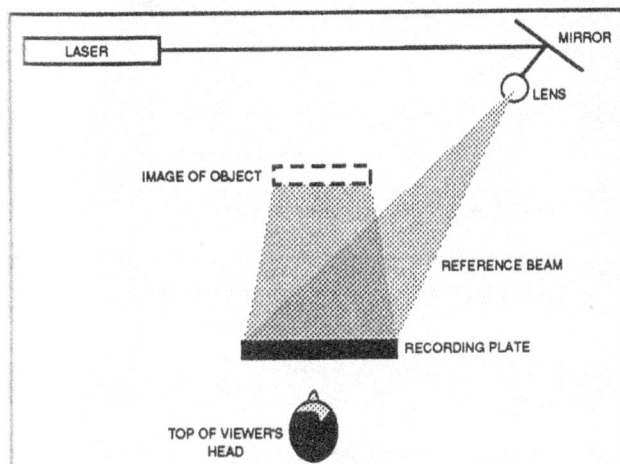

Figure 1.2. Viewing a transmission hologram

The laser is turned on again. When one looks through the plate, an image is seen of the original object, in its original place and at its original depth.

A detailed explanation of why this happens would be exhaustive. A simple explanation might go like this: the object and reference beam strike the photo-sensitive recording material at the same time and naturally create a pattern as they expose the chemicals in the emulsion. Since both beams strike the plate at the same time, the light waves in the beams naturally interfere and combine with each other. therefore the pattern we get when we develop the plate is called an interference pattern.

After development, we aim only the reference beam at the plate, at exactly the same angle that originally exposed the plate. As the reference beam goes through the plate, the interference pattern naturally causes part of the reference beam to change direction. This phenomenon is called diffraction. The interference pattern, because it was created by light from the original object, diffracts the reference beam passing through it back in the direction of the original object. This deflected beam has exactly the same form as the bearn that was originally reflected from the object because the object beam made the patterns in the emulsion. Now, the reference beam travels through those patterns to recreate an image of the object in the same place. In other words, a hologram mimics the way light reflects from an object, without the object being there.

In order to record a clean and clear interference pattern of our object and reference beam on the plate we use a laser which has a single beam of light at one wavelength. We cannot use just any light as our source of light, because the light from common light bulbs contains many constantly changing wavelengths of light. If we make the exposure using regular light, the interference patterns would be completely blurred and useless because of the changing wavelengths and multitude of interference patterns that would be created.

Reflection Holograms

The hologram just discussed is called a transmission hologram because the light passes throu~h the plate to us.

It is also possible to make a hologram where the light reflects off the surface and back to our eyes for us to see the image. This is called a reflection hologram.

How are reflection holograms made? Look at Figure 1.1 that we showed you earlier. If we transfer the reference beam around with mirrors so it illuminates the recording plate from the back instead of from the same side as the object beam, we create a reflection hologram, as shown below. It is that simple.

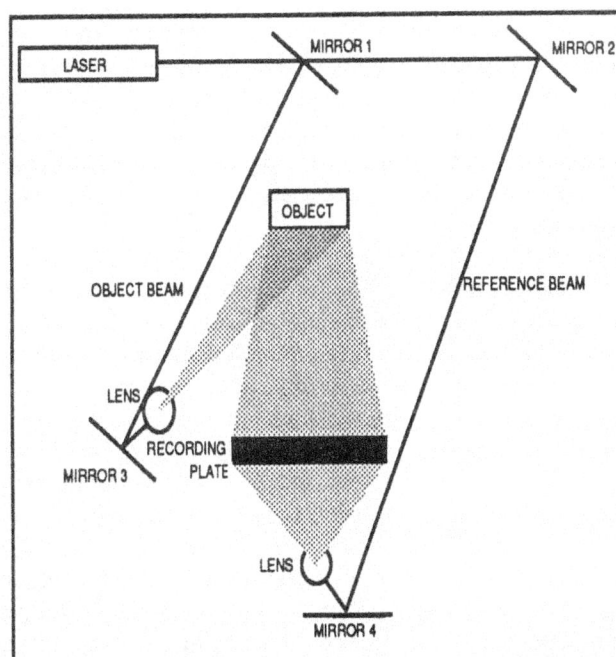

Figure 1.3. Reflection Hologram

As their name implies, all reflection holograms are illuminated from the same side as the viewer. In other words, when viewing a reflection hologram light comes from your side of the plate, strikes the plate and reflects back to your eyes. It should be pointed out that all reflection holograms can be viewed in sunlight (frequently called daylight-viewable or white-light viewable holograms) whereas transmission holograms, depending on how they are made, can be either daylight viewable or only viewable with a laser light (called laser-viewable holograms). We will discuss these categories shortly.

Within the two major divisions of holograms (reflection and transmission), there are many variations. Like any other specialized field, holography has its own lingo, and in some cases the same hologram can be described using more than one name.

We will try to clarify the sub-divisions of holograms next - but remember, each of these subsets can always be classified into the larger families of transmission or reflection holograms.

Through Thick and Thin

Another popular way to broadly classify holograms is by calling them a "thick" (sometimes called "volume") holograms or "thin" holograms. One of the reasons the words "thick" and "thin" are used in conversation is that they allow one to instantly get an idea of some of the properties of the hologram. Very thin holograms provide little depth to their object upon reconstruction. Embossed holograms, such as the images on bank charge cards, are examples of thin holograms. Thick holograms have the ability to replay or reconstruct the image with considerable depth or projection.

How do you decide if a hologram is a thick or a thin hologram? The classification works like this: If you look closely at the surface of the hologram you see the fringes caused by the interference pattern. A hologram is considered to be "thick" if the thickness of the recording medium is considerably greater than the spacing between the interference fringes. Otherwise the hologram is considered a "thin" hologram.

The distance between interference fringes will depend on a number of things, such as the wavelength of light being used, and the density of particles in the emulsion.

These interference fringes are called Bragg planes after Sir William Henry Bragg and Sir William Lawrence Bragg, who did much of the early work on the subject. As you would expect, Bragg planes actually

go all the way through the medium, but are visible to our eye only where they meet the surface.

It should be pointed out that in a reflection hologram, where the reference beam and object beam strike the plate from opposite sides, the Bragg planes slice through the medium at very shallow angles.

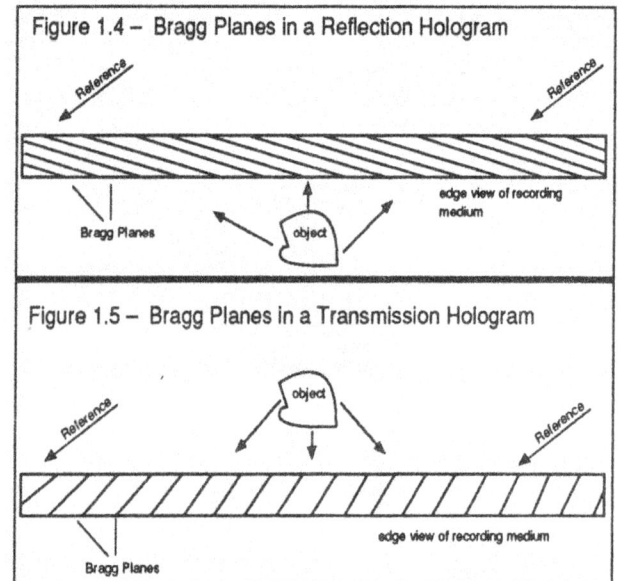

Figure 1.4 – Bragg Planes in a Reflection Hologram

Figure 1.5 – Bragg Planes in a Transmission Hologram

Conversely, in a transmission hologram, where the reference beam and object beam strike the plate from the same side, the Bragg planes cut the emulsion at much sharper angles and thus are further apart.

Although transmission holograms seem to be more naturally designed to create a hologram with considerable projection, one can make reflection holograms that have a great deal of projection as well. In fact, reflection holograms with considerable projection are a favorite among artistic holographers and the buying public. They are favored because they can be hung on the wall and illuminated just like a painting, whereas transmission holograms need to be lit from behind, often requiring a much larger viewing area.

Laser-viewable transmission holograms demonstrate amazing depth and projection when the correct equipment is used. It should be noted that the depth of the holographic image is not so much a function of the power of the laser as it is the coherence length of laser light. You can read more about the coherence length in the laser chapter (Chapter 4), or in several of the books listed in the bibliography.

Theoretically, the maximum image projection in front of the hologram plate can be as great as the projection in back of the plate (depth of the image). Unfortunately, it is difficult for our brains to make sense of greatly projected images. Because of this and

A.H. PRISMATIC is the World's leading Holographic gift manufacturer and distributor. Why? Because we offer the most comprehensive range of Holographic gifts, Toys, Jewelry and Stickers in the World. At all price points, we have something for everyone; including . . .

HOLOGRAMS. The best selling and widest range of Film Holograms on the market, now including our exclusive "Red Beam" collection, the lowest-cost brightest images available today.

HOLOGRAPHIC GIFTS. Our Laser Jewelry, Stickers, Watches, Laser Spex, Laser Discs and Laser Boxes are among the best selling Holographic gifts on the market. Our Executive Accessories are fascinating and innovative new products.

DISPLAYS. In addition to our industry-leading high quality packaging, we offer a full range of point of purchase Display options, from counter-top units to fully lighted floor displays.

A complete Catalogue is available upon request.

Pictured above: Counter-Top Hologram Centre

NEW YORK
A.H. Prismatic Inc.
Head Office
285 West Broadway
New York, NY 10013
Tel: (212) 219 0440
Fax: (212) 219 0443

SAN FRANCISCO
A.H. Prismatic Inc.
Sales Office
1792 Haight Street
San Francisco, CA 94117
Tel: (415) 221 4717
Fax: (415) 221 4815

UNITED KINGDOM
Enquiries outside the U.S.A. to:
A.H. Prismatic Ltd.
New England House,
New England Street,
Brighton, Sussex BN1 4GH, England
Tel: (0273) 686966
Fax: (0273) 676692

RETAIL LOCATIONS
Museum of Holography
11 Mercer Street
New York, NY 10013

Holos Gallery
1792 Haight Street
San Francisco, CA 94117

A.H. PRISMATIC
Incorporating Holos Gallery

the fact that there are some optical distortions in the image planing process, projected distances in transmission holography are usually kept under four feet.

Laser transmission holograms have the widest parallax (the ability to see around an object side to side) and resolve the greatest depth of objects. There are laser transmission holograms, for example, ofpeople and objects in a 4000 cubic foot room exposed with a pulsed laser.

Not surprisingly, projected hologram images like this generate one of the highest shock and thrill responses from viewers. A good percentage of first time viewers respond by waving their hand back and forth through projected images in disbelief.

The laser viewable transmission hologram is most often used as a master hologram. Transfer copies (making another hologram using the image on the master as the subject) can be made in quantity from the master. These transfer holograms can either be other laser-viewable transmission holograms, white-light viewable transmission holograms or reflection holograms (always white light viewable).

The H-1 And Master Hologram

It is important that we cover the topic of the H-1 and master in this introduction because it is a fundamental procedure in the making of almost every commercial hologram.

H-1 stands for "hologram one", which simply means it is the first hologram you make on the path to your desired final hologram. Sometimes the H-1 is the master hologram from which you make multiple copies. Frequently, though, there is more than one hologram that needs to be made before you get the desired master hologram from which you will make copies. If this is the case, the next hologram in the sequence is called the H-2, and then H-3, and so forth.

A question that immediately comes to mind is why would anyone want to make an H-2? Well, historically one of the big problems that holographers used to have was placing the object to be holographed exactly where they wanted it. Suppose, for example, you want the object in the final hologram to appear half in front and half behind the recording plate. How would you do it? You obviously can't do it on your first shot because the object would have to be going right through your photographic plate.

This problem was solved by the following procedure:

• Make a transmission hologram. We call this H-1 because it is our first hologram.

• Since the H-1 hologram creates an image of the object, why not use the image made by our H-1 as our subject and make a hologram (H-2) of the image made by the H-1? In other words, make a hologram of our hologram. This H-2 hologram can be a transmission or reflection hologram depending on your need.

• It sounds strange, because you are making a hologram of an image and not an object. But it works.

• Now, since you can make a hologram of the H-1's image, take time to move the image around to wherever you want it positioned. In this case, adjust the H-2 recording plate so that the image of the object is half in front and half behind the plate and then make your H-2. The problem of getting our object to be half in front of our plate and half behind our plate is solved.

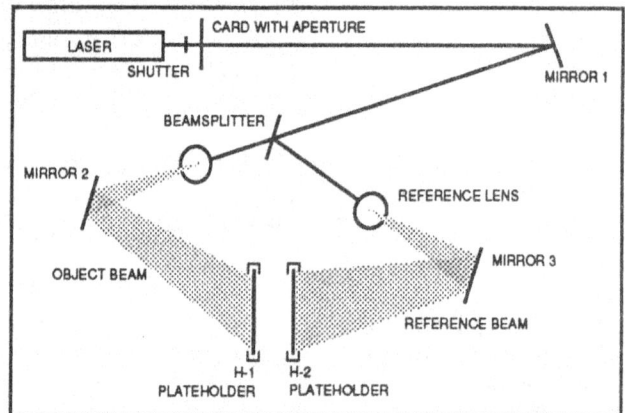

Figure 1.6 – Reflection H-2 being made from H-1

In short, there are at least three good reasons why an H-2 should be made:

1) The H-2 allows you to reposition the image of your subject. When you reposition your image from the H-1, you may make your subject focus out in front of the recording plate, behind the plate or anywhere within the limits of your equipment (you are usually limited by the laser's ability and the quality of the optics). The creative potential here is enormous because you are able to move solid objects around like they are ghosts. You can have two objects occupying the same space, etc. The process of moving the image around to make the H-2 is called image planing.

2) It gives the holographer a chance to brighten up the

image. Since you may move your image anywhere, you can focus the image right at the recording plate. This concentrates the light directly on the recording material and brightens up the image considerably. This is commonly done in silver halide reflection holograms.

3) It saves time on remakes. That is, after you develop the H-2 and decide you don't like the position of your subject astride the recording plate, you don't have to find the original subject and set it up again. This can be important if there are large costs in arranging the H-1 shot.

Going through the pains of making H-1, H-2, etc. to get a good master is necessary for creating most holograms. It is technically possible to get some desired holograms by skipping this process but the results are generally very inferior. A master is almost always used for commercial jobs of value.

Sunlight-viewable And Laser-viewable Holograms

We mentioned earlier that although it is necessary to use a laser to make a transmission hologram, it is not always necessary to use a laser to see a transmission hologram. In fact, most transmission holograms can be seen in sunlight.

This may seem confusing because we have said that in order to see a holographic image you have to shine the reference beam that made the hologram on the plate. This is true, but sunlight contains a multitude of wavelengths, including the one we used to make our exposure. The sun is far enough away that it appears to be a single beam of light shining on our plate. It would seem that we have only to position the plate so we are at the proper angle and we should see our image.

This is logical, but it also stands to reason that if sunlight passed through a transmission hologram you should also get your image being formed by all of the other wavelengths that are somewhat close to the wavelength of your reference beam. These other frequencies of light would diffract at a somewhat different angle than your original reference beam because they are a different frequency. The result would be a whole multitude of images forming right next to each other, thus causing a blur instead of a clear, crisp image.

That's exactly what does happen and it took a

while for a solution to be developed. Around 1969 Dr. Seven Benton came up with a solution. The resulting hologram is sometimes referred to as a Benton hologram or, more frequently, a rainbow hologram.

Rainbow Holograms

Benton reasoned that since our problem was that we were getting too much in the way of imagery at the point where we want to reconstruct our object, why not block off some of the imagery? In other words, suppose we had an opaque mask that we put up against the transmission hologram with a long narrow horizontal slit in the mask through which we viewed our transmission hologram. This would certainly clean out a lot of the annoying secondary images that are blurring our ability to see our primary image reconstruct. It would, of course, come at a price because the mask would make it so we would lose our vertical parallax (the ability to be able to see over and under our object). We would, however, still have our horizontal parallax (ability to see side to side around the object) and humans, with feet fixed on the ground and eyes on a horizontal plane, are actually more accustomed to the horizontal parallax of images than vertical parallax.

The procedure to produce this masked hologram is as follows:

1) First a normal transmission hologram is made.

2) Next, a transfer copy of the transmission hologram is made but an opaque card with a horizontal slit in it is placed between H-1 and H-2 .

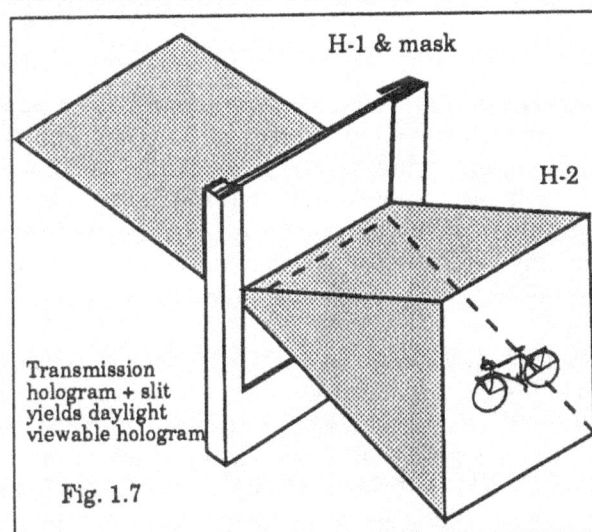

H-1 & mask

H-2

Transmission hologram + slit yields daylight viewable hologram

Fig. 1.7

If the copy hologram is viewed in the frequency of our laser light, the eyes must be located at the real image of the slit to see the image.

Now, imagine viewing this H-2 hologram in two different colors of light. A hologram of the image made through the slit will be played back, but each of the two wavelengths of light will diffract through the hologram fringes at a slightly different angles. There will be two different images of the object, each a different color and each at a slightly different vertical position.

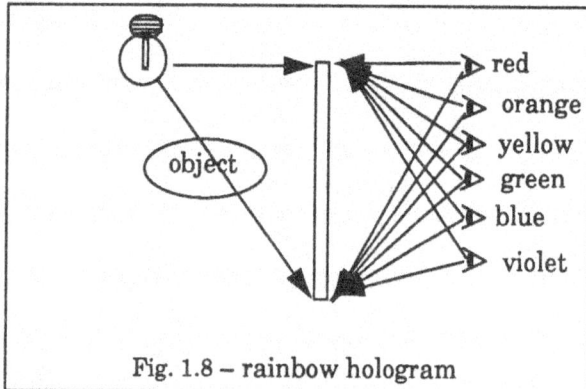

Fig. 1.8 – rainbow hologram

Now imagine viewing the image in white light or sunlight. All of the wavelengths will reconstruct their own image, all slightly displaced vertically with respect to one another. We are faced with the same problem we had with the original transmission hologram except the images being recreated have no vertical parallax. As you move up and down in front of the plate the color of light will change but the image will be the same (hence the name "rainbow hologram"). As you move from side to side you will have horizontal parallax because nothing has been done to destroy it.

As you can see, rainbow holograms get their name from the fact that as you shift your viewing angle up and down, instead of the image changing its perspective, it changes colors, shifting through the colors of the rainbow. By careful planning, the image may be made any desired color, or even a combination of colors (a multi-color rainbow hologram).

In effect, the hologram is filtering the white light while all that is sacrificed is vertical parallax, which, as we mentioned, our two horizontal eyes usually don't miss anyway. Rainbow images are often extremely bright and this is because all of the frequencies in white light are being used to form the image. So the rainbow hologram technique is a way of making a transmission hologram sunlight-viewable. Other names for this are daylight-viewable or whitelight viewable. They all mean the same-- a hologram you

can see without the need of a laser.

Some rainbow transmission holograms are displayed in art galleries on glass plates and on film. However, they are much more popular in two other forms. The two most popular forms in which rainbow transmission holograms are seen are as embossed holograms and holographic stereograms.

In an embossed hologram, the light goes through a rainbow transmission hologram that has been embossed in clear plastic, strikes a mirrored backing and reflects back through the rainbow transmission hologram to the viewer's eyes. Chapter Nine of this book concerns itself entirely with embossed holography.

Lighting

Sunlight is not the only source of light that works with white-light viewable holograms. There are a whole range of lights with which to view a white light hologram; some light sources are better then others. Just remember that white light holograms require a light source which contains the original exposure wavelength and enough intensity to replay the hologram. Ideal light sources cast sharp shadows, like a spotlight or an average clear light bulb with a single filament. Some very bad light sources, such as fluorescent lights, are extremely diffuse sources which in some circumstances render white light holograms unviewable.

Whether or not a white light hologram is viewable in bad lighting depends on how the hologram was made. Especially vulnerable are holograms which project an image far out in front of the plate or have great depth. If a hologram like this is illuminated with spotlights coming from multiple sources at different angles, the hologram forms projected images at all the different angles dictated by the spotlights. The mixture of projected images "blurs out" the image the holographer is attempting to recreate. Hence, holograms made to be viewed in a wide range of lighting, including light from multiple sources, use subjects that have very shallow depth. This is because if there is little or no depth to the object, all the images being created by the different sources appear to be focusing in the same place.

Thus, if you go into a shop that specializes in holograms, one usually finds that the shop has subdued, overhead lighting with spotlights focused on the holograms. This serves the dual purpose of creating a lighting environment that is pleasant to be in as well

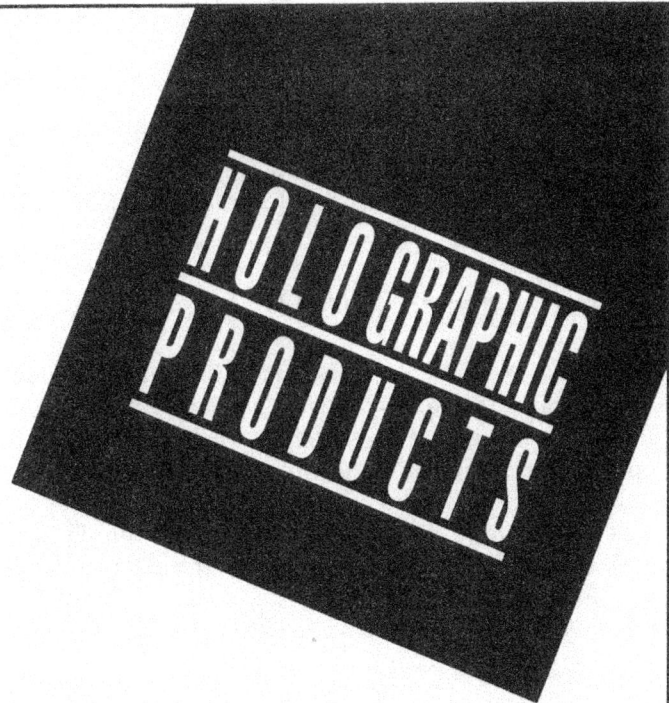

as providing a single spotlight for holograms with considerable projection.

People who display holograms in their homes find an inexpensive way to illuminate them is with a clear light bulb having a single filament. These bulbs are available at any shop with a large selection of light bulbs.

Pseudoscopic and Orthoscopic

Both laser viewable and white light transmission holograms share a fascinating property. Remember in the beginning of this chapter how, after developing, the hologram was put it back in the plateholder and, with no object on the table, the viewer was then able to look through the holographic plate and see the original object appearing deep within the plate at its original position on the table? Well, now comes the interesting feature. Take the transmission hologram out of its plateholder, flip it over, and put it back in the plateholder. Step back and look at the plate. You see the image forming out in front of the plateholder (between you and the plateholder). It focuses in air the same distance in front of the plateholder as it originally sat behind the plateholder. You also see that the image is a pseudo scopic image.

What is pseudoscopic? An image as normally seen in everyday life is an orthoscopic image. A pseudoscopic image is the opposite of this. For example, if one's viewpoint moves to the right, you do not see more of the right view of the image but the left view, and when the viewer raises his viewpoint, the lower part of the subject comes into view instead of the upper part.

A pseudoscopic image yields an exciting effect, but it can be confusing to the viewer. Some artists, however, have produced exciting pseudo scopic work using geometric shapes like wide spirals, pyramids and cones.

Recording materials

Although we have discussed how holograms are made, we have not discussed the material that they are recorded on. In photography, the most common item used to capture images is a silver halide emulsion on a film base. In holography, there are a number of items used to record your image. The most common recording media are:

1) silver halide
2) dichromated gelatin
3) photopolymer
4) photoresist

Note that we use the phrase "recording materials" instead of emulsions. That is because not all the items used to capture holographic images are emulsions. In an embossed hologram, for example, the holographic image is literally stamped into clear plastic. A discussion of the different recording materials requires a chapter of its own and we dedicate Chapter Three to that discussion.

Lasers

We will discuss lasers a little more in depth later in this publication (Chapter Four) but it is important that we touch on the subject in the introduction. The type of laser you use affects what subjects you can holograph and we want to discuss that next.

There are two major kinds of lasers; the Continuous Wave (CW) laser and the pulsed laser. The CW laser emits a continuous wave of laser light whereas the pulsed laser emits laser light in bursts.

Continuous Wave Laser: The power of a CW laser is measured in watts (w). The CW laser is by far the most common laser used in holography. In holography labs, most of these lasers fall in the 5 to 50-mw (milliwatt) range. One of the great problems with CW lasers is that they cannot make the extremely short exposures necessary to capture a live subject. Consequently, there must be absolutely no motion at all during the exposure with a CW laser. An exposure with a CW laser can take a fraction of a second to many seconds. Because there cannot be any motion at all during the exposure, we need eliminate any vibration coming from the ground. To do this we make or buy a vibration isolation table on which to put our laser, optics, and objects. Since it is absolutely critical that we have no motion at all, the subjects that we holograph with CW lasers have to be "dead" objects. Feathers might move in the breeze and living things simply move too much.

Remember that what we are recording on the plate is the reference beam and the object beam converging (or interfering) with each other at the plate. If the object moves even a microscopic amount (on the order of wavelengths) from one moment to the next, we will record two different interference patterns and the hologram fuzzes out or doesn't even show. The effect is like some photographic daguerreotypes of the 1800's -

only much less forgiving.

Pulsed Laser: Pulsed lasers, on the other hand, emit extremely quick bursts of very powerful laser light. Consequently, the exposure time is much shorter than a CW laser. Exposures can be made in nanoseconds. One nanosecond is one billionth of a second. You don't need a vibration isolation table for the pulsed laser. What can you shoot? Anything you want. You can shoot an entire room of people belly dancing in costumes of paper with feathers in their hair, and birds flying around the room. Why such freedom? Because your subject can't move significantly in a nanosecond.

What are the drawbacks of pulsed lasers? Why doesn't everyone buy one? The answer is money. They cost about $60,000 and require a lot of extra overhead and care. Lasers don't last forever and when a pulsed laser burns out it is expensive to fix. Holographers are anxiously awaiting, with cash in hand, a low-cost, easily maintained pulsed laser.

The reason this discussion on lasers is important is that who you go to for making your hologram depends on what you want holographed. If you have a corporate logo or some other object that does not move, a CW laser can do just as good a job as a pulsed laser. There is no need to pay extra to have your hologram made by pulsed laser if you do not need it. On the other hand, if you have a live subject, you must use a pulsed laser.

Choosing A Subject

One last topic we should cover in this introduction is what kind of images you can make a hologram of.

As with any creative art form there are a whole multitude of choices available for you and everything depends on what you want to accomplish. To simplify things, we will list some of the most common items that are used for holographic subjects and then comment on each item. The most common items are:

1) 3-D model Oike a sculpture).

2) 2-D model - just like a graphic arts layout you would make, complete with overlays, for a printer.

3) Movies - specially-shot motion pictures for a hologram that moves.

Although some of the above topics are covered in more detail later in the book, we will give an overview

of each of them now.

3-D model: This is the most common way to create a subject for a hologram. What your model is, of course, depends on the type of laser you are going to use for your exposure. With a pulsed laser, as we have already mentioned, you can shoot live subjects and just about anything you wish. Most holograms, however, are made with a CW laser and the models used have a dramatically different look depending on the type of paint used to cover the item. In order to get some idea of what your hologram will look like, many artists paint the subject and then look at it under the light of the laser they are going to use for exposure. Several repaints may be necessary to get exactly what you want. It is possible, with some of the latest techniques, to shoot a hologram of a subject and then reduce the subject in the copying process. Consequently you can have portraits of people reduced and still preserve much of the parallax that holograms have to offer.

2-D model: You will see this method being used with great abundance in embossed holograms. You have the ability to create camera-ready art for an embossed hologram much the same way you would create artwork for a printer with the different overlays designated to be different colors and at different depths in the final hologram. You can also have a combination of photos, line art and 3-D objects in your final embossed hologram, although the depth of the 3-D object is limited because we are dealing with a thin hologram. Colorization for this process has come a long way in the last few years. Enclosed in this book is an ad featuring a sample of the type of 2D hologram we are discussing. This topic will be discussed in much more depth in the embossed holography chapter.

Movies: Motion picture footage is used in the making of holographic stereograms, which will be covered in Chapter Two. Suffice it to say that you make holograms of motion picture footage, if it is shot correctly (with the view in mind of using it in a holographic stereogram). Popular usage of this format includes plastic cylinders which rotate and, as you look into the plastic, you see the image you filmed suspended in mid air in the center of the cylinder performing whatever was done in the movie. It is also possible to make flat stereograms and even emboss them so that you are looking at an embossed hologram and as you move the hologram with your hands, the image moves as well. This will be discussed more in the next chapter.

We have now had a look at some ofthe basic principals behind making holograms. Obviously these principals are applied by practicing holographers in a multitude of ways. Let's go on to Chapter Two and

examine some of the different types of holograms that
are made.

RICHARD BRUCK HOLOGRAPHY
3312 WEST BELLE PLAINE
CHICAGO, IL. 60618
312.267.9288
USA

QUALITY SILVER HALIDE PRODUCTS
LIMITED EDITIONS   FINE ART   COMMERCIAL RUNS

2

Varieties of Holograms

In Chapter One we covered some of the principals of how holograms are made. The bulk of artistic holograms sold in public are holograms that are simply mass-produced from a master, packaged and sold. There are, however, some unusual applications of the holography principals discussed in Chapter One which give us a variety of holograms beyond the film or plate repoduction of a single image. In this chapter we will cover some of these varieties.

Embossed Holograms

Probably the most widely-seen example of holograms are embossed holograms. They appear on everything from money to breakfast food and are one of the largest money makers in the field of holography. Creating an embossed hologram is a very involved process. Due to the popularity ofthis process, we devote a whole chapter (Chapter 9) to an explicit discussion of exactly how embossed holograms are made and we refer you to Chapter 9 for a detailed discussion.

Holographic Stereograms

Creating holographic stereograms is one of the most exciting fields of holography. If you have seen "moving holograms" you most likely have seen one of these. A stereogram is defined as "a diagram or picture representing objects with an impression of solidity or relief." Consequently, a holographic stereogram is a hologram of pictures or diagrams which gives the impression of solidity or relief.

Several techniques are used to make holographic stereograms. Names are given to the various methods and one will hear names like:

· Integram
· Cross hologram
· Multiplex hologram
· Benton stereogram
· Embossed stereogram
· Alcove hologram
· Holodisk®

As with any trade, one of the problems in holography is understanding the jargon. Sometimes several names are used to describe the same item and there are many special nicknames used to describe special techniques. The names you see above are all holographic stereograms and the first five are really the same type of hologram. To give a clearer idea about stereograms, following is a brief synopsis of how a Cross holographic stereogram is made.

Making A Holographic Stereogram

Generally, this is the way a 360-degree Cross holographic stereogram is made:

1) Make a small stage that rotates 360 degrees.

2) Put the subject on the stage.

3) Set up a regular movie camera on the stationary floor.

4) Film the subject as it turns the full 360 degrees. Slight motion is possible but the subject cannot make radical or jerky moves. This latter kind of movement creates what are known as time smears in the hologram---places where the subject looks jagged. Slow, even movements when filming yield the best results.

5) Develop the movie film.

6) Make a hologram of the image in each frame of the movie, making sure to shoot at least 3 frames for each degree of rotation.

7) The holograms will be' on a roll of holographic film. Each hologram, as one can see, will be as tall as the width of the roll but of very narrow width.

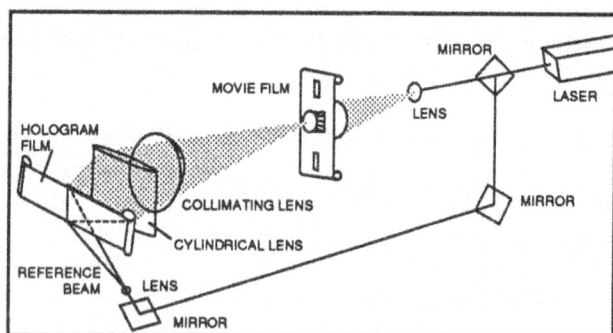

Fig. 2.1 – Optical set-up for a white light holographic stereogram

8) Set up a mask in front of the roll of holographic film to get one narrow hologram (you will be shooting a white light rainbow transmission hologram).

9) Shoot the first exposure, advance the holographic film one frame, advance the movie film one frame, expose again and so forth for the entire movie film.

Viewing A Holographic Stereogram

After development, take the roll of holograms and wrap it around a strong cylinder of clear plastic that is mounted on a display which is able to rotate 360 degrees. Place a clear (unfrosted) light bulb with a single filament in the center just below the holographic film.

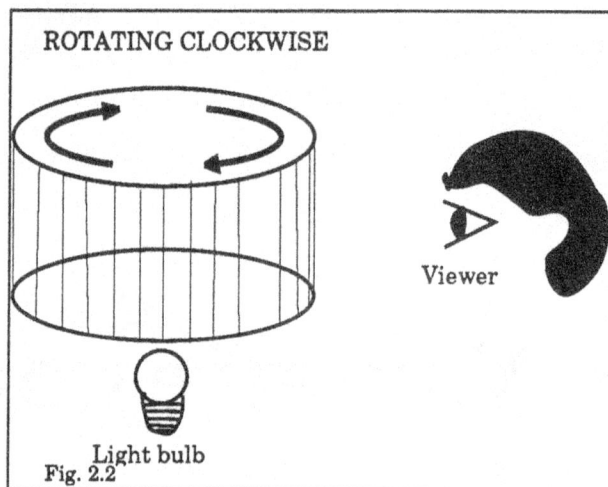

Turn the light on and rotate the cylinder. Several things are happening here:

• The movie frames are moving past the eyes.

• Each eye sees different images at the same time, thus creating a stereo view.

• Since this is a rainbow white light transmission hologram, the image forms in the same position it was when originally shot (the image usually forms in the center of the cylinder).

What we have is a moving holographic stereogram, in which the viewer sees a three-dimensional "movie" of the image moving around in the center of the cylinder.

Other Holographic Stereograms

Less-Than-360-degree-Curved-Cross-Holographic Stereogram: The holographic stereograms like the example above can be made in a curved format less than 360 degrees. On the market today are 90-, 120-

180-, and 360-degree curved holographic stereograms. The ones that are smaller than 360 degrees generally stay fixed and the viewer walks past them to see the motion. The smaller curved stereograms are inexpensive compared to the 360-degree stereograms.

Flat Holographic Stereograms

Instead of a curved stereogram as described above, one can make a flat holographic stereogram. The procedure for shooting is a little different.

The subject is not on a moving stage this time but on the ground at a distance from the camera. Now we build a straight railroad track on which to put the movie camera. Without aiming the camera directly at the subject, but facing the camera in the subject's direction, the camera moves along the track and takes photos at equal distances. We then develop the film and holograph it much the way we did in the Cross hologram described above.

The result is a flat sheet which, when held up to a light and tilted from side to side, displays an image in motion. With some changes in the holographic process, you may also make this a reflection hologram. The reflection holographic stereogram is viewed by standing under a light, holding the flat sheet and tilting it from side to side. The image can display above the surface of the hologram and move about as it is rocked from side to side.

Embossed Holographic Stereogram

Earlier it was mentioned that rainbow transmission holograms are used as embossed holograms. Since a holographic stereogram is usually a transmission hologram, why not take a flat holographic stereogram and emboss it?

It turns out that this can be done, and it is a popular application. Other techniques along this line include shooting live subjects with a pulsed laser. Very recent developments allow holographers to educe the size of the image in pulsed shots. In theory one should be able to make a series of live shots, using a pulsed laser, and reduce them to fit an embossed holographic stereogram. Perhaps someday there will be curved embossed holographic stereograms as labels for canned products.

Alcove Holographic Stereograms

What has been done in the stereograms just discussed is to take flat art, such as movie film, in which the subject moves from frame to frame and make a hologram of the film. Computers, of course, can make original flat art. Computer-generated images are obviously easier to work with, particularly when making corrections. Why not generate all the art one needs to make a holographic stereogram using the computer?

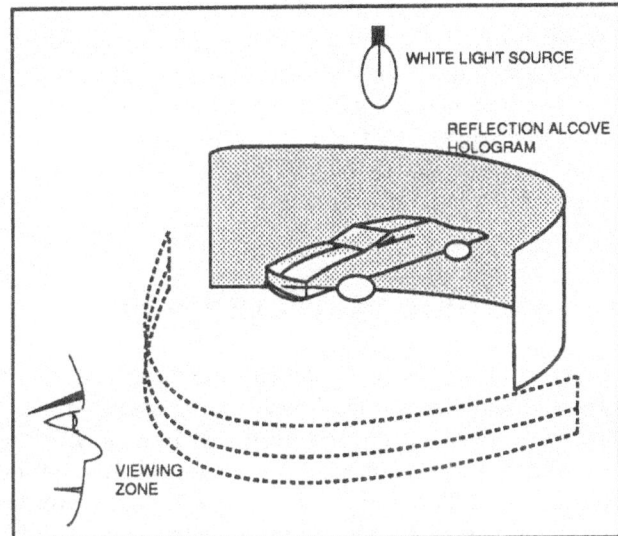
Fig. 2.3- Oblique view of reflection alcove hologram

In a very broad way, this is the concept behind computer-generated holograms. This is a field of holography that has a great deal of interest at present and we devote a chapter (Chapter 7) to discussing the subject in more detail.

One example of CGH is the alcove holographic stereogram. The group led by Steve Benton at the Massachusetts Institute of Technology's Media Laboratory is working hard on this concept and they have already produced some remarkable results. The computer, in this case, creates the flat art. The flat art is then filmed from the video display terminal. The film is then used to make the holographic stereogram. What also sets this hologram apart from the others is the way it is viewed. The image is seen within a clear, concave half-cylinder (an alcove), with nearly 180 degrees of horizontal parallax. In addition to the wide parallax, the image in the hologram can have depth going back through the hologram to infinity. Image distortions are corrected before transfer within the computer.

The alcove hologram is probably the closest holography has corne to a totally synthetic, three-dimensional subject to date.

Holodisk®:

This is a fascinating application of the reflection holographic stereogram. The process takes 360-degrees' worth of photographic images and holographically integrates them onto a flat Holodisk®. It is lit from above and the subject can appear below, straddling, on, or above the disk surface within the limitations of the reflection hologram.

A slowly rotating turntable brings 360-degrees of view to a stationary viewer. The obvious application would be to put the Holodisk~ on phonograph records. As one listens to a favorite song, an image could dance before one's eyes. Unfortunately the revolutions per minute are too high for the Holodisk~ to be used on records, but research continues.

Holographic Stereogram Sizes

Sizes vary with each holographic stereogram. The standard curved Cross holographic stereogram is about ten inches high and eighteen inches in diameter with a curved face in 120-, 180- and full 360-degree formats. The widest film roll one can purchase is a standard 42 inches wide. Attaching one end of a roll to another, these holograms can theoretically be made in indefinite lengths. Using motion picture footage, one could conceivably run hundreds of feet of continuous holographic film. Current technical restrictions, however, need to be overcome before this can be a reality.

Color Variations in Holograms

An obvious question that most people ask is, how close can holograms come to full-color pictures? The answer is that each of the different recording materials are having different degrees of success. Let's look at each of them:

Silver Halide: Full-color holograms in silver halide are not available as far as we know but multi-colored silver halides are abundant.

How are multi-colored holograms made? Let us say you want to have two subj'ects in the hologram. You want each one to be a different color and you want them to be seen by the viewer at the same time (i.e. not multi-channel where you have to move the hologram from side to side to get another "channel" and see the other image). For two colors, the exposure would go something like this:

1) The first exposure is made.

2) The subjects are exchanged.

3) The recording material is pre-swelled to another color's percentage.

4) The reference angle IS changed and the second exposure is made.

The pre-swelling makes the whole recording surface fatter for the exposure. When the hologram is processed the swelling agent is washed out and the thickness of the hologram returns to normal. At normal thickness, the hologram does not reflect the red wavelengths by which it was made, but instead reflects shorter wavelengths - ranging from an orange-red to violet, depending on how much it was pre-swelled.

For registration purposes the reference angles also must be changed with each pre-swelled color exposure. If you make a two-color hologram by pre-swelling the recording material but do not change the reference angle, the result will be a two-channel hologram. In this kind of hologram, as stated above, only one image can be seen at a time.

There are many complications with the above procedure: swelling agent ratios, changing of reference angles, and multiple exposures. Added to this is the problem that the hologram cannot be copied without going through the whole process again (it cannot be mass manufactured). Consequently, this method is used for a single hologram or for small-run, custom work. It should be noted that some film suppliers, like Agfa-Gevaert, make pre-swelled emulsions for just this effect.

In sum, full-color holograms, with colors like those seen in a color photograph, are still very difficult to produce in silver halide.

Photopolymers: There has been some progress in photopolymers with color holography and we have seen multicolored photopolymer holograms. Photopolymers still have a way to go before they get to full color (DuPont says they will have an improved emulsion in 1991 that is much easier for color work but we will have to wait and see what that brings).

Embossed holography: There has been good success in producing results close to full-color pictures in embossed holography. An example of what can be done is the Light Impressions advertisement bound in this book. A flat color stereogram has also been pro duced ("The Clown" by Sharon McCormack) where the figure is in color and also moves as you move the hologram from side to side.

Bird's Eye View Of Holograms

Below is a chart showing the most common types of holograms which we have briefly introduced here.

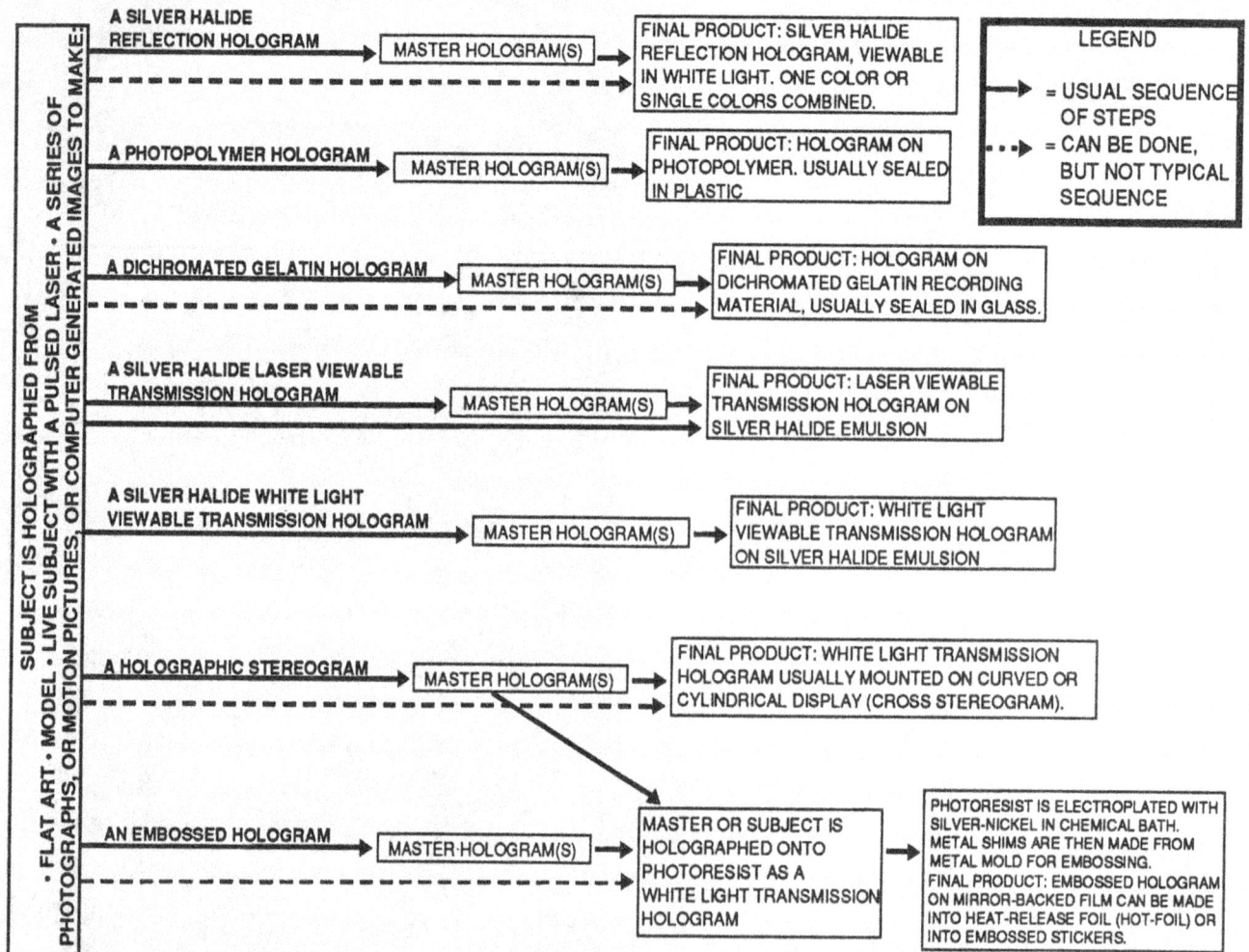

SUBJECT IS HOLOGRAPHED FROM

• FLAT ART • MODEL • LIVE SUBJECT WITH A PULSED LASER • A SERIES OF PHOTOGRAPHS, OR MOTION PICTURES, OR COMPUTER GENERATED IMAGES TO MAKE:

A SILVER HALIDE REFLECTION HOLOGRAM → MASTER HOLOGRAM(S) → FINAL PRODUCT: SILVER HALIDE REFLECTION HOLOGRAM, VIEWABLE IN WHITE LIGHT. ONE COLOR OR SINGLE COLORS COMBINED.

A PHOTOPOLYMER HOLOGRAM → MASTER HOLOGRAM(S) → FINAL PRODUCT: HOLOGRAM ON PHOTOPOLYMER. USUALLY SEALED IN PLASTIC

A DICHROMATED GELATIN HOLOGRAM → MASTER HOLOGRAM(S) → FINAL PRODUCT: HOLOGRAM ON DICHROMATED GELATIN RECORDING MATERIAL, USUALLY SEALED IN GLASS.

A SILVER HALIDE LASER VIEWABLE TRANSMISSION HOLOGRAM → MASTER HOLOGRAM(S) → FINAL PRODUCT: LASER VIEWABLE TRANSMISSION HOLOGRAM ON SILVER HALIDE EMULSION

A SILVER HALIDE WHITE LIGHT VIEWABLE TRANSMISSION HOLOGRAM → MASTER HOLOGRAM(S) → FINAL PRODUCT: WHITE LIGHT VIEWABLE TRANSMISSION HOLOGRAM ON SILVER HALIDE EMULSION

A HOLOGRAPHIC STEREOGRAM → MASTER HOLOGRAM(S) → FINAL PRODUCT: WHITE LIGHT TRANSMISSION HOLOGRAM USUALLY MOUNTED ON CURVED OR CYLINDRICAL DISPLAY (CROSS STEREOGRAM).

AN EMBOSSED HOLOGRAM → MASTER HOLOGRAM(S) → MASTER OR SUBJECT IS HOLOGRAPHED ONTO PHOTORESIST AS A WHITE LIGHT TRANSMISSION HOLOGRAM → PHOTORESIST IS ELECTROPLATED WITH SILVER-NICKEL IN CHEMICAL BATH. METAL SHIMS ARE THEN MADE FROM METAL MOLD FOR EMBOSSING. FINAL PRODUCT: EMBOSSED HOLOGRAM ON MIRROR-BACKED FILM CAN BE MADE INTO HEAT-RELEASE FOIL (HOT-FOIL) OR INTO EMBOSSED STICKERS.

LEGEND

→ = USUAL SEQUENCE OF STEPS

⇢ = CAN BE DONE, BUT NOT TYPICAL SEQUENCE

3

Recording Materials

In this chapter, we'll focus on the photosensitive materials that actually record the hologram patterns. It is tempting to call them emulsions, but the word emulsion is reserved for particles suspended in a gelatin and this word really only applies to the silver halide material.

Therefore, the broader term "recording material" will be used, which simply means something that captures or records our interference patterns. How many recording materials are there? In what format are they available?

There are four popular recording materials:

1) silver halide emulsion

2) dichromated gelatin

3) photopolymer

4) photoresist

No material is perfect and here is a chart of the advantages and disadvantages of each type of material:

MATERIAL	ADVANTAGES	DISADVANTAGES
Silver Halide (Bleached)	High Speed	Only Fair Quality Complex Processing
Dichromated Gelatin	Highest Quality	Poor Shelf Life Complex Processing Poor Image Stability
Photoresist	Easily Reproduced (Embossed)	Low Speed, Limited Spectral Range Transmission only
Photopolymer	High Quality	Short Shelf Life Complex processing

Popular hologram recording materials
Fig. 3.1

These photosensitive recording materials can be applied to either glass or film; both are used in the industry. Film is obviously less expensive and easier to mass-manufacture, but glass in many cases is more popular for original masters because of its stability.

This chapter gives an overview of the different types

of recording materials and details their availability, advantages and disadvantages, commercial manufacture, chemical composition, proper exposure and processing, and finally, manufacturer's tips for best results.

Silver Halide Emulsion

Photography has utilized silver halide emulsions for decades. As you would expect, some major manufacturers of photographic film have adapted their formulae to provide film and plates for holographers.

Availability: There are three primary manufacturers of silver halide films and plates: Ilford, AgfaGevaert, and Kodak. Each company has several different materials available; for a complete listing, see the chart at the end of this chapter.

Advantages: Silver halide emulsions require less light for exposure than most of the other recording materials, so the primary reason for its popularity is its speed. Fast emulsions reduce stability problems. This allows people to buy less powerful, hence less expensive lasers and still make good holograms. It is also convenient to buy silver halide emulsions in precoated film and plates that have a reasonable shelf life, something which cannot be said of some of the other recording materials.

Drawbacks: The silver halide emulsion is granular, so it scatters more light than the other recording materials. Commercial Manufacturing Costs using Silver

Halide: Before mass ma'nufacturing a hologram you must make a model and have the master hologram made. This cost is one of the things that is holding the industry back. Model costs range from $1,000 to $4,000, depending on the detail, and masters cost between $2,000 and $10,000. Extremely short runs are generally not done because the cost for model and master makes the unit price too high. Although most of the mass manufacturing of silver halide holograms on film and plates is not fully automated, there are fully automated machines for silver halide film reproduction which are capable of runs in the tens of thousands. The mass manufacturing quality, which at first was poor, has improved considerably.

Silver Halide Chemistry: There are five atoms which, because of their atomic similarity, are called the halides. They are chlorine, bromine, iodine, fluorine and astatine. Silver halide emulsions are made using either silver chloride, silver bromide, or silver

iodide. The other two halides are not used because silver fluoride is insoluble in water and astatine is radioactive.

A typical silver halide emulsion is made by adding a solution of silver nitrate to a solution of potassium bromide and gelatin. Silver bromide crystals form in the emulsion. The emulsion is heated for a certain amount of time, which is called the ripening process.

During the ripening process, the grain size increases and the speed of the emulsion is increased. Some doping agents may be added to the emulsion at this time to foster proper crystal growth. Afterwards, the gelatin is allowed to cool. It is then shredded, and the soluble potassium nitrate is washed out of the emulsion.

The emulsion is heated again, with more gelatin added; then it is cooled and applied to a base. The thickness and hardness of the emulsion is important in holography because emulsions too thick tend to deform during development. Emulsions that are too hard can either retard chemical reactions or create vacuoles in the emulsion left by migrating atoms. These vacuoles tend to scatter light.

Exposure Photochemistry: Let's assume our emulsion is made and we now want to expose our emulsion to light. It sounds surprising, but a perfectly structured crystal of silver bromide does not react to light in any appreciable way. A crystal with defects, however does react with light. Fortunately, most silver bromide crystals will have defects which consist of some interstitial (out of order) silver ions displaced in the crystal structure.

The process of the photochemical reaction is not known in exact detail but it is believed that when light strikes a silver bromide crystal, enough energy is available to remove an electron from an occasional bromide ion. The electron produced is able to migrate through the crystal until it comes in contact with an interstitial silver ion. The silver ion takes the electron and becomes silver metal. Silver atoms formed by this mechanism apparently act as a nucleus for the formation of aggregates of 10 to 500 silver atoms, known as latent images because they are too small to be seen by the naked eye.

After exposure, the emulsion is developed. The developer goes to the site of any silver bromide crystal with a latent image and causes all the silver in that particular silver bromide crystal to be reduced to silver metal and deposited on the already existing latent image of silver metal. This causes a worm-like grain

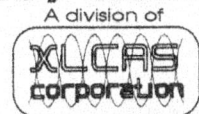

of silver metal to form which is limited in size by the amount of silver available in the silver bromide crystal. This growth is considerable, amplifying the size of the latent image silver metal by a factor on the order of 106· If the developer is left in contact with the emulsion long enough it eventually attacks all the silver in the emulsion. The speed of development is slow enough, though, that you can use a timer to take the emulsion from the developer just after the latent image, but not the unexposed silver bromide crystals, have been developed. At this point the developer has converted silver ions to silver metal if and only if they belong to a silver bromide crystal that was exposed to light. We now place the emulsion in a fixer solution which attacks all silver bromide crystals that were not exposed to light. The fixer makes these silver bromide ions soluble and removes them from the emulsion.

The result is an emulsion with black spots where light has struck, and clear spots where no light struck.

Ideal Silver Halide Emulsion: An ideal silver halide emulsion depends somewhat on its use but there are three main factors to consider in any emulsion: thickness of emulsion, grain size of silver halide crystals, and sensitivity (or density of silver halide crystals) in the emulsion. We can generally state the following:

Thickness of emulsion: It is generally agreed that emulsions of more than 10μm are neither practical or theoretically necessary to produce most volume holograms. It is pointed out that thicknesses above this size cause problems in development.

Grain size of crystals: Grain size becomes an important issue in holography because we are recording fringe patterns that are wavelengths apart. Too large a grain size may create excessive scatter which may fog or destroy your hologram and too small a grain size makes the emulsion have no usable sensitivity. It is generally agreed that the most ideal grain size is in the range of .01μm to .035μm.

Sensitivity of emulsion: The ideal exposure would probably be 100 - 300μJ/cm2 to give a useful density (D=2-3). If exposures are much longer than this, the main attraction of silver halide emulsion, its speed, comes into question and other emulsions become more attractive.

Some of the more common brands of silver halide emulsions used in holography are listed below and after the chart we present some of the recommenda-

PLATES	FILM	SPECTRAL SENSITIVITY	EMULSION THICKNESS		GRAIN SIZE
			PLATE	FILM	
Agfa-Gevaert					
10E75	10E75	Red	7μm	5μm	.090μm
8E56	8E56	Blue-Green	7μm	5μm	.044μm
8E75HD	8E75HD	Red	7μm	5μm	.044μm
Eastman Kodak					
649F	649F	All	15μm	6μm	.060μm
120-02	SO-173	Red	6μm	6μm	.050μm
	SO-253	Red	9μm	9μm	.070μm
	SO-424	Blue-Green	7μm	3μm	.065μm
Ilford					
SP673	SP673	Blue/Red	6μm	6μm	.040μm
SP675T	SP672T	Blue/Green	6μm	6μm	.040μm
HOTECR	HOTECR	Red	6μm	6μm	

Note: Soviet and Bulgarian emulsions exist with a grain size of .015μm but they are not readily available to the market.

Common silver halide emulsions
Fig. 3.2

tions by the manufacturers.

Notes From Manufacturers For Silver Halide Emulsions

1) Ilford Recommendations

Processing: "Careful attention should be given to proper processing techniques, regardless of the material to be processed.

When preparing processing solutions, ensure that mixing vessels and processing dishes have been thoroughly cleaned before use. Discard processing solutions at the end of their working life. Do not attempt to economize by keeping solutions from one working period to the next if there is any risk that the solutions will not perform in the recommended way upon reuse. Mix fresh chemicals if there is doubt about the condition of any processing solution.

In general, it is satisfactory to mix chemicals with ordinary tap water. Care should be taken with bleach baths and the final rinse solution: de-ionized or distilled water is strongly recommended for making up these solutions to prevent the formation of precipitates.

For highest quality holograms, it is important to keep all processing solutions, including the wash water, at about the same temperature (+/- 2 C or 5F). In this way, image distortion due to random shifts in the emulsion layer, as the gelatin alternately swells and shrinks during processing, will be minimized.

While exposure conditions can be varied to achieve good holographic performance over a wide range of development times and temperatures, it is generally advantageous to standardize on processing parameters such as time, temperature and agitation, and thereby minimize the effects of processing variability. In the same way, while it may be tempting to 'develop by inspection' to obtain the required result, for consistently good results, it is always best to process for the standard times. The 'hologram should then be examined after processing and the appropriate revised exposure or development time determined to produce a satisfactory hologram.

Stop Bath: Stop bath is recommended between the development and bleach stages, to prevent premature exhaustion of the bleach bath. The stop bath solution should be replaced whenever developer is replaced.

Treatment with potassium iodide: Phase holograms, consisting of silver halide, are inherently susceptible to photo reduction (printout). Amplitude holograms, where the fringes consist exclusively of metallic silver, are not. The light stability of phase holograms can be significantly improved by the use of a bath of potassium iodide. This should be employed after the hologram has been washed following the bleach bath.

Standard developers: When working with a standard developer such as Kodak D-19 or Tetenal Dolumol, it is important to bleach the film using a ferric nitrate bleach. The formula for this is given below.

Ferric nitrate: 100g
Potassium bromide: 30g
Water to make: 1 liter

ILford SP679C holographic bleach

Recommended for use with Ilford developer; should be diluted 1 + 4 with water to give a working strength solution.

SP679C bleach has a good processing capacity and for normal work a large number of holograms may be processed before there will be a noticeable loss of quality. For the best working conditions, no more than 20 8" x 10" holograms should be processed in each liter of working-strength bleach.

2) Agfa-Gevaert

Transmission holography: "While theoretically the maximum diffraction efficiency that may be achieved with an amplitude hologram will be 6.25% at the utmost, theoretically 100% may be achieved with phase holograms.

Technical literature describes a great number of processing systems that, with the highest possible diffraction efficiency enable the noise to be kept as low as possible.

The exposure doses that are required for making a good phase hologram amount to -50 μ J/cm2 for the emulsion 8 E 56 HD, and to -25 μ J/cm2 for the emulsion 8 E 75 HD, as a relatively high density (between D=1.5 and D=2.5) proves to be necessary. As the HOLOTEST 10 E emulsions after bleaching produce more noise than the 8 E types, they are not recommended for phase holography."

Reflection holography: Though theoretically emulsion layers of a thickness of 20 μm are necessary for reflection holography so as to achieve reflection holograms of top quality, it is still recommended to use the materials 8 E 56 HD and 8 E 75 HD with thickness of the emulsion layer of 7 μm, as with these materials the distortion of the Bragg planes after processing will be smaller. This is why it is also possible to achieve high-quality reflection holograms on thinner emulsion layers.

3) Kodak Silver Halide Recommendations:

"When selecting a silver halide material, it is important to match peak sensitivity as closely as possible to the wavelength emission of the laser being used. Early attempts to produce holograms also demonstrated the need for emulsions with lowest possible graininess characteristics, and highest possible resolution.

"Another factor to consider is latent image fading, a propensity for microfine-grain emulsions to lose density after exposure; that is, for a given exposure, developed density is likely to decrease with increased time between exposure and processing. Storage conditions between exposure and processing can also contribute to the rate of fading. When conditions do not permit prompt processing, adoption of a uniform protocol for exposure and processing is necessary, especially when each hologram in a series must have the same density."

Dichromated Gelatin:

The dichromate hologram is very popular in artistic holography. It is usually sold in a sealed glass sandwich. The most common DCG hologram sizes range from small, inch-and-a-half circles, up to standard sizes of 4 x 5 inches.

Availability: DCG holograms are produced by holographers who make the batches of dichromate gelatin on site because of its inherent instability. For obvious trade reasons, no company would want to divulge its own formula for making dichromates.

Advantages: Dichromate holograms (also known as DCG, or dichromated gelatin) are well liked by holographers because they give a very sharp, almost grainless image and they reflect light extremely well. Due to this large refractive index, dichromate holograms are among the brightest holograms available. The DCG hologram's images are usually close to the hologram surface and display the image in almost any available light, including fluorescent light. The DCG recording material holds metallic colors very well, e.g. images of the inside of a watch look almost identical to the original watch.

Drawbacks: The dichromate chemicals do not have a long shelf life and the exposures are not as fast as silver halide due to its complex chemistry. DCG is also very sensitive to the environment: the image disappears if it comes in contact with moisture.

Commercial Manufacturing Costs For Dichromates: DCG holograms can be made in very short runs. Orders for as small a run as 1 to 100 are done commercially. Some short runs are done by shooting the final DCG hologram directly from the object, thus skipping any mastering. This shortcut is used commercially because it allows businesses to test-market their product in short runs without the expensive mastering costs involved in a mass manufacturing run. Serious commercial runs, however, go through the whole mastering expense.

DCG Chemistry: The dichromated gelatin emulsion takes a completely different approach to image recording. In this process the substance you suspend in the gelatin reacts, upon exposure to light, with the gelatin itself. The exact cp.emical constituents of gelatin cannot be listed because gelatin is a biological product made by boiling animal parts such as hooves, bone, etc. and processing it chemically. It has many uses, such as medicine, food (jelly, etc.) and in photography.

In sum, gelatin is mass-produced for a variety of uses, has a molecular structure that cannot be fixed with any certainty, and its properties vary from batch to batch. This makes scientific control difficult.

Dca Exposure Photochemistry: It was discovered that gelatin with a small amount of dichromate, such as $(NH_4)_2Cr_2O_7$, becomes progressively harder upon exposure to light. After investigation, it was found that the Cr_2 element of the molecule absorbs light to become the ion Cr^3+. This ion then stimulates crosslinking of molecular chains in the gelatin. The crosslinked parts of the gelatin then have a different refractive index than the unlinked gelatin. This difference produces the holographic fringes.

Photopolymer

Photopolymer is a relatively recent recording material. It can be made in three different formats:

1) A dry film composition containing photinitiator system, monomer, and a polymeric film-forming binder.

2) A liquid coating containing photinitiator and monomers.

3) A dry film composition containing crosslinkable polymer and a photosensitizer or initiator. According to one of the manufacturers of photopolymer, DuPont, the first format is the only one which has enjoyed commercial success (as of 1990).

Availability: There are two major manufacturers of photopolymer, Polaroid and DuPont.

DuPont's Omnidex 352 film is the only photopolymer that is available for sale to the public at present; however, several others are in R&D and one, HRF 700 is expected to be available in mid-1991. For information on theses, contact E.I. DuPont in Wilmington, Delaware at (302) 695-4893. DuPont also plans to make available a limited number of replication equipment systems for their film sometime in 1991. This way holographers can mass produce photopolymer holograms.

Polaroid's photopolymer is not commercially available--if you want to produce a hologram with Polaroid's material, you should contact their "Mirage" Holography Group in Cambridge, Massachusetts at (800) 237-5519.

Advantages: Photopolymer has some of the same attributes that dichromated gelatin has. It has a very sharp, almost grain less clarity and reflects light extremely well. The famous "skull" hologram on our cover, made by Polaroid, is an example of what can be done with photopolymer.

Disadvantages: Previously, photopolymers didn't last long: Polaroid and DuPont claim that their newer materials have a shelf life of up to one year. However, complex processing is involved to produce a hologram.

Commercial Manufacturing Costs For Photopolymer: Due to the model and mastering expense, it is not cost-effective to do an extremely short run of photopolymer holograms, but the price is competitive with dichromate and silver halide in long runs.

Polaroid's Mirage holograms are furnished to the customer with either a black or clear backing. The clear backing allows you to place the Mirage hologram right over a printed area allowing type or graphics to show through. A black backing gives an image high brightness and contrast. On average it costs about $0.10 per square inch (2.54 cm) for a minimum volume production run. The price may drop as low as $0.05 per square inch for a large volume application. The master may cost from $8,000 to $15,000.

Polaroid offers a library of stock holographic images and prices vary depending on size and quantity. It takes between 8 and 10 weeks to produce a Mirage hologram, from artwork to shipping, and the sizes range from .5 x .5 inches to 10 x 14.5 inches.

Photopolymer chemistry: DuPont's photopolymer is composed of the following materials: photosensitizing dye (0.1-0.2%); initiator (1-3%); chain transfer agent (2-3%); plasticizer (0-15%); acrylic monomer (28-46%); and polymeric binder (45-65%). Except for dyes and initiators, all ingredients are obtained from regular commercial sources and used as received.

DuPont films are manufactured on a clean room web coater with on-line filtration, static control and web cleaning systems. The resulting films consist of a three-layer laminate with the photopolymer layer (5-50 microns thick) sandwiched between a 2-mil Mylar base and a 1-mil Mylar cover sheet.

Photopolymer Exposure Photochemistry: The following is part of a report by SPIE (Practical Holography IV, 1212-04, Jan 14-19, 1990) on tests done using DuPont's photopolymer:

In these tests, the photinitiator system used consists of three separate components: a visible-light absorbing dye; a UV-light absorbing hexaarylbiimidazole (HABI) initiator, and a chain transfer agent, 2-mercaptobenzoxazole (MBa). The dye makes the composition sensitive to blue-green light (400-560 nm with ideal wavelength of 480 nm).

Prior to exposure, photopolymers are highly plasticized polymeric binder compositions in which monomer and other low molecular weight components act as plasticizing agents. Because of this the initiator, chain transfer agent and monomer diffuse freely through the composition during the early stages of photopolymerization.

During the imaging, electronically excited dye forms and sensitizes decomposition of HABI to triarylimidazoyl radicals; these oxidize the chain transfer agent to thiol radicals. Addition of thiol radicals to acrylic monomer initiates polymerization.

This chemical reaction causes the composition, once highly viscous, to gel and harden. The monomer is converted to photopolymer, and fresh monomer diffuses in from neighboring dark regions, thus setting up concentration and density gradients that result in refractive index modulation.

At this point, the hologram image consists of photopolymer-rich regions that monomer diffused into and binder-rich regions that it diffused away from, probably with some residual unreacted monomer distributed throughout.

It was found that the presence of plasticizers allows more efficient diffusion to occur at the later stages of polymerization, and it is believed that this is the principle reason for higher diffraction efficiency. It was also found that index modulation is significantly enhanced with little change in playback wavelength by simply heating the hologram in a conventional oven while sealed between glass and/or Mylar.

Notes From Manufacturers For Photopolymer Emulsions

1) DuPont

Recommendations from DuPont for producing an OmniDex 352 reflection hologram:

1. Remove cover sheet and laminate photopolymer film directly down onto glass surface of holographic reflection master using a soft rubber roller.

2. Holographically expose the blue-green sensitive film using the proper reference beam for the master.

3. Use ultraviolet light to cure the film.

4. Delaminate the film from the master.

5. Heat treat film in a vented forced-air oven at 100 degrees C for 1 hour.

2) Polaroid:

Specifications for masters used to produce Polaroid Mirage Holograms: An H1 is a transmission laser viewable hologram of an object or stereogram composite. An H2 is an image plane reflection hologram using the above H1. Both H1 and H2 are required for copying into a photopolymer. Generally, the H2 is used to determine image plane and object framing. However, a direct contact copy of this H2 is often made for proofing.

H1 Specifications:

1. HI size: 8" x 10" or 30 x 40 cm glass plate.

2. Object size=hologram size. If object is larger than 5" x 5" contact technical representative.

3. Object position- 7"-11" from proposed image plane to H1, centered on H1, image plane must be parallel to H1.

4. Depth: 3/8", with most important information on the image plane, gives fairly sharp results in most lighting conditions.

5. Efficiency value (diffracted laser power/incident laser power) must be more than 0.35. Examples of acceptable efficiencies found in 8" x 10" H1s:

Incident laser power: 256 μw/cm2
Diffracted laser power: 104 μw/cm2
Efficiency: 0.4

Incident laser power: 256 uw/cm2
Diffracted laser power: 200uw/cm2
Efficiency: 0.78

6. Image contrast: The diffacted laser power of the brightest spot measured in the real image should be approximately 200 or 300 times the diffracted laser power of a non-image point at the image plane. Do not use a mirror-like reflection as your brightest spot for determination of contrast or efficiency.

7. Collimation of the reference: Highly recommended and mandatory in the majority of cases.

8. Reconstruction at 647.1 nm, 45 degrees from bottom, landscape format (longest side of plate parallel to horizon).

H-2 Specifications:

1. Make it look like the hologram you want: proper framing, proper image plane, illuminated at 45 degrees from above, on a glass plate or film.

2. Reconstruction: Approximately 45 degrees from above, 647.1 nm, 632.8 nm.

Photoresist

This recording material is used exclusively for preparing embossed holograms. Photoresist is used extensively in the computer industry to make circuit boards. A similar procedure is used in holography to produce a stamping die that is then used to emboss the holographic fringes in plastic to produce embossed holograms.

Photoresist is a photosensitive recording material that records the hologram unlike the other three recording materials. In a conventional silver halide hologram the interference patterns that create the holographic image are contained within the light-sensitive emulsion; in a photoresist hologram, the pattern is in relief.

The relief allows us to make a replica of the surface pattern in hard metal, and stamp out holograms in hot plastic in much the same way as we make audio discs. Embossed holograms can be delivered to you from the manufacturers as adhesive stickers or as hot stamping foil for stamping on books or magazines. Stickers can also be applied to books and magazines by machines.

Availability: Shipley (tel. 215-820-9777) sells bottled liquid photoresist if you want to coat your own plates. Towne Laboratories (tel. 908-722-9500) and Resist Masters, Inc. (tel. 313-481-1980) sell pre-coated plates.

Advantages: Photoresist holograms are widely used on bank cards, magazine covers, food containers, and many other products. Their primary uses are for identification (security) and as point-of-purchase attention-getters. Their advantage is that they can be mass-manufactured and machine-applied to products by the millions. The ability, to apply large quantities of holograms at a comparatively low unit cost greatly

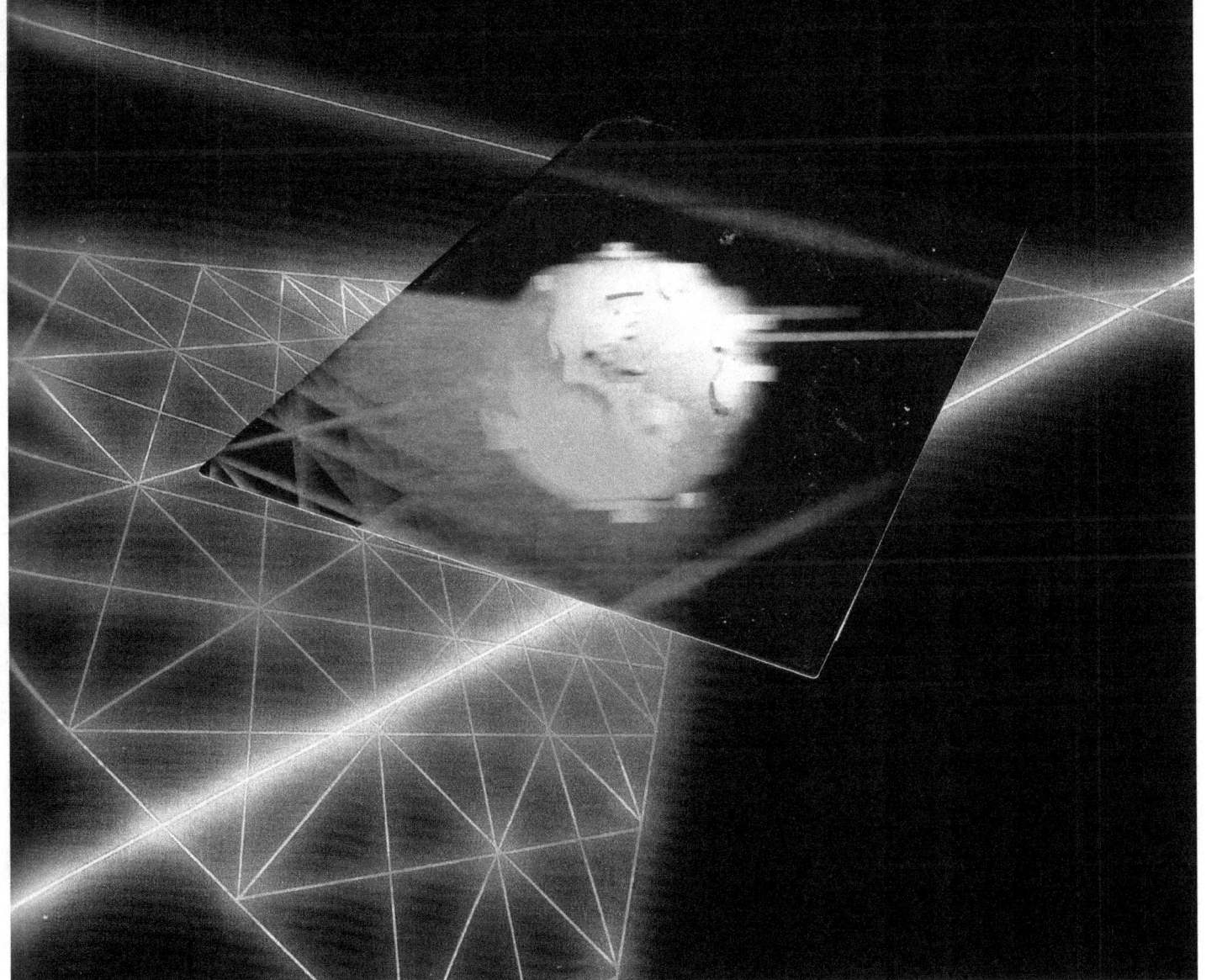

HOLOGRAPHY

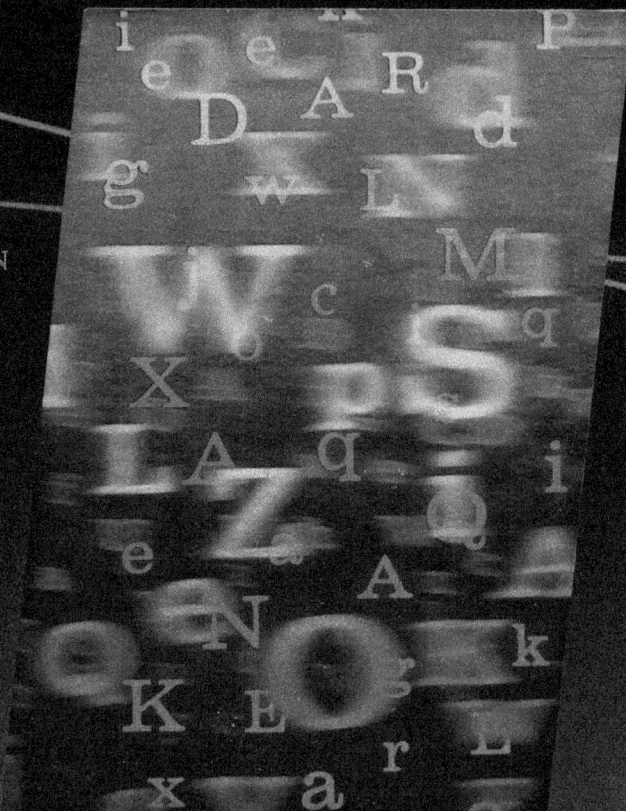

Harness the power of American Bank Note Holographics' "laser perfect" holography

American Bank Note Holographics, Inc. is the world's most experienced and innovative holographer. Our mass-produced holograms first appeared on MasterCard® and VISA® credit cards as an anti-counterfeiting device—a logical outgrowth of our 200 year history in the security printing business. For almost two centuries, we have printed stocks, bonds, foreign currency, stamps, travelers checks, and other documents of value—using delicate engravings, fluorescent inks, special paper, and numerous other security devices to make the counterfeiter's job more difficult.

Today, the miracle of holography has been performed over 2 billion times by American Bank Note Holographics on a wide range of products. Discriminating companies in all disciplines of industry and business have discovered the many applications and benefits of holography. ABNH's eagle hologram appeared on the cover of the March 1984 National Geographic Magazine, signaling the "coming of age" of the hologram as a 21st century commercial graphic medium. Today's holography is used in America's aeronautical and space programs, in medicine, for satellite communications, for protection of valuable documents against duplication, and for high profile product and packaging enhancement and recognition. Holograms can be used in thousands of ways on thousands of products, and ABNH is at the forefront of inventive holographic ideas in advertising, publishing, direct marketing, product design, promotion, and security/anti-counterfeiting applications.

Holograms used commercially have been proven to:

- set all-time records in reader interest and recall—crucial first steps in the selling process—when used in direct response advertisements;

- increase sales for a publishing company by 40% when shown on book covers;

- increase sales of a brand name cereal by over 300% when used as a primary packaging label; and

- all but eliminate counterfeit bank cards, and reduce the banks' losses by $50 million or more a year.

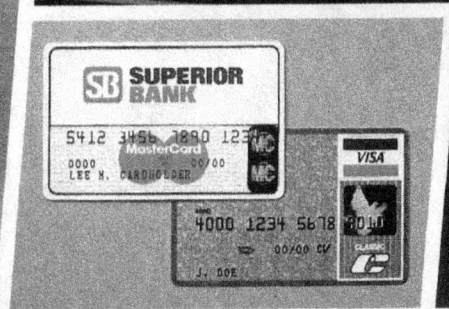

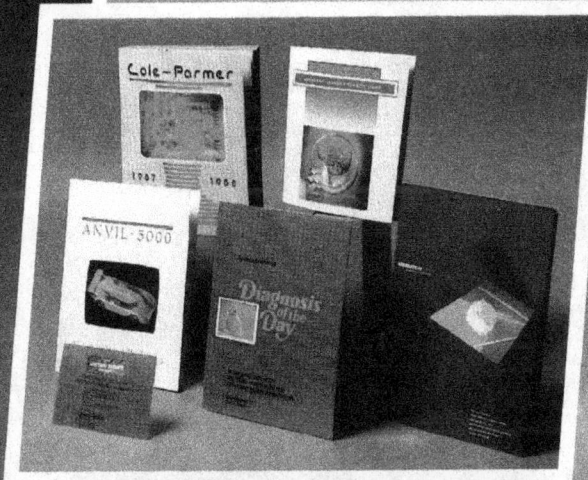

ABNH scientists, engineers, technicians and designers are eager to work together with you on holographic projects for your organization. Our staff is comprised of the largest private multi-disciplined holographic organization in the free world. Our laboratories and equipment are the most extensive and up-to-date available in holography. Research and development are serious, full-time occupations at American Bank Note Holographics, Inc.

Call us today, and let us show you how to put our "laser perfect" technology to work for you!

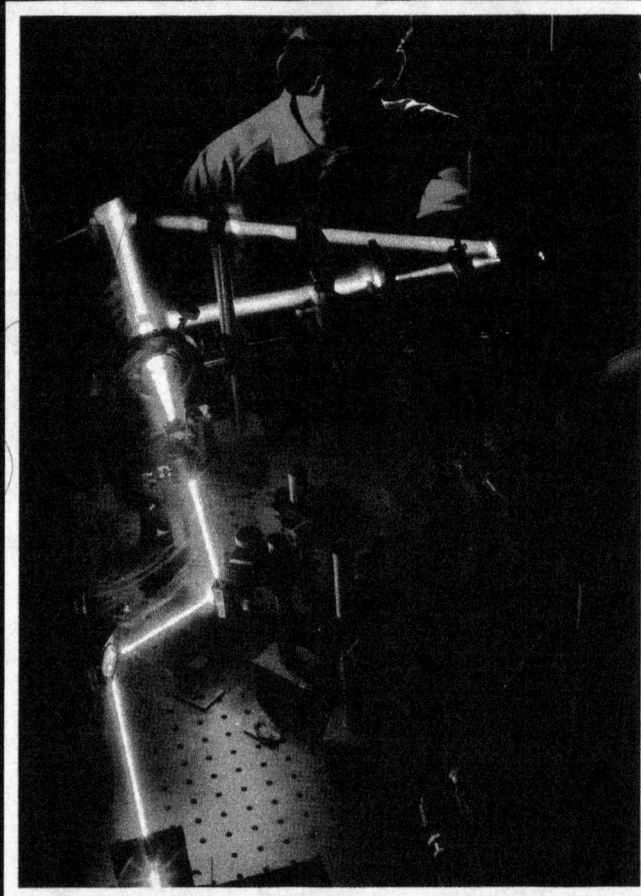

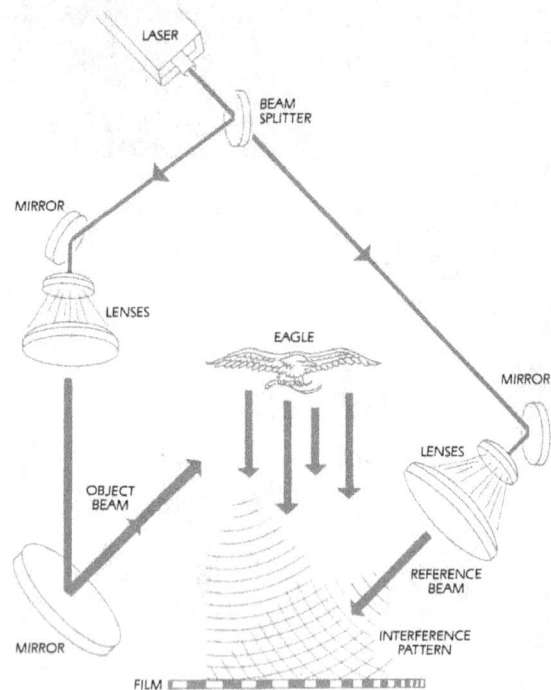

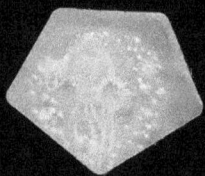

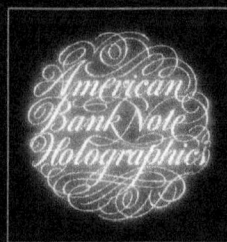

favors the embossed hologram over DCG's and silver alides at present.

Commercial Manufacturing Costs Of Photoresist Holograms: Once again, the model and mastering fees make this method appropriate for manufacturing in large quantities. Embossed holograms are literally stamped into plastic on a modified commercial printing press and consequently millions of holograms can be produced very quickly. To reiterate, the unit cost of embossed holograms is far lower than other types of recording materials if made in large quantities.

The model and mastering fee for the typical 5 x 7 inch hologram is about $5,500. Stock images can be purchased from major manufacturers of embossed holograms, which eliminates the model and mastering fee.

To give you an idea of production prices, one major manufacturer of embossed holograms gives this scenario: If the mastering cost is paid and you plan a 2 x 2 inch embossed hologram to be delivered as stickers in a quantity between 5,000 and 200,000, your price ranges between $0.04 to $0.02 per hologram.

Photoresist Chemistry: According to Shipley, their photoresist "consists of, essentially, three main components: 1. Etch-resistant, film-forming, base-soluble polymer; 2. a photoactive compound which confers dissolution-inhibiting properties to the unexposed resist film and dissolution rate enhancing characteristics to the exposed resist; 3. a carrier solvent employed as a casting agent for the resist material. Minor additives, such as leveling agents, absorption-enhancing dyes and adhesion promoters, may also constitute part of the photopolymer resist formulation."

The photoresist normally is packaged in a precleaned, chemically-inert bottle which is opaque to white light, thus ensuring photochemical stability. Refrigerated storage is recommended; absolutely no greater than room temperature (ambient) conditions (45-65 degrees Fahrenheit) are allowable. Depending on the lithographic requirements of the customer, Shipley quotes 6 months to 1 year shelf life.

The photoresist should be applied by spin coat, spray or dip, usually by spinning. Some people coat their own plates, although the emulsion is available on plates from some manufacturers. If you do coat your own plates, it is recommended to have an anti-halation backing of some kind on your plate. The layer of emulsion is around 1 to 211m thick.

It should then be soft-baked at 90-100 degrees Celsius for 30 minutes in a convection oven. After exposure, it is developed for 60 seconds in an immersion tank, and rinsed with water.

The most widely used photoresist is Shipley AZ-1350. The sensitivity of this emulsion is greatest in the ultraviolet and drops off rapidly toward the blue end of the spectrum. Therefore holograms are best recorded with either an Ar+ laser at 458nm or with a HeCd laser at 442 nm. A typical exposure on factory supplied emulsion is 250 mJ/cm2 at 458 nm. A commonly used developer is Shipley AZ-303 A. It is usually diluted 4:1 or more. The exposure can also be controlled by pre- or post- exposing the resist to an incoherent source such as a fluorescent lamp.

Photoresist Exposure Tips From Shipley: The following are a few points to remember:

1) The characteristics necessary in the H-1, relative to exposure of the photopolymer resist, is that the master must be opaque to the exposing wavelength of the resist in the unexposed areas; i.e., the master must absorb or reflect light, in those areas where you don't want the photopolymer exposed to provide the necessary edge sharpness for high fidelity pattern transfer.

2) Depending on the isomer type of photoactive compound (PAC), the spectral sensitivity of the resist material may occur at either 345-375nm for the 214 PAC or 345-440nm for the 215 PAC. Recommended exposures depend on the type of images one chooses to implant into the resist film; grating patterns usually require less energy than interconnect holes, isolated patterns tend to need more energy than patterns in neighboring positions.

3) At 436 nm (G-line) wavelengths, reciprocity failure is not normally observed for most positive-tone resists. However, at 365 nm (I-line) higher intensity exposure sources do tend to be more efficient in that there is less chance for competing chemical reactions to occur than for the longer exposure times required for less intense light sources. Therefore, depending on exposing wavelengths, reciprocity failure mayor may not occur.

4) Dose is much more important than time. However, with very high power and short time (i.e. less than 100 m sec) it is possible to "pop" the resist when N_2 is evolved too quickly.

REFERENCES AND READING:

Silver Halide Emulsion

Phillips, N.J., (1985). The role of silver halide materials in the formation of holographic images, Proceedings of the SPIE, 532, 29-38.

Ilford, Inc., West 70 Century Road, P.O. Box 288, Paramus, New Jersey, 07653

Agfa Gevaert, NDT Technical Information, NDT & Scientific Systems, 100 Challenger Road, Ridgefield Park, New Jersey 07660

Eastman Kodak Company, 343 State Street, Rochester, New York 14650

Photopolymer

Hay, W.C. & Gurenther, B.D. (1988). Characterization of Polaroid's DMP-128 hoographic recording photopolymer, Proceedings of the SPIE,883, 102-105.

Ingwall, R.T., Stuck, A. & Vetterling, W.T. (1986). Proceedings of the SPIE, 615, 81-7.

Ingwall, R.T. & Fielding, H.L. (1985), Optical Engineering, 24,808.

Ingwall, R.T. & Troll, M. (1988). The mechanism of hologram formation in DMP-l28 photo polymer, Proceedings of the SPIE, 883, 94-10l.

Smothers, Monroe et al. (1990). Photopolymers for holography, SPIE/OE Lase Proceedings, 1212-03.

Weber, Smothers et al. (1990). Recording in DuPont's new photopolymer material, SPIE/OE Lase Proceedings, 1212-04.

Emulsion And Exposure

Hariharan, P. (1987). Optical Holography, Cambridge University Press, 88.

Larimore, L., (1965). Introduction to Photographic Principles, Dover, 10 1-2.

Photoresist

Bartolini, R.A. (1974). Characteristics of relief phase holograms recorded in photoresists. Applied Optics, 13, 129-39.

Bartolini, R.A. (1972). Improved development for holograms recorded in photoresist. Applied Optics, 11, 1275-6.

Burns, Joseph R.(1985). Large-Format Embossed Holograms, SPIE Volume 523, Applications of Holography.

Hariharan, P. (1987). Optical Holography, Cambridge University Press, 106.

Saxby, G., (1988). Practical Holography, Prentice Hall, 285.

Norman, S.L. & Singh, M.P. (1975). Spectral sensitivity and linearity of Shipley AZ-1350J photoresist. Applied Optics, 14,818-20.

Shipley Co., Inc.; Santa Clara, Calif.

Livanos, A.C., Katzir, A., Shellan, J.B. & Yariv, A. (1977). Linearity and enhanced sensitivity of the Shipley AZ-1350 B photoresist. Applied Optics, 16, 1633-5.

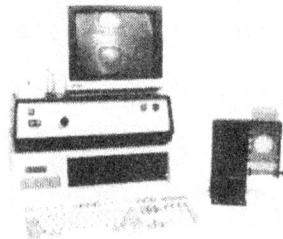

4

Lasers

Since the laser is at the heart of making holograms, a basic understanding of laser principles and operation is essential for creative applications as well as successbuying. In this chapter, we will first trace some of the important developments that led to the modern laser, consider what lasers do and how they work, discuss the features oflasers important to holography and finally, list costs and suppliers of various lasers.

A Brief Historical Background

Today, the laser's widespread applications range from playing audio compact discs to pinpointing targets for bombs. Lasers have become indispensable tools for such diverse fields as construction, pollution detection, medical surgery and of course, holography.

Few people realize that this technological triumph started as a rather esoteric paper written by Albert Einstein in 1917. In his paper, Einstein claimed that atoms could generate light and other forms of radiation by a mechanism he called stimulated emission. Previously it was thought that atoms and molecules could generate light only by a mechanism called spontaneous emission.

To understand Einstein's distinction we need to define two important ideas that have shaped much of modern physics: the idea of the photon and the concept of distinct energy levels in the atom. The photon is an irreducible unit of light energy. Light beams are thought to be composed of a finite number of photons. The theory of the photon shattered the belief that light could be infinitely subdivided into smaller and smaller units of energy. The theory of distinct energy levels in the atom, or quantum levels, upset a similar belief, the belief that any amount of energy could be injected into an atom. In fact, as quantum theory suggests, the energy of atoms changes in "quantum jumps" from one energy level to another. An atom cannot have an amount of energy between two levels.

If we put these two ideas side by side, we reach an interesting conclusion: photons that travel past atoms may be absorbed by the atom if the photon energy equals the transition energy to excite the atom to a higher quantum level. If a photon passes an atom with one half the energy needed to put the atom to the next energy level, the photon cannot be absorbed by the atom. Only if the transition energy and that of the photon are the same will absorption occur. Conversely, if an atom moves from a higher to a lower energy level, then a photon will be emitted. The energy of this photon will, again, equal the atom's transition energy.

In nature, atoms are frequently excited to energy levels above their lowest state, ground state. However, in a tiny fraction of a second these atoms emit a photon and return to their ground states. This emission occurs without any outside influence and is therefore called spontaneous emission.

Einstein suggested that light might also be produced by a process he called stimulated emission. Previously, it was thought that if a photon passed by an excited atom, the photon might be absorbed if its energy matched the atom's transition energy or the photon simply would pass the atom.

Stimulated emission was Einstein's alternative to this thinking: the photon passing by the excited atom would "stimulate" the emission of a photon from the atom. Thus two photons would leave the atom simultaneously.

The main point of Einstein's argument was that two processes, not one, explain the generation of light in nature. But this theoretical point raised a technological challenge: building a source that would generate radiation such as visible light primarily by the mechanism of stimulated emission rather than spontaneous emission.

The problem was in establishing a large number of excited atoms since atoms tend to remain in their ground state or return very quickly if they are excited. The first success came in 1953 when Charles H. Townes created the maser (an acronym for Microwave Amplification by the Stimulated Emission of Radiation). Townes found a way to separate excited ammonia atoms from ammonia atoms in the ground state. The maser generated microwaves, a form of radiation less energetic than light, by stimulated emission.

The difficulties of generating visible light were only overcome about seven years later when Theodore Maiman made his first laser (an acronym for Light Amplification by the Stimulated Emission of Radiation) by putting mirrors on the two ends of a synthetic ruby and illuminating it with a powerful flashlamp. Within a few years, Emmett Leith and Juris Upatienks made the first laser 3-dimensional holograms.

Laser research over the past 30 years has multiplied the different types of lasers available and their range of features. Most lasers fall into three classes: (1) solid state lasers, like Maiman's ruby laser, (2) semiconductor lasers, which are typically found in CD players and (3) gas lasers which are the type commonly used in holography. This classification excludes some of the more exotic lasers like the chemical dye lasers.

Lasers also generate electromagnetic waves of many different types, one of which is visible light (actually, the laser by original definition produces only visible light, but since laser principles are used in other devices like the maser, we will loosely call all these devices "lasers"). Electromagnetic waves generated by lasers are conceptualized as "ripples" in the electromagnetic field that permeates all of space. These waves are characterized by their wavelength, which is the distance between successive wave crests, and by their amplitude, which is half the height between a wave crest and a wave trough. The wavelengths generated by lasers range from the long wavelength microwaves to the short wavelength X-rays. Laser power, which is proportional to the amplitude squared, can vary from a fraction of a milliwatt to trillions of watts.

Common to this diverse array of lasers are a few basic principles which we consider next.

How a Laser Works

The road between Einstein's paper on stimulated emission and Theodore Maiman's ruby laser was filled with technical obstacles. The major problem was in energizing the atoms so that a majority were in an excited state. Remember that stimulated emission will occur only when a photon passes an excited atom and the excited atom's transition energy from excited state to ground state equals the photon's energy. The problem is that excited atoms decay to their ground states in extremely short times. The net result is that a fraction of atoms are excited at any given time, leaving little chance for stimulated emission to occur. Only if there is a population inversion, or a predominance of excited atoms with transition energies matching the photon energy, will laser action occur.

To see how population inversion is achieved, let us look at the Helium-Neon laser, the laser most commonly used by beginning holographers. Figure 4.1 illustrates three of the important energy levels in neon. If a neon atom is excited to the second energy

No More Mode-Hops

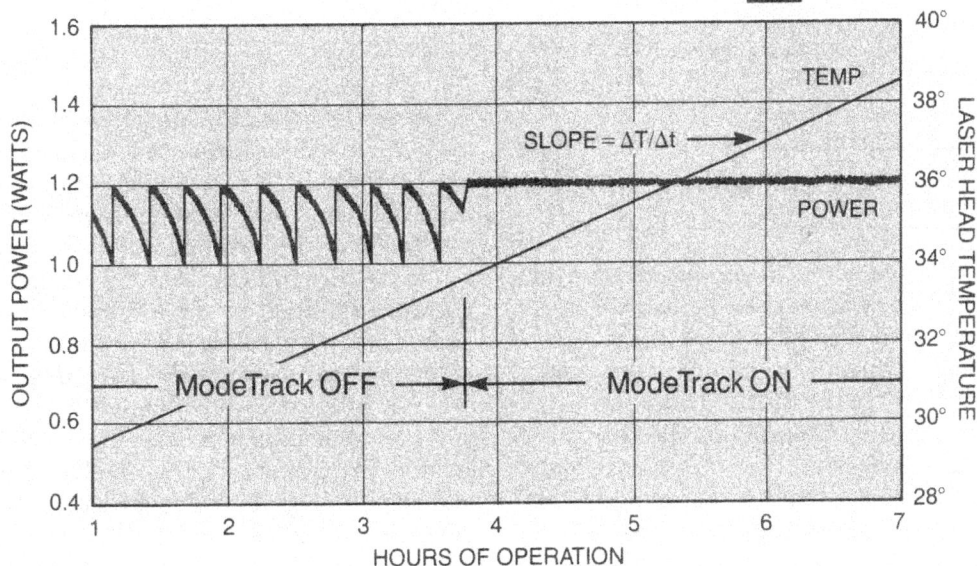

Stable Operation Guaranteed

The Innova® 300 Ion Laser System with ModeTrack™ eliminates mode-hops. It's that simple. Now you can make long exposure holograms on the first attempt. And you'll experience a lower reject rate in your high volume holographic applications.

Changes in ambient air or cooling water temperature are the main causes of mode-hops in ion lasers. ModeTrack utilizes the Innova 300's system CPU to continuously monitor the single-frequency stability and controls the etalon to prevent mode-hops.

The Innova 300 also features PowerTrack™ an actively stabilized optical cavity. PowerTrack automatically provides both optimum output power and unmatched long-term stability. PowerTrack—coupled with ModeTrack—provides you with hours of hands-off operation, without a single mode-hop.

No other laser offers either of these capabilities. No other laser can increase your productivity and quality like the Innova 300.

Innova 300. The Laser for Holography.

To learn more about Innova 300 Mode-Track, call (800) 527-3786, ext. 336; U.K. (0223) 420501; W. Germany (06074) 9140; Japan (03) 648-8115, or call your local Coherent representative.

COHERENT LASER PRODUCTS

✳®COHERENT®

level, it will decay quickly back to the first energy level. Atoms excited to the third energy level decay slowly back to the second and are therefore called metastable. Remember that each time an atom decays to a lower energy level, a photon is emitted. For example, the decays from the third to the second energy level emit photons with a wavelength of 632 nanometers (632 billionths of a meter), which is the characteristic red light of the He-N e laser.

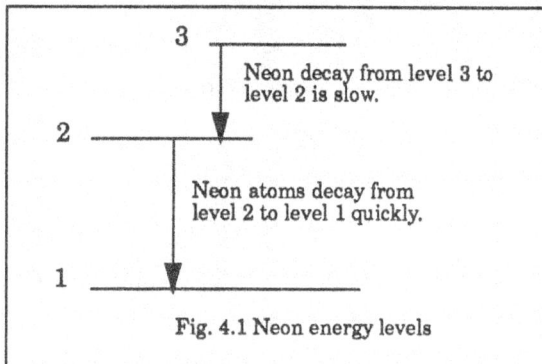

Fig. 4.1 Neon energy levels

This configuration of energy levels suggests a promising possibility: if neon atoms can be excited to the third energy level, then their slow decay to the second energy level together with the rapid decay of atoms in the second energy level to the first or ground level will result in a predominance of atoms in the third level over atoms in the second level. In other words, population inversion occurs between the second and third level, resulting in stimulated emission.

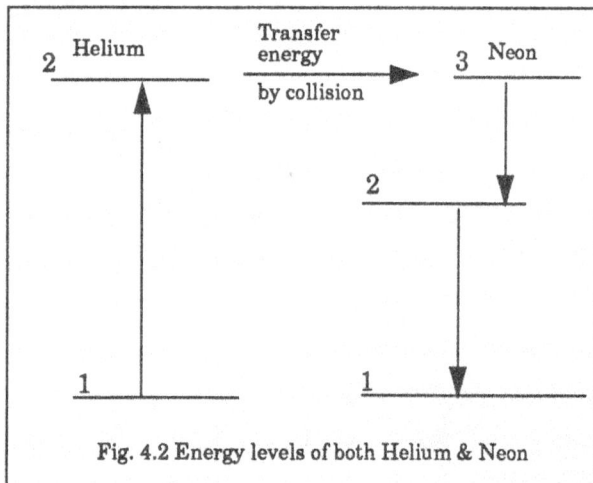

Fig. 4.2 Energy levels of both Helium & Neon

The trick to exciting neon atoms to their third level is illustrated in figure 4.2. It turns out that it takes the same amount of energy to excite helium to its second level as it does to excite neon to the third or metastable level. When excited helium atoms collide with ground state neon atoms, the energy is transferred, exciting the neons to the third energy level. Once neon atoms are excited to the third energy level, population inversion between the third and second occurs.

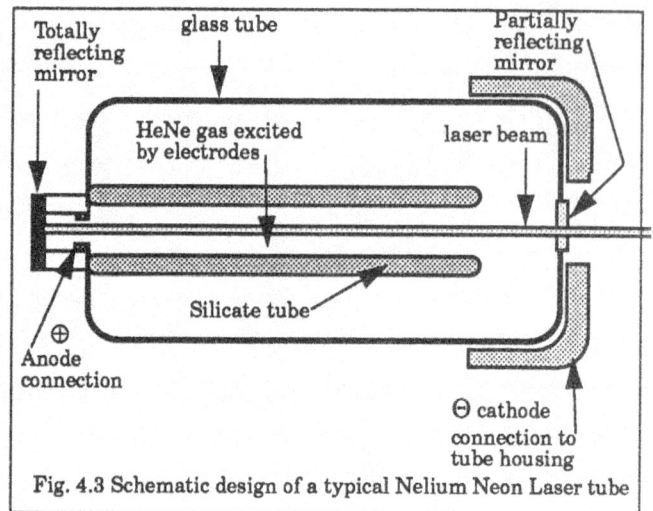

Fig. 4.3 Schematic design of a typical Nelium Neon Laser tube

Figure 4.3 schematically illustrates a typical HeNe laser tube. While the entire glass tube is filled with low pressure helium-neon gas, only the central region near the axis of the tube is excited. The excitation is accomplished by two electrodes that send high-speed electrons through the gas. An anode (positive charge plate) is attached to the left end of the tube and the cathode (negative charge plate) is attached to the tube housing. Charged particles traveling between the anode and the cathode collide with and thereby excite helium atoms mostly in the silicate tube region. These excited helium atoms then collide with neon atoms and create a population inversion between level 3 and level 2 neon atoms.

Decays from the third to the second level set off a chain reaction of stimulated emissions. The cascade of photons is directed through the narrow silicate tube toward the mirrors at the opposite ends of the laser. Most of the light bounces back and forth between the mirrors in order to induce more stimulated emissions. This increases the number of photons in the light beam. A small fraction of this beam is allowed to escape through the partially reflecting mirror on the right side of the tube. This light that leaks out is the characteristic red beam of light generated by a laser.

Now that we have explored basic laser operation, we can tum to the laser features important in holography.

Laser Applications In Holography

The three common lasers used in holography are the Helium-Neon laser, Krypton laser and Argon laser. All three are gas-filled lasers. Unless you are not constrained by cost, the best starter laser is the

Helium-Neon (He-Ne) laser. The major distinction among the three is power. Krypton and Argon lasers typically generate 2-4 W, thousands of times more power than the weaker HeNe lasers. This intensity will allow you to make larger holograms and work with a greater variety of photographic film or emulsions. Expect costs to start in the tens of thousands of dollars.

The best starter laser is the He-Ne laser, which is reliable and has a very long lifetime. A bottom-of-theline model sells for as little as $200-$300. This starter model will probably have a power rating of 0.5 to 1 mW - about 100,000 times less intense than an average light bulb. You will have no problem seeing the characteristic red dot if you shine the laser on a wall, but creating holograms may be difficult because of the limited light intensity. For a few hundred dollars more, a laser in the 2 to 5 m W range can be purchased. Serious holographers may want to purchase HeNe lasers as powerful as 50 m W. Although holograms will be easier to make, you will still be limited by the photographic film or emulsion you can use. Most HeNe holograms will require a silver halide emulsion.

To conserve costs, you may consider purchasing a used laser, perhaps reconditioning some old tubes, or building your own power supply.

In addition to power, other important laser features are continuous-wave/pulsed emission and coherence. The helium neon laser just described operates by continuous emission. The light emitted by the laser is a steady beam of photons that is generated by the continuous build-up of level 3 electrons. Another mode of operation is pulsed emissions. If the atoms in a laser are excited very suddenly and then decay all at once, the result will be a short flash of laser light. This type of laser is more expensive.

An example of a pulsed laser is the Q-switched laser. A typical Q-switch basically controls whether the cavity mirrors in the laser will absorb or transmit light (the mirror switches from a high "Q" or "quality" of transmission to a low quality of transmission). While the mirrors are in the low Q mode, they absorb light and prevent a cascade of stimulated emission by keeping the light waves moving between the two mirrors at a low intensity. This gives the atoms ample time to build a large population inversion. When the mirror switches back to a transmitting mode (high Q), a cascade of stimulated emissions produces a short, intense burst of light.

Holograms can be made either with pulsed or con-

tinuous-wave lasers. Pulsed lasers have the advantage of freezing motion like a high-speed camera shutter. The exposure time is merely the time it takes for the light pulse to sweep across the object. Continuous lasers have lower light intensities that usually require a longer exposure time for the image to form. Therefore high-speed holography is not possible with a continuous laser system. Furthermore, the holography must be done on a vibration-free table because any motion would blur the hologram. Most holograms are made with continuous rather than pulsed lasers simply because they are cheaper.

For both continuous and pulsed lasers, coherence is the feature most important in creating a hologram. Two waves are coherent if they move in phase with each other, meaning the crests and the troughs of the two waves continue to line up as the waves move. Laser light is coherent because the waves generated by all the different atoms are in phase or very nearly in phase. This is one of the basic properties of stimulated emission. Coherence is what makes threedimensional holography possible. Ordinary twodimensional pictures merely capture light intensity. This intensity is represented by the relative brightness and darkness on a photograph.

Holograms actually capture the interference pattern between the light from the object and a reference beam of light (intensity and phase). Thus when the hologram is developed, instead of having an actual image, the photographic plate has small microscopic interference fringes. When a reference beam is shone through these fringes, the light diffracts. The resulting light that reaches our eyes in an exact replica of the light that bounced off the object that was holographed. This creates the illusion that we can see the object from different angles.

There are two types of coherence holographers need to understand: temporal coherence and spatial coherence. To understand temporal coherence, imagine that we split a laser beam into two halves and let each half run side by side separated by a divider. If the waves are perfectly coherent, we can tell from one side of the divider what the phase was on the other side, next to us as well as very far up the beam path. However, with an imperfect coherence, we can only predict the phase a limited the distance up the beam path. Beyond that point, the phases of the two beams do not correspond in any predictable way. This maximum distance is called the coherence length. Thus temporal coherence or coherence time is simply the coherence length divided by the speed of light.

When making a hologram, it is important that the

dimensions of the object are less than the coherence length. A good rule of thumb is that the laser's coherence length should be about two or more times the length of the object. This is because holograms are constructed by dividing the beam of light into two beams, an object beam and a reference beam. The object beam reflects off the object onto the photographic plate while the reference beam simply illuminates the photographic plate.

When the two beams rejoin at the photographic plate, they must still have a predictable phase correspondence in order to record phases from the different points of the object (remember, it is by recording this information that holograms achieve their three-dimensional effect). If the dimensions of the object are greater than the coherence length, then those parts of the object beam that out travel the reference beam by more than a coherence length will have no way of comparing their phases with the reference beam when they reach the photographic plate. By losing phase information, the three-dimensional effect will be lost.

An inexpensive Helium-Neon laser typically has a coherence length of about 10 cm, sufficient for holographing small objects. If you want to make holograms of larger objects, make sure to find out the coherence length before you make your purchase.

Another type of coherence is spatial coherence. One way of testing the spatial coherence of your laser is to shine it on a wall. If you are fortunate, the pattern will be a bright dot that gradually fades towards the edges. When this occurs, a laser is operating in the TE-Moo mode. This means that there are no interference patterns across the beam caused by stray transverse electro-magnetic (TEM) light waves. The modes refer to the way the light moves at angles to the beam line inside the laser cavity. Certain mixtures of straight moving and angling beams (represented by the two subscripted numbers after "TEM") create characteristic patterns. If the pattern were a bright ring with a dark hole in the middle, it would be operating in the TEM10 mode. For a laser to create holograms, it is essential for it to operate in the TEMoo mode. Otherwise the object will not be illuminated with an even phase front, and the hologram will be distorted.

Finally, if you have decided to purchase a laser, you may wish to contact any of the following manufacturers to find out more about the latest costs and features (for more complete information, refer to the Laser heading in our charts):

(1) Hughes
(2) Aerotech
(3) Metrologic
(4) Melles Griot
(5) Ealing
(6) Jodon
(7) Coherent
(8) Spectra Physics

This chapter was contributed by Gabe Paulson of Berkeley, California

5

Holographic
Optical Elements

One of the most financially rewarding fields of holography has been the making of Holographic Optical Elements, commonly referred to as HOEs.

As the name implies, HOEs are optical elements such as lenses, mirrors, etc. that are made holographically. Although the fabrication can be quite involved tethnically, the concept of the HOE is relatively simple. The best way to explain the concept is in the following way.

We know from our earlier discussion that in making a transmission hologram the light from the object beam reflects off our object at innumerable points and goes on to strike the photosensitive plate at the same time that the reference beam does. We then develop the plate to get our hologram. Whenever we shine a reference beam on our hologram at the same angle used to create the hologram, the light from the reference beam is reflected off the hologram and forced to focus in space to visibly recreate our original object.

Although it sounds awkward, we could say that there are innumerable points of light coming from the object which together form the object beam. After the hologram is developed and we illuminate it, the transmission hologram refocuses the reference beam in such a way that it recreates the points of light that were reflected off the object. What we actually see, then, is a compilation of countless points of light, each with its own focal length, being refocused in space by the hologram.

Let's consider a much simpler transmission hologram. Instead of an object which reflects many points oflight, suppose our object has just one point of light. If you stop to think about it, the transmission hologram of a single point source of light functions in the same way as a traditional concave lens (illustrated in fig. 5.1).

You will also find that a reflection hologram functions in much the same way as a convex mirror, as shown in fig. 5.2.

These types of holograms are Holographic Optical Elements (HOEs). They are generally made without

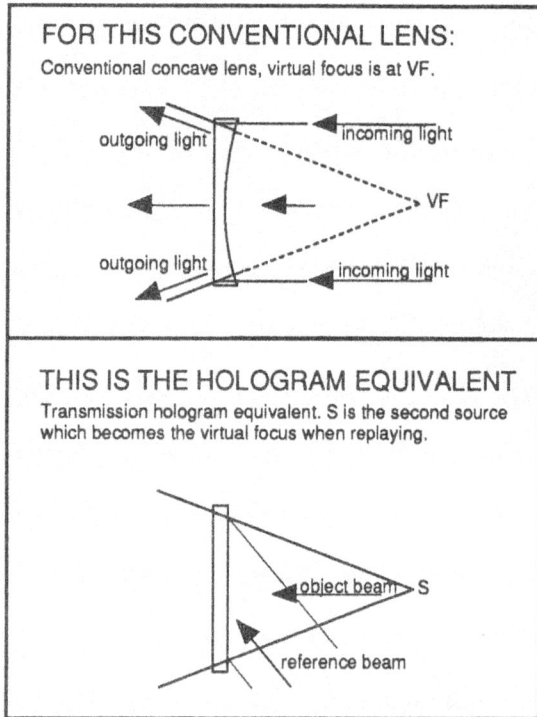

Conventional concave lens, and holographic equivalent.
Fig. 5.1

objects and with the intention of mimicking a conventional optical lens or a variation of conventional optics. With an HOE, the method for producing the image differs from conventional optics but HOEs do obey basic principles of conventional optics. These include such fundamentals as the equation for determining the focal length of the lens.

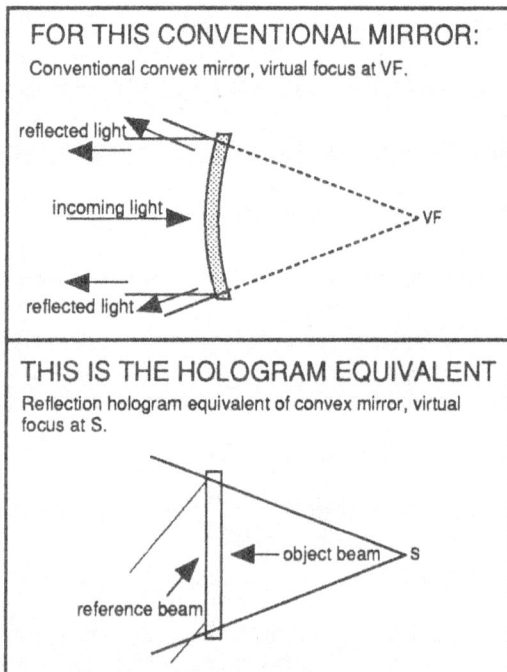

Conventional mirror and holographic equivalent.
Fig. 5.2

As you can imagine, there are differences between conventional lenses and HOEs. Here are a few of them:

- The angle of light used by a holographic lens is very selective since it is the reference angle .

- The frequency of light used by holographic lens can be very selective since it is the frequency used to make the hologram.

- Unlike a conventional mirror, a holographic mirror can be rotated 180 degrees to form a real image with a positive focal length.

Advantages of HOEs:

- Since you can create an HOE lens that, unlike conventional lenses, is very angle-and frequency-specific it makes possible the construction of unique and very selective optical system configurations. One example is an HOE mirror that reflects light coming in from an angle greater than 45 degrees, or a mirror that reflects a given percentage of a pre-designated, specific wavelength.

- Unlike conventional optical elements, the function of HOEs is relatively independent of the substrate geometry. Consequently, HOEs can be produced on thin substrates that are relatively light even for large apertures.

- Since HOEs are holograms, spatially overlapping elements are possible because several holograms can be recorded in the same layer.

- HOEs can correct system aberrations, so that separate corrector elements are not required.

- HOEs offer the possibility of mass production at a significantly cheaper unit cost than conventional optics.

- Some HOEs can be produced from computer ge erated holograms. When the waveforms needed to · form the HOE are calculated and generated by a computer, an HOE that has little noise and imperfection can be produced.

In order to make an HOE that reproduces a desired optical setup, there are two approaches to take. One is to physically set up the optical path that you wish to reproduce, using conventional optics, and make a hologram of it. In this case, the setup itself serves as a master for a hologram which can be mass produced for a far lower unit cost. For many holograms, this is still the easiest and most accurate method. The disadvantage of this method is that all of the optical components still need to be purchased for he master setup.

The other way to make a wide variety of optical components holographically is to reproduce the optical properties desired and follow the same optical laws. There are many different types of HOEs. In general they are either gratings, lenses, mirrors, beam splitters or combiners, but almost anything is possiale.

Diffraction Gratings

Diffraction gratings are among the simplest HOEs :0 make. For a very simple diffraction grating, you nly need the interference of two or more beams of aser light, such as in this general setup below.

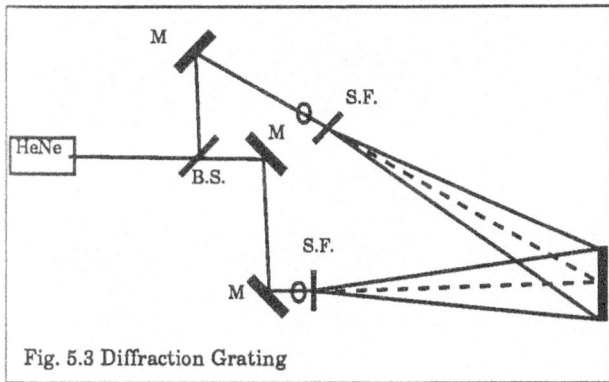

Fig. 5.3 Diffraction Grating

A diffraction grating is the holographic equivalent of a prism. If a diffraction grating is shot with a HeNe laser, when it is reconstructed with the same laser at the same angle, the light will diffract at that original angle. If, on the other hand, it is reconstructed with white light, only the red light with which it was made will diffract at the original angle. The shorter wavelengths of light will diffract progressively less until we are out of the visible spectrum (see fig. 5.4).

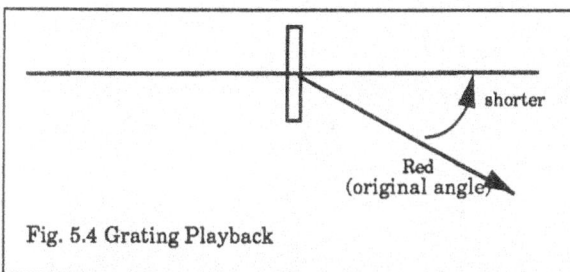

Fig. 5.4 Grating Playback

Holographic Lenses

As mentioned earlier, a transmission hologram can be much like a conventional lens in form and function. It can be mounted in much the same way and placed as a conventional lens, yet can cover a far greater area at a much lower cost. A holographic lens is called a Gabor Zone (see fig. 5.5). There are two simple techniques for constructing holographic lenses, both of which derive from the lens equation:

$$1/F = 1/U + 1/V$$

where F is the desired focal length, U is the object distance and V is the image distance.

The first technique is to use a collimated beam for one of the beams. Since $1/oo = 0$, therefore $F=V$. In other words, the focal length of the lens is the distance of the spatial filter from the holographic plate. (see Fig. 5.5).

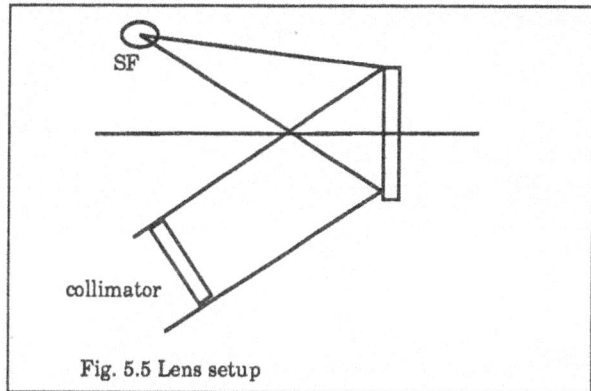

Fig. 5.5 Lens setup

A more general setup is to use two finite sources at different distances from the plate such that:

$$1/F = 1/D1 - 1/D2$$

The minus sign here comes from the fact that, on reconstruction, the reference beam distance is reversed. An algebraic trick here is to use D2 = 2D1, so that:

$$1/F = 1/D1 - 1/2D1 = 1/2D1 = 1/D2$$

i.e. F = 02, but this is not essential. (See Fig. 5.6)

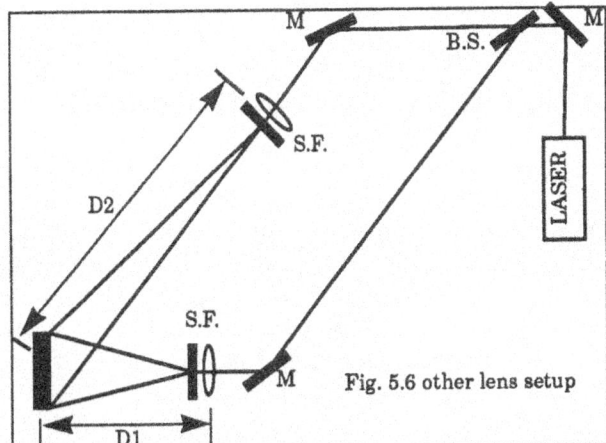

Fig. 5.6 other lens setup

Holographic Mirrors

Both the above formula and its concept are also true of holographic mirrors. One only has to convert the setup to a reflection hologram setup, by illuminating the back of the plate with the reference beam to produce mirrors.

Again, if D2 = 2D1, then the focal length is going to be equal to D2. You can also keep in mind that using different powered microscope objectives at appropriate distances can also create the desired effect. (Fig 5.7)

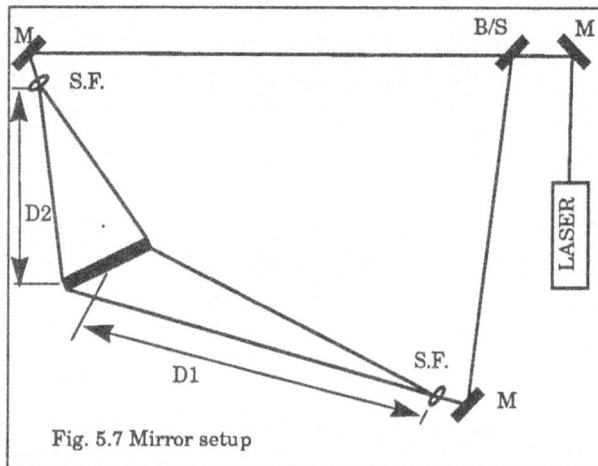

Fig. 5.7 Mirror setup

Holographic Collimators

It is worthwhile for us to notice at this point that a collimated beam has the property D = 00. What can be derived from this is that it is possible to approximate a collimated beam holographically, by moving one of the beams very far away' in relation to the other beams. At this distance, the spherical waves coming from the spatial filter are almost parallel and approximate a plane wave front, or collimated beam. This kind of geometric approximation is often enough for a given application. This can be done with either the lens setup or the mirror setup, but since the distance is great and therefore the power is low, the higher efficiency of the mirror collimator is recommended.

Holographic Mirror Beamsplitter

Another essential component of a two-beam holographic setup is the beam splitter. Usually this is a half-silver or a specially coated mirror, whose functions can be mimicked holographically. The procedure is to make an HOE mirror with beams that enter the plate perpendicular to one another. This can be done using only a single beam in this simple configuration,

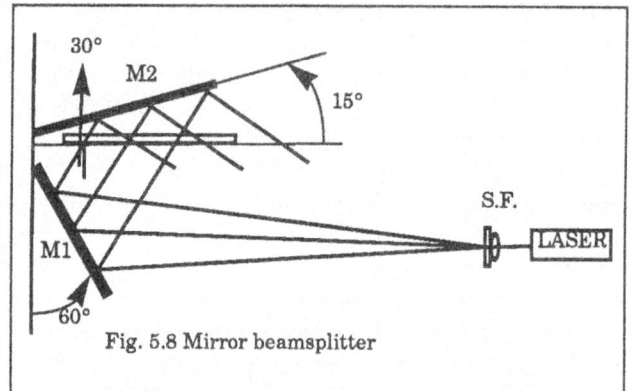

Fig. 5.8 Mirror beamsplitter

a usetul trick since, without a beamsplitter, the beginning holographer has only one beam.,

In Fig. 5.8, Mirror 1 is a front surface mirror at 60° to the holographic plane. It is far enough away from the source that a collimated beam is approximated. This means that the waves passing through the holographic plate are plane waves. Mirror 2 is at 15° to the holographic plate, so that the incident rays at the plane are perpendicular to one another.

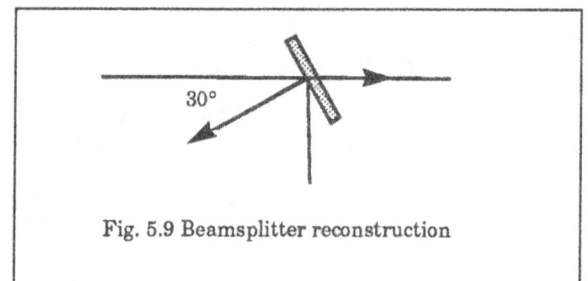

Fig. 5.9 Beamsplitter reconstruction

As you can see in fig. 5.9, when reconstructed at the shooting angle the HOE reflects a portion of the beam at 90° to the incident.

A variable beam splitter is often useful and can be made by moving the source off-axis, so that the strength of the beam across the holographic plane reduces as it gets further away from the source. Then you have a beam splitter that will transmit and reflect more or less of an incoming beam as it is moved along it own axis.

The beam splitter provides an excellent example of the cost-effectiveness of HOEs. In this case, what you would have to purchase, other than your table and your laser is two front surface mirrors of a few square inches each (approximate cost $20). These mirrors are re-usable in many other applications, so the total cost of our beam splitter (without a mount) is negligible. Beamsplitters from large supply houses cost between $500 and $2000 and can, of course be broken or damaged.

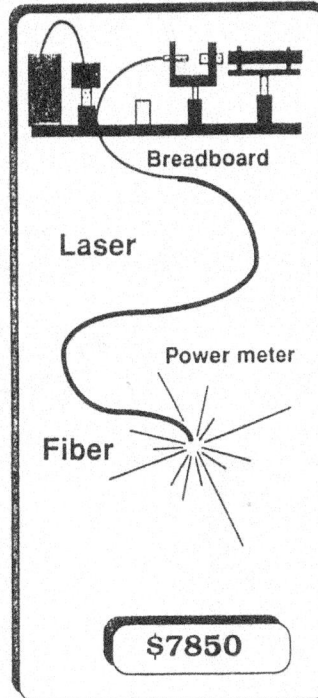

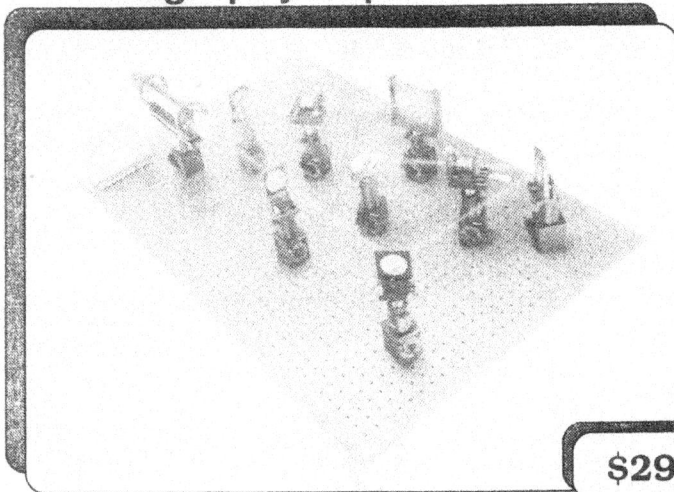

Other HOE Applications

HOEs have many applications other than just simple table components in our modern marketplace. Among others, there are automobile and defense applications, artistic applications and industrial applications. We will present a few examples.

Tandem Optics.: In many situations it is desirable to be able to look at a volume of material interferometrically. One way to do this is to make a set of gratings that diffract light at precise, controlled angles and efficiencies and position them in such a setup as fig. 5.10.

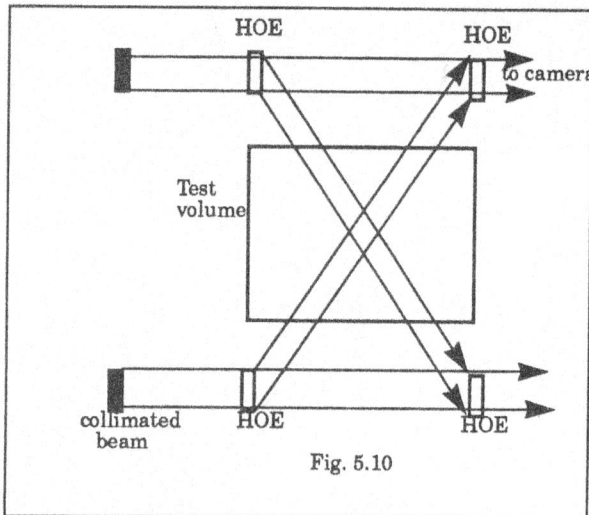

Fig. 5.10

The purpose of this is to establish a dual object beam passing through the material and a dual control (reference beam) to provide the interference.

Another application of the same procedure is to provide what is called .the Perfect Shuffle optical system, whereby irrelevant information can be eliminated and relevant data stored in the correct place by a series of HOE diffraction gratings that shunt the light at different angles in the system (Fig. 5.11).

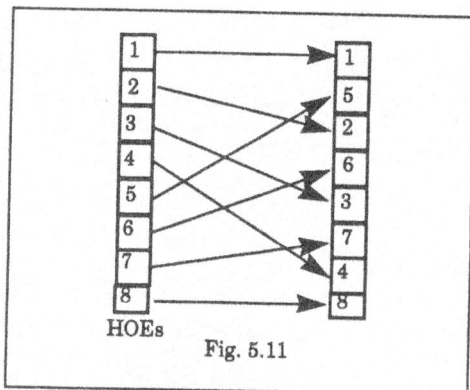

Fig. 5.11

Artistic applications: In holographic artwork one of

the problems that arises is that the holograms with the best depth, silver halide laser viewable holograms, are also the hardest to display, because of the amount of space needed behind them. Again, a grating is a possible solution; in this case, a grating designed to receive an incoming beam at a very shallow angle and diffract it to the reference angle of the art hologram. This HOE can be sandwiched to the original hologram and sold as a package (fig. 5.12).

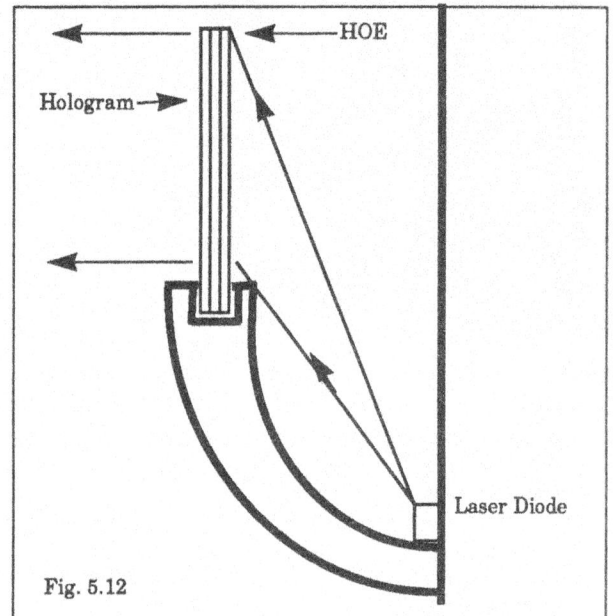

Fig. 5.12

Military and Automobile Applications: Heads-lip Displays (HUDs) have been common in aircraft for quite a few years, and they have now worked their way into high-end automobiles. The idea is that a CRT display can be projected up onto a glass substrate at eye level, which will neither require the driver (or pilot) to take his eyes off the road nor interfere with his view of the road. Calls to five major automobile manufacturers indicated that none are using holographic HUDs at present, however.

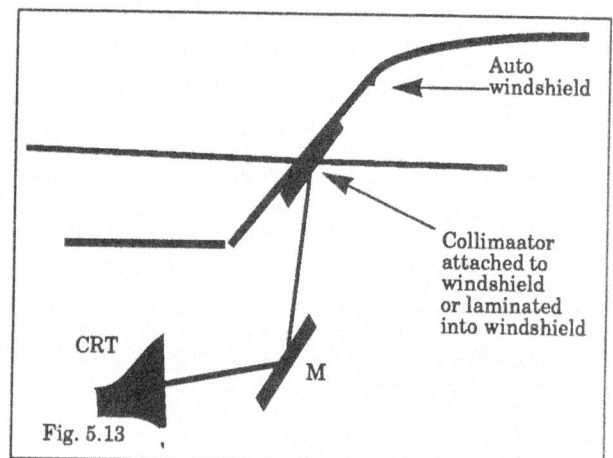

Fig. 5.13

This chapter was contributed by Phil Burfoot of El Cerrito, California.

References

Saxby G. Practical Holography. (1988) Prentice Hall International. 228-234.

Stojanhoff, C.G:, & Windeln, W., Development of high efficiency holographic optics. SPIE Proceedings #I 812 (1987)

Kobolla, Harald., Sauer, Frank., & Volkel, Reinhard. Holographic Tandem Arrays. SPIE Proceedings #I 1136 (1989)

Jeong, Tung & Wesly, Edward. HOE for Holography. SPIE Proceedings #I 1183 (1989)

Del Vo, P., Rizzi, M.L., & Stibelli, S. Holographic Head-up Display for the Automotive Industry. SPIE Proceedings#l1136 (1989).

6

NDT: Holographic Non-Destructive Testing

NDT is a rapidly-growing field of holography. Although the techniques and analysis can be very involved, the general concept is not difficult to understand.

Suppose you make a transmission hologram of a still object. Then, without disturbing the setup, you make another transmission hologram of the object. If you were to take the two holograms and put them on top of each other, you would simply see one hologram pattern reinforced by the other, since both holograms are identical. Ifthere is a small amount of movement in the object between exposures, however, the two patterns would be slightly different at the site of the dislocation.

When you put the two patterns on top of each other you then see a Moire pattern where they differ. These are called interference patterns. Since the holograms do not match at the site of the dislocation, the interference patterns show us where and by how much the object is dislocated. Investigations that started in 1965 showed that in some cases, very exact measurements can be obtained (measurements down to the order of

light waves in some cases) by studying the interference patterns.

The immediate applications are clear. You can take one holographic exposure of an object, subject the object to stress and then take a second holographic exposure. Comparing and analyzing the interference patterns that result when the two patterns are superimposed tells you where the object deformed under stress. Analyzing this has become a whole industry in itself and is variously called "Holometry", "Holographic Interferometry", or by its much more sales-oriented name of "Holographic Non Destructive Testing" (NDT for short).

Some applications of holographic interferometry include:

- Locating the presence of a structural weakness: the object is stressed by the application of a load or change in pressure or temperature.

- Detecting cracks and the location of areas of poor bonding in composite structures.

• Medical and dental research.

• Aerodynamics, heat transfer, and plasma diagnostics.

• Solid mechanics such as measuring the changes in shape due to absorption of water and corrosion.

There are a number of techniques for holographic interferometry testing and each has its advantages and disadvantages. Some of the major classifications follow.

Real Time Holographic Interferometry

This test is done to obtain immediate results. A hologram is shot, developed and then re-placed in the holder it was in for the exposure. When the object is illuminated as if to expose it for another shot, there are two holographic patterns on the plate. One pattern will be the one you developed and the second pattern is created by the light you just turned on reflecting from the object still in its original position.

The two patterns should match exactly if nothing moved and the plate is replaced exactly in its original spot. Therefore, if you look through the plate at the illuminated object, you should see just one, reinforced, holographic pattern.

Suppose, however, that there was some small movement in your object while you were developing the plate. If there was movement, then after the plate is put back in the holder .and the laser is turned on, the observer viewing the object sees it covered with a pattern of interference fringes around the areas where the object has changed shape. If you move the object or put it under stress while viewing it, you see interference fringes change to map the area where the object is being deformed.

It is awkward to have one person at a time viewing the interference fringes. Capturing the image being viewed is also a problem. These problems may be solved by using a closed circuit television camera. A Polaroid camera can photograph any desired scene. It is then possible to color code the fringes to identify the direction of the displacement by using filters and a double exposure. Another method is to have the video signals read digitally and input directly to a computer for analysis.

Since pulsed lasers are not subject to vibration re-

strictions, it is also possible to mount a pulsed laser on a truck, drive to an object that cannot be moved, such as a water tower, and from the truck perform a real time NDT stress analysis. This also applies to buildings, bridges, etc. for applications like earthquake analysis.

Although the above procedure for real time holographic interferometry sounds very simple and straightforward, there are many problems with it. Precise positioning of the hologram after processing is necessary, and uniform drying of the emulsion to avoid deformations in the hologram is another potential problem. One method of solving these problems is to develop the hologram on the spot using a "liquid gate" arrangement.

Another alternative for processing the hologram, which eliminates the need for wet processing altogether, is to use thermoplastic recording. There are commercial units available in which a hologram can be recorded and viewed in less than a minute.

Advantage: The advantage of real time NDT is its instant results. There are numerous cases where you must have instant analysis so that work on a project can continue. You are also able to view the interference patterns as the object is undergoing stress. In other words, you see a real time "movie" of the stress points developing as the experiment is conducted, and you may photograph or videotape what you see.

Disadvantage: One disadvantage is that there is a serious drop in the visibility of fringes because the light diffracted by the hologram remains linearly polarized while the light scattered by the object is largely depolarized. To avoid the decline in fringe visibility, it is necessary to use a polarizer when viewing or photographing the fringes. A second disadvantage is exact registration of a developed hologram with the original object. A third disadvantage is that you do not wind up with a precise, permanent record of the interference fringes except for what appeared on the television monitor. Recording on photographic emulsion offers more precision.

Double Exposure Holographic Interferometry

In this case a double exposure of the hologram is made on the same plate. The first exposure is usually made with the object in an unstressed condition, and the second with stress applied to the object.

Advantage: Double exposure interferometry is much easier to perform than real time holography because the two holograms are in exact register. Any distortions to the emulsion affect both holograms equally. Both holograms have the same polarization. Therefore, the fringes are much clearer and no special care needs to be taken when illuminating the holo- gram. It is also possible to use rainbow holograms for double exposure interferograms. Fringe patterns due to different effects can then be displayed in different colors.

Disadvantage: The double exposure tends to brighten the image of the object, which makes it difficult to see small displacements. You can help correct this by shifting the phase of the reference beam or tilting the object beam between exposures. Another disadvantage is that with a double exposure you have the "before" and "after" snapshot but you do not see what happens in between. This can be overcome, to some extent, by multiplexing techniques. Another problem is that you cannot compensate for and control the fringes if the object tilts between exposures. This problem can be solved by using two holograms, recorded on the same plate, with two different, angularly separated reference waves.

Sandwich
Holographic Interferometry

The sandwich hologram takes advantage of the fact that multiple exposures can be made on the same holographic plate. This procedure provides a much more elegant solution to the problem of viewing varying stages of stress on an object between exposures.

Two plates (with no anti-halation backing) are set in the same plate holder with their emulsion surface toward the object. An exposure is made of the object under no stress. The object is then tilted, stress is applied, and a second exposure is made. This procedure is repeated as many times as desired.

To view the hologram, you put the plate that recorded the undisturbed object in its original plateholder, and add to the plateholder one of the holograms of the object under stress. You see interference patterns at the points of deformation. You can see the incremental deformations of your object by going through all the plates in sequence. Using any combination of two plates, the incremental changes between any two stages of deformation can be seen.

Advantage: This technique allows you to see a wide

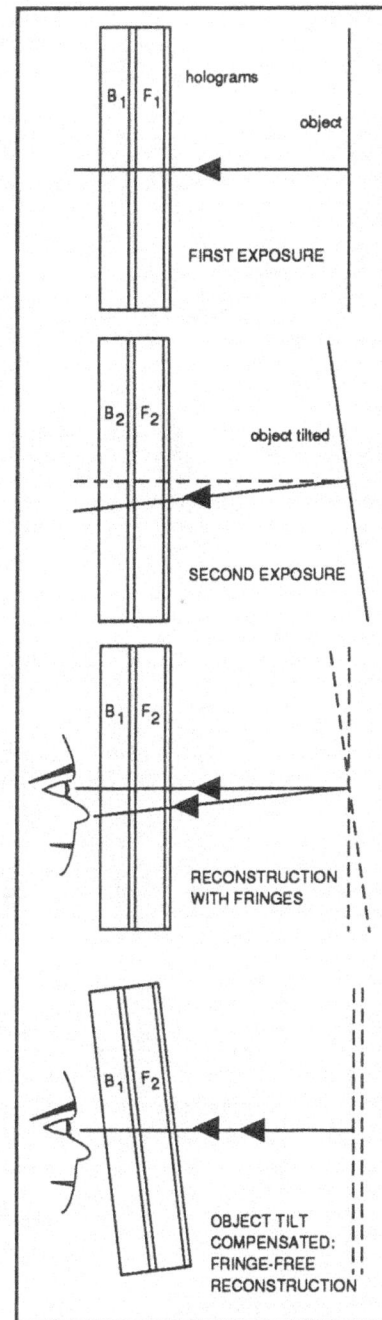

Sandwich hologram interferometry

latitude or combination of stress loads. You could, for example, select the plate B1 and F2 to view what happens when the first stress load is applied. Or, you could select B2 and F3, which show the deformations between stress load 1 and 2.

Disadvantage: You are looking through two different plates to see the image and there may be some parallax problems. It is also time- and material-consuming to catalogue and handle all these different plates.

Holographic Interferometry Through Dense Media

One example of this procedure begins when a hologram is taken of an undisturbed chamber filled with a gas. The chamber is disturbed in some way and a second hologram is taken. Any variation in density of the gas alters the path length of the laser light and hence creates a hologram which produces interference patterns when compared to the undisturbed chamber's hologram.

Another example of dense-media interferometry is plasma diagnostics, in which measurements of the light's refraction at two wavelengths make it possible to determine the electron density directly.

Time-Averaged Holographic Interferometry

There are numerous instances in which the object can only be properly tested while operating. To see clearly all the uneven stress in a hi-fi speaker or a rotating fan, for example, you need to test it while it is in operation. In order to make interferometer tests on these subjects a number of clever methods have been developed. Time-averaged holographic interferometry is one of them.

If you take a flexible metal ruler and whip it back and forth in the air while holding the end of it tightly in your fist you see the image of the ruler at the two most extreme points of the flex, but it is a blur in between. If you were to calculate it, you would find that the amount of time spent at the two extremes of the cycle is a substantial percentage of the time in the entire cycle.

Suppose we look directly into a hi-fi speaker. When it is operating, the speaker membrane or cone spends most of its time at the two ends of its vibration (i.e. the cone is either fully extended toward us or fully extended away' from us). We simply make a hologram of the speaker while it is operating and extend the exposure so that it is long compared to the cycle of the speaker membrane moving in and out. What is the result? Since the time spent between the two extremes is a blur, the result is a hologram of the speaker membrane fully extended toward us, superimposed on a hologram of the speaker membrane fully extended away from us. In other words, we seem to obtain a "double exposure hologram" even though it is only one exposure.

If the speaker is not in perfect resonance and the distance the speaker membrane extends towards us is not exactly equal to the distance the speaker membrane moves away from us, the two superimposed holographic patterns create an interference pattern. You can calculate how out of resonance the speaker is by counting the fringes. The distance between fringes is slightly greater than one quarter of a wavelength.

The above technique of time-averaged holographic interferometry works well for studying objects vibrating in a stable manner. It was first reported in 1965 and is usually used to identify the resonances of a test object. It does this by monitoring the object's response while varying the excitation frequency and the point of excitation on the object.

Typical applications of the process are testing of musical instruments and electromechanical devices such as loudspeakers, turbine blades and aircraft structures.

Advantage: In time-averaged holographic interferometry you are allowed to view and adjust your object to obtain an almost perfect resonance with little effort.

Disadvantage: Time-averaged holographic interferometry is good only for objects vibrating in a stable manner, and compares only the two extremes of the vibration cycle. This procedure does not show the relative phases of the various modes of vibration. Also, the contrast of the fringes falls off rapidly as the fringe order increases.

Strobed Holographic Interferometry

Suppose that you have an object that vibrates in a stable manner, but the point you want to inspect in the vibrating cycle is not at one of the two extremes of the cycle. In 1968, articles were written describing a solution to this problem using a sort of strobe light effect. In this process, a sequence of light pulses are triggered at desired times during the cycle. The resulting hologram is equivalent to a double exposure hologram (or sandwich hologram if you use that method) recorded while the object is in the desired state of deformation at any time in the cycle.

Advantage: Allows you to study the object at any time in the vibration, in relation to the object at any other time in the vibration.

Disadvantage: This is a much more involved set up than the time-averaged technique.

References & reading:

General:
Hariharan, P. (1984) Optical Holography. Cambridge: Cambridge University Press.

References for applications of holographic interferometry:
Greguss, P. (1975). Holography in Medicine. London: IPC Press

____ . (1976). Holographic interferometry in biomedical sciences. Optics & Laser Technology, 8, 153-9.

Sciammarella, C.A., (1987). Advances in the application of holography for NDE, SPIE Proceedings, 523, 137-44.

Von Bally, G. ed. (1979). Holography in Medicine and Biology. Berlin: Springer-Verlag.

Willenbourg, G.C., (1987). Holography and Holometry Applictions in Dental Research, SPIE Proceedings, 747, 51-6.

References for Interference Fringes:
Powel, L. & Stetson, K.A. (1965). Interferometric Vibration Analyill by Wavefront reconstruction, Journal of the Optical Society of America, 55, 1543-8.

References for real time holographic interferometry:
Dandliker, R., Ineichen, B. & Mottier, F.M. (1973). High resolution hologram interferometry by electronic phase measurement. Optics Communictions, 9, 412-16.

Hariharan, Oreb & Brown, (1982). A digital phase-measurement system for real time holographic interferometry. Optics Communications, 41, 393-6.

____ . (1982). Real-time holographic interferometry: a microcomputer system for the measurement of vector displacements. Applied Optics, 22, 876-80.

References for liquid gate recording:
Hariharan, P. (1977). Hologram interferometry: identification of the sign of surface displacements. Optica Acta, 24, 989-90.

Rariharan, P. & Ramprasad, B.S. (1973). Rapid in situ processing for real-time holographic interferometry. Journal of Physics E: Scientific Instruments, 6, 699-701.

Van Deelen, W. & Nisenson, P. (1969). Mirror blank testing by real time holographic interferometry. Applied Optics, 8, 951-5.

References for thermoplastic recording:
Harihalran, P. & Hegedus, Z.S. (1975). Relative phase shift of images reconstructed by phase and amplitude holograms. Applied Optics, 14, 273-4.

Saito, T., Imamura, T. & Tsujiuchi, J . (1980). Solvent vapor method in thermoplastic photoconductor media. Journal of Optics (Paris), 11, 285-92.

Thinh, V N. & Tanaka, S. (1973). Real time interferometry using thermoplastic hologram. Japanese Journal of Applied Physics, 12, 1954-5.

References for double exposure holographic interferometry:
Yu, F.T.S. & Chen, H. (1978). Rainbow Holographic interferometry. Optics communications, 25, 173-5.

Yu, F.T.S., Tai, A. & Chen, H. (1979). Multiwavelength rainbow holographic interferometry. Applied Optics, 18, 212-18.

____ . (1979). Multislit one-step rainbow holographic interferometry. Applied Optics, 18, 6-7.

References for shifting reference beams:
Collins, L.F. (1968). Difference holography. Applied Optics, 7, 203-5.

Hariharan, P. & Ramprasad, B.S. (1972). Simplified optical system for holographic subtraction. JOUI'}lal of Physics E: Scientific Instruments, 5, 976-8 .

____ . (1973). Wavefront tilter for double-exposure holographic interferometry. Journal of Physics E: Scientific Instruments, 6, 173-5.

Jahoda, F.C., Jeffries, R.A. & Sswyer, G.A. (1967). Fractional fringe holographic plasma interferometry. Applied Optics, 6, 107-10.

References for multiplexing techniques:
Caulfield, H.J. (1972). Multiplexing double-exposure holographic interferograms. Applied Optics, 11, 2711-12.

Hariharan, P. & Hegedus, Z.S. (1973). Simple multiplexing technique for double-exposure hologram interferometry. Optics Communications, 9, 152-5.

Parker, R.J. (1978). A new method of frozen-fringe holographic interferometry using thermoplastic recording media. Optica Acta, 25, 787-92.

Two holograms on the same plate:
Ballard, G.S. (1968). Double exposure interferometry with separate reference beams. Journal of Applied Physics, 39, 4846-8.

Gates, J.W.C. (1968). Holographic phase recording by interference between reconstructed wavefronts from separate holograms. Nature, 220, 473-4.

Tsuruta, T., Shiotake, N. & ltol, Y. (1968). Hologram interferometry using two reference beams. Japanese Journal of Applied Physics, 7, 1092-100.

References for sandwich hologram interferometry:
Abramson, N. (1974). Sandwich hologram interferometry: a new dimension in holographic comparison. Applied Optics, 13, 2019-25.

____ . (1975). Sandwich hologram interferometry: 2. Some practical calculations. Applied Optics, 14, 981-4.

____ . (1976). Sandwich hologram interferometry: 3: Contouring Applied Optics, 15, 200-5.

__ . (1976). Holographic contouring by translation. Applied Optics, 15, 1018-22.

__ . (1977). Sandwich hologram interferometry: 4: Holographic studies of two milling machines. Applied Optics, 16,2521-31.

Abramson, N. & Bjelkhagen, H. (1978). Pulsed sandwich holography. 2. Pratical application. Applied Optics, 17, 187-91.

Hariharan, P. & Hedgedus, Z.S. (1976). Two-hologram interferometry: a simplified sandwich technique. Applied Optics, 15, 848-9.

References for holographic interferometry through dense media:

Ostrovskaya, G.V. & Ostrovskii, Yu. I. (1971). Two-wavelength hologram method for studying the dispersion properties of phase objects. Soviet Physics - Technical Physics, 15, 1890-2.

Radley Jr, R.J. (1975). Two-wavelength holography for measuring plasma electron density. Physics of Fluids, 18, 175-9.

Zaidel, AN., Ostrovskaya, G.V. & Ostrovskil, Yu. 1.(1969). Plasma diagnostics by holography. Soviet Physics - Technical Physics, 13, 1153-64.

References for time averaged holographic interferometry:

Agren, C.H. & Stetson, K.A. (1972). Measuring the resonances of treble viol plates by hologram interferometry and designing an improved instrument. Journal of the Acoustical Society of America, 51, 1971-83.

Bjelkhagen, H. (1974). Holographic time average vibration study of a structure dynamic model of an airplane fin. Optics & Laser Technology, 6,117-23.

Chomat, M. & Miller, M. (1973). Application of holography to the analysis of mechanical vibration in electronic components. TESLA Eletronics, 3, 83-93.

Saxby, G., (1988). Practical Holography, Prentice Hall, 321.

Stetson K.A. & Powell, R.L. (1965). Interferometric hologram evaluation and real-time vibration analysis of diffuse objects. Journal of the Optical Society of America, 55, 1694-5.

References for strobed holographic interferometry:

Archbold, E. & Ennos, A. E. (1968). Observation of surface vibration modes by stroboscopic hologram interferometry. Nature, 217, 942-3.

Shajenko, P. & Johnson, C.D. (1968). Stroboscopic holographic interferometry. Applied Physics Letters, 13,44-6.

Watrasiewicz, B.M. & Spicer, P. (1968). Vibration analysis by stroboscopic analysis. Nature, 217,1142-3.

7

Computer-Generated Holograms (CGH)

As we explained in the introduction, some types of holograms sacrifice their depth so they may be seen in a wide variety of commonly encountered lighting conditions. For the sake of convenience, authors sometimes divide holograms into two broad categories: "thin" holograms and "volume" holograms!. Basically, holograms are considered "thin" or 2D when the thickness of the recording material is small in relation to the average spacing of the interference fringes (less than 1:1). Volume holograms have a ratio greater than 1:1.

"Thin" holograms have little or no depth. "Volume" holograms can have great depth. Although volume holograms have all the glamorous 3D qualities, thin holograms have their own reasons for being popular. Thin holograms are:

· Easy to view even under harsh lighting conditions.

· Easy to make since we are only dealing with 2D subjects.

· Marketable holograms because many are HOEs like gratings, mirrors, and lenses.

Let us consider only the thin holograms, since CGH are thin holograms. With a fairly thin hologram, we can look at the emulsion and think of it as a flat, two dimensional object like a piece of paper. We can store all the information for this flat hologram in a computer by simply recording, on an (x,y) coordinate graph, the size and location of all the holes in the emulsion.

Following is an overview of how a computer-generated hologram (CGH) is made.

1) Input Hologram
We either have a real physical object or an imaginary, digitized object.

Real Object: If we have a real object, we make a "thin" hologram at this point. Then we have the holographic patterns on the plate digitized by scanning the plate using a microdensitometer. A microdensitometer measures the percentage of light that goes through a material in an extremely small area.

Imaginary Object: If we want to make a CGH of an imaginary object, we calculate the wave patterns at the plate surface by using theories appropriate for Fraunhofer, Fresnel, or near-field diffraction. It should be pointed out that the more complex our object is, the more difficult this task will be.

2) Edit Hologram

If we need to edit or change our hologram in any way, we do so in the computer at this point.

3) Output Computer Generated Hologram

We now output or reproduce our hologram onto some media. Eventually we want to wind up with a transparency that has our holographic pattern on it. There are several techniques used to map the hologram pattern and we cover the techniques later in this section. The most common hardware used to output CGHs are:

- Computer-driven plotter
- Computer-driven laser beam
- Computer-driven electron beam
- Computer monitor, displays hologram (a photograph is made from the monitor)

4) Reconstruct and Detect Hologram

Once the hologram is output, we copy the hologram and mass-produce it, or illuminate the hologram and record it using means that suit our purpose. Detectors may include:

- Eye-simply view the hologram if that is the purpose
- Charged-couple devices
- Photographic film
- Small antennae and receivers have been used for detecting microwaves

Advantages: The advantages of the computer generated hologram are clear. Here are six main reasons:

1) Objects: No objects are necessary in the simple cases where we can calculate the wave patterns by computer.

2) HOEs: Since we don't need to use real objects, idealized waveforms can be made. This eliminates unwanted extraneous light and "noise" one would get from a normal hologram. This is of great value if we are making HOEs (see also Chapter 5, HOEs).

3) Editing.: We can digitally edit our hologram in the computer.

4) Storage: We store our holograms electronically instead of on time sensitive emulsions.

5) Transmittal: We can send our holograms electronically worldwide in a matter of moments.

6) Translation: As we look into a Fourier Transform hologram we see a unique feature. We see a central spot of light and two images, one on each side of the spot. One object is an inverted and pseudoscopic image of the other. If we rotate the plate clockwise, the images rotate with the plate. If we simply turn the plate so back becomes front and vice versa, like we tum the pages of a book, the images stay fixed in space. This unique feature is used in some specialized cases.

Disadvantages: Here are two major disadvantages of computer-generated holograms:

1) Hologram thickness: The major disadvantage to computer-generated holograms is that they currently only work for "thin" hologram. One of the great attractions to holography is the "volume" or three dimensional qualities of a hologram. It is true, however, that there are clever ways to create depth using thin holograms such as stereograms. One of the most notable current items is the computer-generated alcove hologram being done by Dr. Stephen Benton at MIT's Media Lab.

2) Computer Time And Storage: Generating the holographic fringe structure of a simple object such as a toy car can require tremendous amounts of computer memory and time, making it difficult and not cost-effective for many applications.

Techniques for Outputting Computer-Generated Holograms

As mentioned earlier, there are several techniques for plotting or mapping the holographic pattern we send to our output hardware. Four of the most commonly-used methods are:

- Binary Detour Phase Holograms
- Kinoform
- ROACH (Referenceless On-Axis Complex Hologram)
- Computer Generated Interferograms

Binary Detour Phase Holograms

This method was first reported in 1966. The final product is an opaque mask with transparent holes or apertures that represent the hologram.

To output this, the computer calculates the image

of the hologram mathematically using a Fourier transform. The paper or media on which the computer prints the hologram is subdivided into miniature cells. In each cell, the computer prints a dot which later becomes an aperture. The magnitude of the Fourier transform at the center of the cell is calculated and that calculation determines the height and width of the aperture. The lateral position of the aperture within each cell is proportional to the transform's phase at the center of the cell.

On photoreduction, the black dot becomes a clear aperture on a black background. Representing the phase by using a lateral shift of the aperture within the cell led to the name "Detour Phase Hologram", an analogy to diffraction gratings with unequally spaced rulings. Another term for these holograms is "binary hologram" because any point on the hologram we create has a transmittance value of zero or unity. There are several variations of the Detour Phase Hologram which allow for a more refined output.

Advantage: It is possible to use a simple pen-and-ink plotter to prepare the binary master and problems of linearity do not arise in the photographic reduction process.

Disadvantage: This method is very wasteful of plotter resolution, since the number of addressable plotter points in each cell must be large to minimize the noise.

The Kinoform

If the object is diffusely illuminated, the magnitudes of the Fourier coefficients are relatively unimportant, and the object can be reconstructed using only the value of their phases. This gave rise to the Kinoform. To record a Kinoform, the computed values of the phase are recorded on a multilevel gray scale, which in turn is used to control a photographic plotter that exposes a piece of film. The master is then photographed again, to reduce it to final size, and bleached to convert the gray levels to corresponding changes in optical thickness.

Advantage: Kinoforms can diffract all the incident light into the final image.

Disadvantage: Less information is available since only the values of the phases are used. Also, if there is any error in the recorded phase shift, light is diffracted into the zero order which can spoil the image by creating an extremely bright light in the center of the hologram.

ROACH
(Referenceless On-Axis Complex Hologram)

This method uses multilayer colored film as a recording medium to obtain most of the advantages of the Kinoform without its major disadvantages. Both the phase and amplitude are recorded. Using Kodachrome film, the intensity-variation pattern is exposed through a red filter and the phase-variation pattern is exposed through a cyan filter. If the processed film is illuminated with a HeNe laser, the cyan layer modulates the amplitude and the other two layers modulate the phase.

Advantage: Since all the light is diffracted into a single image, the diffraction efficiency of the ROACH is very high. In addition, because both the amplitude and phase information is recorded, the image quality is superior to that of the Kinoform. The ROACH is superior to the Detour Phase Hologram because only one display spot is required for each Fourier coefficient and quantization noise is negligible.

Disadvantage: The steps required to produce this hologram are much more involved than for the Kinoform.

Computer-Generated Interferograms

Problems were encountered with the Detour Phase Holograms when encoding wavefronts with large phase variations. These problems were attributable to the fact that the apertures in the cells overlapped in cases where the phase of the wavefront moved through a multiple of 27t radians or more. To solve this, an alternative approach was taken by noting that the case of a wavefront which has no amplitude variations is essentially similar to an interferogram. The non-linearity of the emulsion can then be exploited to produce a hologram that is approximately binary. There are methods that can then be used to record the amplitude variations in the binary fringe pattern.

Advantage: Computer-generated interferograms are an improvement to the Detour Phase Hologram where large phase variations are encountered.

Disadvantage: Since amplitude variation is not recorded initially, it has to be calculated or derived later by one of several means.

Three Dimensional Computer-Generated Holograms

Computer-generated holograms were first general

ized to a three-dimensional object in 1968 by Waters. The process involved approximating the 3D object by making a number of equally spaced cross sections perpendicular to the z axis. Then a number of holograms from different angles were produced showing the resulting changes in parallax. Due to the fact that there were lines "hidden" from the front view of the object encountered as successive slices through the object were made, one had to add the contributions to the object wave arising from the hidden lines as one went along.

In 1970, King, Knoll and Berry took a different approach. Their technique makes a tall, thin, holographic stereogram and then "multiplexes" or joins large quantities of these stereograms together on a single plate. The computer produces a series of perspective projections of an object by either programming the holograms or by filming the subject from a number of different perspectives along a horizontal axis, and inputting the holograms to the computer by microdensitometer readings. This is a nice theory, but reports are that it is difficult in practice.

The computer then outputs a series of thin vertical strips, each of which are holograms, on a single plate. Since this is a thin hologram, it can be illuminated with white light to construct a bright, almost achromatic image. When the hologram is illuminated by the reference beam, or any white light, we see the real image, which is two-dimensional and located in the plane of the final hologram. However, since our eyes see a number of different holograms and each one is from a slightly different perspective, the viewer has a stereoscopic view and sees a three-dimensional image. If the plate is large and there are numerous images, we can tilt the plate from' side to side and see the image move about.

The alcove hologram is the latest (Benton, 1987) in the evolution of 3D computer-generated holograms. This is a multiplexed reflection stereogram in which the multiplexed holograms are arranged in a semicircular alcove into which the viewer looks.

The advantage is that as the viewer walks from side to side, the three-dimensional hologram rotates left to right, allowing the viewer to see around the object. To date, a viewing angle of about 30 degrees has been accomplished. Benton suggests enlarging and extending the arrangement so as to have close to 180 degrees of viewing using 900 slit holograms that are 300 mm high and 1mm wide (the interior of the concave would be 600 mm across).

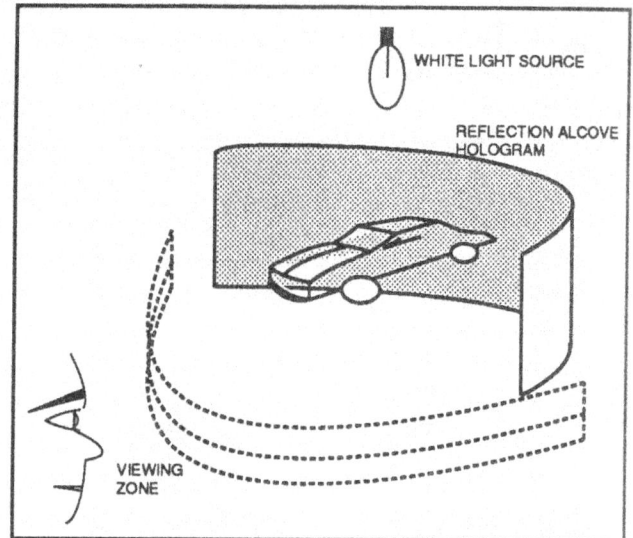

Fig. 7.1 Alcove hologram

Advantage:, Although a volume hologram is the only true three-dimensional recreation of an object, it has a problem with the restricted angle of view and subject matter. Holographic stereograms use any movie film that is filmed with the proper perspectives and, recently, color has been used in holographic stereograms. Industry benefits include being able to move around any desired object at will without having the object present.

Disadvantage: One of the big problems with the CGH is getting the holographic image into the computer. Using a microdensitometer to read 900 holograms into the computer is more of a challenge than anyone wants to take on. Generating the hologram internally by programming the wave patterns is very difficult for all but the simplest objects. Progress is being made, however, and developing technology is on the side of the multiplexed CGH. It should be pointed out that curved, multiplexed holograms made without the aid of the computer are commercially available, but the computer will offer enormous benefits when the system is perfected.

Uses of Computer-Generated Holograms

The uses for CGHs and HOEs are very similar, since one of the main values of the CGH is its ability to produce almost noise-free HOEs.

Some applications for CGHs:

• *Commercial Embossing:* Used to make simple artistic holograms for later use embossing, etc.

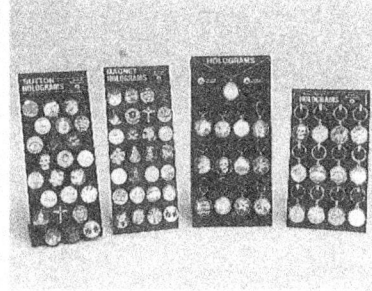

• *Multiplexed Displays:* 3D artistic displays using holographic stereograms such as the alcove hologram.

• *Medicine:* Currently under research at MIT is a process that takes CAT scans and MRI (magnetic resonance imaging), both already digital, and creates a multiplexed CGH from the data.

Footnote:

1. *Thick And Thin Holograms:* There are two popular mathematical models used to describe the pattern of light waves coming through a hologram. One model, the Raman-Nath diffraction pattern, is based on the assumption that the thickness of the emulsion is small compared to the average spacing of the interference fringes. The other model, called the Bragg diffraction pattern, is used to describe results when the thickness of the emulsion is large compared to the average spacing of the interference fringes. There is a "boundary" between the two models where neither model holds up perfectly. Therefore, one has to say that the Raman-Nath "thin hologram" model works when the emulsion thickness is considerably less than the fringe spacing and the "volume hologram" Bragg model works when the thickness of emulsion is considerably greater than the fringe spacing.

References & Reading:

For Binary Detour Phase Holograms:

Brown, R.B. & Lohmann, A.W. (1966). Complex Spatial filtering with binary masks. Applied Optics, 5, 967-9.

____ . (1969). Computer Generated Binary Holograms. IBM Journal of Research & Development, 13,160-7.

Burckhardt, CB. (1970). A simplification of Lee's method of generating holograms by computer. Applied Optics, 9, 1949.

Dallas, W.J. (1971a.) Phase quantization - a compact derivation. Applied Optics, 10,674-6.

____ . (1971b.) Phase quantization in holograms - a few illustrations, Applied Optics, 10,674-6.

Goodman, J.W. & Silvestri, A.M. (1970). Some effects of Fourier domain phase quantization. IBM Journal of Research & Development, 14,478-84.

Haskell, R.E. & Culver, B.C. (1972). New coding technique for computer generated holograms. Applied Optics, 11, 2712-14.

____ . (1973). Computer generated binary holograms with minimum quantization errors. Journal of the Optical Society of America., 63, 504.

Lee, W.H. (1970). Sampled Fourier transform hologram generated by computer. Applied Optics, 9, 639-43.

Lohmann, A.W. & Paris, D.P. (1967). Binary Fraunhofer holograms generated by computer. Applied Optics, 6, 1739-48.

References for The Kinoform:

Kermisch, D. (1970). Image reconstruction from phase information only. Journal of the Optical Society of America, 60, 15-7.

Lesem, L.B., Hirch, P.B., & Jordan, J.A. (1969). The Kinoform: A new wavefront reconstruction device. IBM Journal of Research & Development, 13, 150-5.

Lohmann, et alia. Binary Fraunhofer holograms generated by computer. op.cit., pp.1739-48.

References for the R.O.A.C.H.:

Chu, D.C., Fienup, J.R. & Goodman, J.W. (1973). Multi-emulsion, on-axis, computer generated holograms. Applied Optics, 12, 1386-8.

References for Computer Generated Interferograms:

Bryngdahl O. and Lohmann, A.W. (1968) Interferograms are image holograms. Journal of the Optical Society of America, 58,141-2.

Lee, W.H. (1974). Binary synthetic holograms. Applied Optics, 13, 1677-82.

____ . (1979). Binary computer generated holograms. Applied Optics, 18,3661-9.

References for Three Dimensional CGH:

Benton, S.A. (1982), Survey of holographic stereograms, SPIE Proc., 367, 15-19.

____ .(1987), Alcove Holograms for Computer-Aided Design, SPIE Proc, 761, 53-61.

Brown, et alia. Computer generated binary holograms. op.cit. pp. 160-7.

Holzbach, M., (Sept. 1986), Three Dimensional Image Precessing for Synthetic Holographic Stereograms, M. Sc. thesis, Massachusetts Institute of Technology.

King, M.C., Noll, A.M. & Berry, D.H. (1970). A new approach to computer generated holography. Applied Optics, 7, 1641-2.

Krantz, E .. (Sept. 1987), Optics for Reflection Holographic Stereogram Systems, M. Sc. thesis, Massachusetts Institute of Technology.

Lesem, et alia. The kino form: a new wavefront reconstruction device. op. cit. pp.150-5.

Teitel, M. (Sept. 1986), Anamorphic Ray Tracing for Synthetic Alcove Holographic Stereograms, M. Sc. thesis, Massachusetts Institute of Technology,

Waters, J.P. (1968). Three-Dimensional Fourier transform method for synthesizing binary holograms. Journal of the Optical Society of America, 58, 1284-8.

8

Holography Education

We will be the first to admit that there is only so much knowledge to be garnered from this or any book. Therefore, it will be our aim in this chapter to inform you of the different kinds of holography courses that are available, and the classroom and/or laboratory approach of some.

We may begin by dividing holographic education into broad realms. There are some colleges and universities where holography is implemented in the curriculum as an independent course, but often within another discipline (i.e. Physics, Optics, Lasers, or Art). In some schools it is only included peripherally as part of one of these classes.

There are also a number of holography galleries, museums, and individual artists with labs who teach holography, free from the constraints of a university accredited program. You may learn there what you want or need to know in a day or weekend. The emphasis at these sites tends toward hands-on application 2nd techniques, with brief discourses into the theory of holography. It might be best to do some self-study prior to entering these programs, rather than "going in cold".

Some of the larger annual workshops open to the general public are run at the State University of New York (Buffalo, NY, USA, six years running); Lake Forest College, Lake Forest IL (USA, 19 years running); and Newcastle-Upon-Tyne Polytechnic, England.

Holography education is also expanding into elementary, junior high and high schools worldwide. It has been found to be an effective teaching tool for students with learning disabilities. Among the institutions and persons reaching out to primary and secondary schools (sometimes through state-funded programs, such as the State Education Department of New York's "Holography in the Classroom" program) are the Museum of Holography (NY, USA); Arelene Jurewicz (ME, USA); Doris Vila (NY, USA); Debra Duston (Ottawa, Canada), The Ontario Science Centre (Toronto, Canada), and Linda Law (Huntington, NY, USA).

Questions you should ask before you enroll in a holography class are:

1) How long does this class last? The time spent can vary from a one-hour lecture to a four-year program.

2) How much do you want to spend (always a factor!)? There are free lectures, and $3,000.00 + programs. Some schools charge by the hour, some by the day. You should also check to see if the cost of your chose program includes any texts/materials.

3) Does the class meet your specific needs? Is it designed for professionals or beginning holographers? In some instances, schools can tailor their instruction to your level. For example, the Los Angeles School of Holography offers a corporate seminar level program called "Holography Expo" which they hold in a pre-determined location (hotel, banquet room) for a per-person fee.

We surveyed a wide spectrum of holography schools, and colleges which include holography, in some manner, in their curriculae. Due to the large number of freelance educators, it was decided to make the distinction in the charts following this chapter between schools which have regularly-scheduled programs and those which teach by appointment. For more complete information, please refer to their listings in the Names and Addresses section. In our charts, there is also a heading of Education which is more comprehensive.

If you attend or teach at a college or school which is not listed here, but you feel should be, please send us complete information on your program, including a schedule of classes, and we will be certain to include it in the fourth edition of Holography Marketplace.

Regularly-scheduled Classes
Texas State Technical Institute, Waco, TX 76705 USA (817) 799-3611. Terry Kleypas, ext 2819. .

Saint Mary's College, Notre Dame, IN, USA. (219) 284-4696. Doug Tyler.

Pasadena City College, Pasadena, CA USA. (818) 578-7378. Fred Unterseher.

Royal College of Art, London, England UK. (071) 584-5020. Rob Munday.

Newcastle Upon Tyne Polytechnic, Newcastle Upon Tyne, England, UK. (091) 232-6002 ext. 3660. Graham Rice.

Art, Science and Technology Institute. Washington, D.C. USA. (202) 667-6322. Laurent Bussaut. .

Central Michigan University. Mt. Pleasant, MI USA. Richard Kline.

School of Holography/Chicago. Chicago, IL, USA. (312) 226-1007.

School of the Art Institute of Chicago, Chicago, IL USA. (312) 784-1669. Ed Wesly.

Massachussetts Institute of Technology, Spatial Imaging Group, Cambridge, MA USA. (617) 253-0632. Linda Conte.

Annual Workshops
Lake Forest College, Lake Forest, IL USA. (708) 234-3100 Ext. 340. Dr. T.H. Jeong.

SUNY Buffalo, Buffalo, NY USA. Gerald O'Grady. (716) 831-2426

Newcastle Upon Tyne Polytechnic, Newcastle Upon Tyne, England, UK.

Classes By Appointment
Holography Unit, Wolverhampton Polytechnic, Wolverhampton, UK. (0902) 321966. Graham Saxby.

Holography Institute. Petaluma, CA USA. (707) 778-1497. Patty Pink, Jeff Murray.

Los Angeles School of Holography, Los Angeles, CA USA. (818) 703-1111. Jerry Fox.

Holographic and Media Institute of Quebec. Quebec, Canada. (418) 656-7305. Marie-Andree Cossette.

Doris Vila Holographies. New York, New York USA. (212) 686-5387.

Holographic Studios. New York, New York, USA. (212) 686-9397. Jason Sapon.

New York School of Holography. New York, NY USA. (212) 254-9774. Sam Moree, Dan Schweitzer.

Ontario Science Centre, Ontario, Canada. (416) 429-4100. Peter Harris.

Photon League of Holographers. Toronto, Canada. (416) 531-1224.

Museum of Holography. New York, NY. (212) 925-0581. Sue Cowles.

9

Embossed Holography

Due to the commercial popularity of embossed holography, we treat embossed holography separately and in depth here. In this section we cover the following items:

- General description of some successful uses of embossed holograms.

- Costs involved in buying an embossed hologram on the commercial market.

- Outline the main steps used to create embossed holograms.

- More detailed explanation of each step used to make embossed holograms.

Although embossed holography is used in HOE and other applications, we assume in this chapter that most people are interested in using embossed holography for its commercial artistic effect.

Successful Applications

Clearly the most well-known application of holography, and one of the most lucrative, is embossed holography. As security devices, embossed holograms appear on bank credit cards and general sales merchandise. As product advertising they appear on cereal boxes, books, and magazines. They are also used extensively in laser disks and numerous HOE devices. Here are some commercial examples of how successful embossed holograms can be.

1) Food Boxes: A good example along this line is Ralston-Purina. They developed three distinctive hologram images and embossed them on every box of their "Ghostbusters" cereal. This initiated a brand sales increase of 300%. They are repeating the experience with another campaign .

2) Magazine Tip-ins: Uddelhom Steel Company used a holographic image in its first penetration of the US market. This was the focal point of an insert run in "Iron Age" magazine. A subsequent survey of their readers revealed the ad received a 221% effective

ness rating.

3) *Credit Cards:* Probably the most obvious example. The hologram integrated into each Mastercard significantly reduced the passing of fraudulent credit cards which had previously resulted in counterfeiters charging in excess of $35,000,000 annually.

4) *Magazine Covers:* Applying an eagle hologram to the March 1984 cover of National Geographic magazine resulted in over 400,000 new subscribers to the publication and over 14 months of readership response. Later, an image of the Taung child skull was integrated into the cover of the November 1985 issue, resulting in a 25-year record for advertising revenue in a single issue. Numerous magazines have put holograms on their covers since then.

5) *Labels:* Adhering holographic security labels to Prince tennis racquets forestalled a counterfeiting problem that kept sales at an equivalent level for six successive years. With the protection provided by the holographic image, Prince's sales demonstrated a threefold increase.

6) *Book Covers:* Zebra Books, one of the nation's largest mass paperback publishers, put a hologram on the cover of every book published in one of its romance series. The program was so successful that they are now adding a second line. This means literally millions of covers with embossed holograms on them.

7) *Tickets:* With the price of entertainment constantly going up, tickets to events have become a target to counterfeiters in a mach more serious way than in years past. Holograms on tickets work extremely well at preventing counterfeiting. Tickets for the Superbowl in January of 1991 were printed with holograms, with great success.

8) *Currency:* Everyone is well aware of the fact that color copy machines are quickly getting to the stage of perfection that will make counterfeiting very easy. In the words of a security printer, this means that manufacturers of currency will be forced to take measures to enlarge the expense and time necessary for counterfeiting[1]. Embossed holograms are the primary candidates for filling this need. The Reserve Bank of Australia has already issued the first currency with an embossed hologram (a $10 banknote). Although the first printing was flawed and the image tended to wear off, subsequent research and a second printing has made the banknote durable and success-

ful. Consequently, Australia will begin replacing all notes greater than $10.00 with hologram notes starting in 1991. The Reserve Bank of Australia has expressed interest in printing currency for other countries that are interested in this technique. They have already printed a Singapore $50 bill which has been released (see page one photos). A major likely competitor for them might be American Banknote.

The above embossed hologram examples are some of the best-known to the general public, and have been mass-produced, in some cases, by the millions. We may point out, however, that many clients for embossed holography, working with knowledgable, independent holographers, enjoy similar successful runs of much smaller numbers at much lesser costs than the big projects listed above.

Buying An Embossed Hologram On The Commercial Market

A recent issue of a trade publication for magazine purchasing agents was devoted to embossed holography and its costs (Publishing Technology, July 1989 North American Publishing Co., Philadelphia, PA pp 26-33).

The editors commissioned a hologram and had i hot stamped on the cover of the magazine. They reported the following costs involved:

· 3D models cost in the range of $1,000.00 tc $4,000.00 depending on the amount of detail.

· Masters cost between $2,000.00 and $5,000.OC depending on the complexity.

· The cover of the magazine had a 2.75 x 2.00 cm (7 x 5 inches) hot stamped hologram of a 3D model Modeling and mastering, they reported, cost about $5,500.00. Embossing the hot stamp foil cos: approximately $0.10 per image. Hot stamping cost $0.03 to $0.06 per image.

· Embossed foil cost $7.20 for 1,000 inches. Nonembossed foil sold to embossers cost $1.46 for 1,OOC inches.

· Polaroid has a proprietary process and photopolymer material which it calls "Mirage holograms" On average it costs $0.10 per square inch for production runs and the price can drop to $0.05 per square inch with volume. The mastering would

- ▶ Light Impressions, first in its field worldwide, through its commitment to pioneering research and development, remains at the forefront of embossed holography technology.
- ▲ Creative consultancy in the initial stages of a project leads to the optimum holographic interpretation of our client`s brief, whether a graphics, three-dimensional, easy view, animated or multi-exposure treatment is required.
- ◀ The experience gained over a decade provides the customer with a wealth of proven finishes and applications technology.
- ▼ Setting a standard of excellence at every stage in production ensures total customer satisfaction.

- ▶ Durch ihr ständiges Bemühen in Forschung und Entwicklung verschafft Ihnen Light Impressions einen wichtigen Vorsprung im technologischen Bereich der Prägeholografie.
- ▲ Unsere kreative Beratung gleich am Anfang eines Projekts gewährleistet eine optimale holografische Umsetzung des Entwurfs unserer Kunden, ob eine grafische Darstellung, ein dreidimensionales Motiv, "Easy-view" oder eine Mehrfachbelichtung gewünscht werden.
- ◀ Unsere Kunden profitieren von unseren reichhaltigen Erfahrungen, die wir im Laufe eines Jahrzehnts sammeln konnten. Wir sind in der Lage, unsere Prägehologramme so auszustatten, dass Sie diese in den verschiedensten Verfahren problemlos weiterverarbeiten können.
- ▼ Der hohe Standard in jeder Produktionsphase bedeutet für unsere Kunden Qualität und Sicherheit.

- ▶ Light Impressions, premiers dans leur domaine en Europe grâce à leur programme poussé de recherche et de développement, demeurent en tête dans la technologie de l`holographie estampée.
- ▲ Le service de conseil en création que nous offrons aux stades initiaux d`un projet permet d`obtenir une interprétation holographique optimale des désirs de notre clientèle et de savoir s`il faut choisir une image graphique, une image à 3 dimensions, reproduire un hologramme visible a`l`oeil nu, animé ou exposé sous différents angles.
- ◀ L`expérience de plus de dix ans dont nous bénéficions offre au client une preuve de qualité a la vue de nos produits finis et de notre maitrise d`application. Notre première préoccupation étant de maintenir une qualité excellente à tous les niveaux de production nous assure toujours une satisfaction totale de notre clientèle.

- ▶ ヨーロッパでの先駆者として、ライト・インプレッションは研究と開発を重ねつつ、常にエンボス・ホログラム技術の最前線に立ってきました。
- ▲ まずお客様ときめ細かいクリエイティブな打合せを行います。それに基づき、奥行き感のあるイラスト、立体画像、イージービュー、動く立体画像、多重露光など、いずれかご指定された中に、最大限可能性を追及したホログラムを演出します。
- ◀ ライト・インプレッションは、10年間の経験で得た定評ある仕上がりとアプリケーション技術の豊富さを駆使して、お客様のニーズにお応えします。
- ▼ 製作のあらゆる段階で厳しい基準をもうけていますので、トータルにご満足頂けることをお約束します。

Holography:
The Inevitable Technology
in the Evolution
of Imaging™

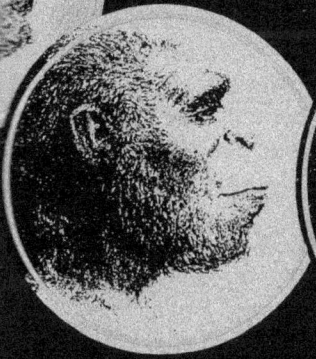

MAXIMAGE™
EMBOSSED HOLOGRAMS

By Bridgestone Graphic Technologies, Inc.

Tear out at perforation.

Bridgestone Announces a Dynamic Consolidation of Talent...

Rich Zucker, General Imaging Corporation, and **Doug Miller, Holographic Design Incorporated,** have joined in partnership with Bridgestone Graphics to form the most comprehensive holographic resource in the industry.

Product design and marketing, holographic development, primary manufacturing and post-manufacturing conversion all take place in our 50,000 square foot facility.

At Bridgestone we also know that technology alone does not make a company. Our group consists of seasoned business professionals who understand the importance of budgets and schedules and we are committed to meeting your requirements.

Evolution doesn't have to take millions of years. Bridgestone can help you evolve in a matter of weeks.

Bridgestone's fully integrated manufacturing facility offers you these advantages:

• Fine quality   • Affordable Prices
• On Time Delivery

So, take the lead over your competition and introduce your customer to the latest in visual communication and introduce yourself to the best in holographic imaging . . . Call Bridgestone today!

1-203-366-1595

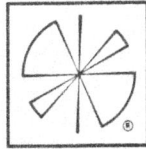

cost $8,000.00 to $15,000.00. The holograms can be produced from .66 x .75 inches to 10 x14.5 inches.

In general, the magazine editors found that the big appeal of embossed holography is that it definitely increases the percentage of returns on direct mail pieces and increases circulation for magazines and books. On the down side, they noted that the mastering cost is the one item that keeps this from being very cost-effective in short runs.

Embossed Holography Step-by-step

Embossed holograms are created in several steps. We will present an overview and then, for those that make holograms, we will cover each step in the making of one type of embossed hologram.

Overview:
The most common steps, once the model (not a simple step) is done, are as follows:

1) The hologram master is made.

2) A rainbow transmission hologram is made on photoresist (photosensitive emulsion) from the master.

3) The photoresist is etched to relieve the hologram patterns.

4) The photoresist is plated with silver and nickel. The photoresist now behaves like a metal mold.

5) The metal mold, or shim, is removed from photoresist and now has holographic patterns on it.

6) This shim is used as a stamping die and stamps the holographic patterns into plastic.

7) The plastic with the holographic pattern stamped in it has a mirror-like backing so light comes through the plastic, strikes the mirror-like backing and, reflecting back out, displays the white light rainbow transmission hologram.

Let's look at these steps in a little more detail and then discuss the model making.

Steps 1-2: There is a possible shortcut here. Sometimes flat art can be directly exposed onto the photoresist material. This is recommended for simple projects only.

Photoresist is a very tricky medium to record on ho-

lographically. It is nowhere near silver halide in responsiveness and requires long exposures even with lasers of several watts of power. A small holography studio would have to be equipped with an expensive laser and heat-resistant optics. New alternatives to this medium of photoresist are being researched.

The typical turn-around time is four to six weeks to receive excellent photoresist plates that are fit for metallizing. The holographer should make several photoresist plates of good quality to cover any problems that might occur in the metallizing phase.

Steps 3-4: After the photoresist hologram is checked for clarity, brightness and overall quality it goes to the metallizing stage of production. A thin layer of silver is deposited on the photoresist. Silver by itself cannot withstand the stamping pressure, so additional coats of a nickel-based material are deposited to reinforce the back of the silver. When it achieves the desired thickness, the nickel-silver shim is pulled from the photoresist plate and this becomes a metal mother die.

Steps 5-7: The first shim must be perfect. This first shim can have several shims made from it and in turn several shims made from those. The heat and pressure of embossing thousands of holograms wears out the shims so extras should be made. Any deterioration becomes very obvious if the shims are not changed regularly.

Producing the final hologram is the job of the embosser. Embossed holograms are made by stamping the shim onto aluminum. Heated polyester material which has a metallized backing is used less often. Aluminum is softer and less destructive of the nickel master. Although not used widely, colored metallized backing is an option. Most often the silver color is selected.

Applying embossed holograms: The press embosses the holograms on rolls of stock which, in the most common commercial work, have an image area of six square inches for each impression. The roll consists of a continuous ribbon of 6 x 6 inch embossed squares separated from one another by about one-half inch. The holograms are often arranged on each 6 x 6 inch square as nine, 2 inch squares of different holograms, but you can use the entire 6 x 6 inch square for one hologram.

There are a number of options for displaying your final embossed product. Embossed holograms can be produced as stickers cut to your size specifications. Typical of the final holograms are individual peel-and stick holograms. If the stickers are to be placed onto some surface be sure the substrate the sticker is on is thick and strong enough not to conform to a textured surface onto which it might be applied. The best surface on which to stick a hologram is a smooth and rigid one so you can be sure that the hologram is flat and consequently able to reconstruct its image properly. There are applicator machines to apply the sticker to your product, but it is often done by hand.

Instead of stickers, you may apply the hot stamp foil directly to a surface like a book cover. If you choose this route, be absolutely certain of the strength and smoothness of the cover material. The embossing foil is very thin---like the foil around individual sticks of chewing gum. If the hologram has ripples in it as a result of a bumpy surface, the image itself might appear rippled.

It is a good idea to have the surface approved by he holographer and the printer, together, before going into production. Also beware of coated paper stock and printed surfaces. These surfaces may present problems for adhesion of holograms.

Hot stamping the hologram onto a paper surface is another popular way to apply your hologram. In general, the hot foil application keeps precise registration better than methods of sticker application. Although hot stamping appears more expensive, when you consider the costs of a very large run using sticker application they both can come out about equal in cost. Smaller runs seem to favor sticker application.

Regardless of which method you choose, it is advisable to be there when the holograms are applied so you can check the quality.

Embossing pulsed holograms and holographic stereograms: Pulsed hologram images are used in embossing and have been very successful. Be sure the pulsed holographer knows that you want this to be an embossed hologram. It is a more complicated approach which may include a step to reduce the pulsed hologram to the six-inch square size of the embossing machine, or whatever limitations your embosser has. Do not use this approach if your turnaround schedule is too tight. Allow enough time to make sure all steps take place with breathing room if a problem occurs. The results of pulsed embossing are stunning and it is worth pursuing.

Preparing Artwork For Embossed Holograms.[2]

Embossed holograms made for advertising or artistic effects have a common problem. Most of them are intended to be seen under any available lighting condition, which includes very harsh lighting like fluorescent lights. This means that the hologram you make cannot have a great deal of projection or depth, because fluorescent light does not display depth well. This means that 3D objects, if they are used, must be very shallow. The most common objects used are either flat artwork or very thin, miniature models.

3D Holograms: A 3D model is frequently made to act as subject in 3D holograms although very lifelike images can be created with photos. Close consultation with the holographer before and during the making of the model is important. In creating a 3D model, the following points need to be kept in mind:

· Get exact specifications on what volume of space your model can occupy before starting. Remember that an embossed hologram has very little depth. Discuss the material. being used to create the model with the holographer.

· The final printed color of the model does matter. Colors that turn out well depend on the laser used to expose the object.

· Once your model is made and painted, you can view it under the light of the laser that will make the master. This will give you a rough idea of what areas of your object the laser light reflects strongly from and from what areas it reflects weakly. You should remember, though, that your hologram may be viewed in unusual lighting conditions and the actual color seen depends on the ambient lighting conditions.

Some of the leading embossers, however, use unpainted models of special white plaster. It is suggested by one holographer that if you insist on painting your model, it is easier if the model maker uses a safelight (available from Roscoe) covered with a nontransparent filter gel of approximately the right color to simulate laser light. This is a simple method of judging appropriate paint colors.

2D3D Holograms: A 2D3D hologram is a hologram made up entirely from flat artwork appearing on two or more levels. If the flat artwork is entirely one level, the hologram is called a "2D" hologram. If it is layered on two or more levels, the hologram is called a "2D3D" hologram. The attraction of a 3D hologram is that you have a model you can see before the hologram is made and, although shallow, there is some 3D depth to the final embossed hologram. Although one might think that a 3D hologram is the "best money can buy", 2D and 2D3D holograms have several advantages.

Advantages of 2D3D holograms from the holographer's point of view:

· 2D3D holograms are all made using exactly the same holography setup, which never has to be changed.

· The "object" for a 2D3D hologram is simply flat artwork, so there are no object motion problems or object mounting challenges.

· All of the image information for a 2D hologram or the 2D part of a 2D3D hologram is contained on the surface of the hologram, so the 2D part of the image can be made much brighter than an ordinary 3D hologram without loss of image detail.

Advantages of 2D3D holograms from the point of view of the graphic artist and the ultimate user:

· Graphic artists are well acquainted with two-dimensional media, so their techniques and skills are easily adapted to designing 2D and 2D3D holograms.

· It is possible to make mock-ups of 2D3D holograms as an aid to making a sale, without going through the hologram mastering process and expense.

· The graphic artist who designs the 2D or 3D hologram can define all aspects of the image: artwork, imagery, colors, and image placement. (The holographer may assist the graphic designer, however, in proper color choices because the former sometimes possesses a better understanding of spectral events, and can therefore produce exciting kinetic effects.)

· 2D3D holograms can be viewed in any kind of lighting without difficulty, making them suited for the many applications in which there is little opportunity to control the lighting, such as packaging, point of purchase displays, advertising, anti-counterfeiting, textiles, clothing, toys, greeting cards, and decorations.

· The greater brightness and sharpness of 2D3D holograms compared to other types of holograms make them excellent eye-catchers.

2D and 2D3D Artwork: The artwork for 2D and 2D3D holograms is essentially the same sort of flat artwork that any ordinary printer would use. The only difference is that, since a hologram allows a person to look around a foreground image, the back-ground should be complete, without "cutouts" for the foreground. Color separations are usually prepared from line art, and the color for each separation is specified. It is also possible to produce 2D3D holograms from color photographs, though it is more complicated.

Line art for 2D and 2D3D holograms should have an unbroken line or boundary for each region of any given color. Each layer in the image should have its own separate artwork. Color overlays should be provided with line art to designate the colors of the various parts of the image in each layer. Each layer should have its own color overlay.

Step-by-step Making Of An Embossed Hologram:

Some people are interested in an explicit step-bystep explanation of how an embossed hologram is made. There are many methods and this is the account of only one method. We list the steps and then cover each one in detail. The steps involved are:

1) Make a laser transmission hologram.

2) Copy hologram onto positive photoresist emulsion.

3) Deposit conductive layer on photoresist.

4) Electroforming of metal masters (1st, 2nd, etc.).

5) Emboss plastic by roller or flat platen.

6) Apply hologram to product at bindery or by hand.

Now we describe each of the above steps in detail.

1) Make a master hologram
The master hologram is generally a rainbow transmission hologram. If you intend to make your own master hologram and have another company do the next step, exposure of the photoresist, then close cooperation is necessary. Some companies require the reference beam to be at a specific angle and perhaps even a specific wavelength. Most likely, the people that do the photoresist step will also want to see a final transfer hologram to use as a guide. Quality, of course, is absolutely essential. Any defects in the master will be magnified in the subsequent steps. If you are creating a multiplex embossed hologram, the Benton method is preferred since the images are truly multiplexed and not slits.

2) Coating and exposing the photoresist

If you intend to coat your own photoresist, there is a good article by Burns that you should read because it describes both photoresist and electroplating in detailS. There are also worthwhile discussions by Saxby4, Bartolini and Clay. The theory of embossed holography has been covered in technical literature as early as 1973[4].

Emulsion: Most holograms to be embossed are recorded in Shipley emulsions because they are commercially available positive photoresist emulsions with sufficient resolving power for holography (other competitors for Shipley are Towne Labs and Resist Masters, Inc.). This photoresist emulsion is designed to be used for the microcomputer industry, however, and consequently does not have the sensitivity that one would wish for making holograms.

Some individuals have tried to make their own resist coating. It is not recommended as resist plates of excellent quality are readily obtainable up to 7 x 7 inches5. Emulsions are generally applied by either spin coating or dipping. Spin coating is the preferred method and you can find suppliers that will coat plates for you, but you should be very cautious about the results. Your coating should be mirror-smooth, and free of defects and striations when viewed in yellow safelight.

Exposure: A frequently used exposure is at 457.9 nm using an argon laserS. This is the argon wavelength to which this emulsion is most sensitive but, unfortunately, it is also one of the weakest argon wavelengths. If you use an etalon the available power is reduced by about 50%: At our desired wavelength of 457.9nm, an 18 watt argon with an etalon can put out about 700mw and a 5 watt argon with an etalon produces 120mw. Burns reports the Shipley emulsion is a factor of about ten times more sensitive to the 441.6nm wavelength, which is available from HeCd lasers, but HeCd lasers are not readily available with more than 40mw of power. In addition, the coherence length of HeCd lasers is poor relative to argon lasers equipped with etalons. Therefore, an argon laser with an etalon is preferred for HIs and H2s; whereas HeCd lasers are preferred only for contact copies.

To give you an idea of the sensitivity, Burns reports that the Shipley emulsion requires a threshold energy to be effectively exposed. At 457.9nm, it generally requires a total exposure energy of 250,000 ergs/cm^2 to 2,500,000 ergs/cm^2 depending on the thickness of the emulsion. Burns found he could achieve this exposure by delivering 150 ergs/cm^2 per

second for a 30 minute exposure or 420 ergs/cm^2 per second for ten minutes. Either way it is a long exposure and system stability could be a problem.

The exposure Burns finally used delivered 1,000 ergs/cm^2 per second and the exposure time was 4 minutes 38 seconds. Development was for 40 seconds with tray agitation at a temperature of 53° Fusing Shipley 303A developer diluted 6: 1 with distilled water.

3) Deposit Electrically Conductive Layer Onto Photoresist

We now coat the photoresist with a layer of electrically conductive silver. Three popular methods for coating and their respective benefits and disadvantages are described below. Regardless of the method, the thickness of the conductive layer only needs to be several hundred angstroms. You want a nice, even thickness with no pinholes or cracks. Watch the temperature to make sure it does not exceed the melting point of the photoresist.

METHOD	COST	SPEED	QUALITY
Silver Spray	Inexpensive	Fast	Possible quality problems
Vacuum Deposited Silver	Expensive	Moderate	Excellent quality
Electroless Nickel	Inexpensive	Moderate	Depends on operator

Figure 9.1. Popular methods for coating photoresist with electrically conductive layer.

Silver Spray: This method offers high production rates with low start up costs. It requires only a spray-gun system, spray booth, and solutions. You can use a two nozzle spray gun to mix the two reagents with a commercially available two part silver solution or mix your own solution from existing formulas.[7,8]

Vacuum-Deposited Silver: To proceed with this method, you need to buy vacuum metallization equipment. In this procedure, the photoresist is affixed at the top of a vacuum bell jar and a small quantity of silver is evaporated from a hot filament in the bottom of the jar9. The benefit of this method is that it requires little skill on the part of the operator, and therefore leaves less room for error.

Electroless Nickel: This is a three step immersion process that is inexpensive. It requires dip tanks (one of which must be heated) and solutions that may be mixed in house or purchased commercially.lo.n The steps are first to sensitize the photoresist with stannous chloride solution, then dip it in palladium chloride. The third and final step is immersion in electroless nickel deposit, which takes place in a heated tank. There is a wash step between each of the three solutions. The electroless nickel process does take some operator experience since the type of agitation, temperature of solution, immersion time, etc. can all affect the outcome .

4) Electroplating The Conductive Photoresist (Making The Stamping Shim)

The metal master or shim, as it is referred to in the plating industry, is the first generation of masters for your embossed hologram. Nickel sulfamate is used by almost everyone for the electroforming bath. It is commercially available and relatively inexpensive. I2 The size of the plating tank is, of course, dependent on the size of your master. The other components of your plating system are:

- DC rectifier
- Nickel anodes
- Filtration pump
- Solution heater
- Solution agitation system

Burns used a 12 volt, 100 amp DC rectifier with less than 3% ripple. The anodes were Inco nickel rounds contained in a titanium basket. A submersible fil tration pump, two quartz immersion heaters and a rocker-arm agitator were also used.

A number of factors must be controlled to have an efficient plating bath. 13 Among the items are:

- Current density
- Bath temperature
- Bath PH
- Bath specific gravity
- Agitation quality
- Bath cleanliness

Organic materials like rubber tubing that go into the bath must first be leached for at least 24 hours in a heated 10% sulfuric acid solution. Metallic materials that might have surface contaminants should also get this treatment. Continuous filtration of the system keeps down the contaminants, agitates the solution and helps stabilize the temperature. A continuous flow of fresh solution should pass over the object being plated.

Sometimes tensile stress causes the nickel deposit to curl away from the photoresist and there are additives for the solution to control this.I4 These additives also tend to increase the grain size and make the master brittle.[15]

Jigging: Before you immerse the photoresist in the electroforming bath, you need to make a conductive frame, or jig, to hold the photoresist. The idea is that the DC current from the rectifier goes through the jig to the conductive layer on the photoresist. The jig, then, should make a complete circuit around the perimeter of the plate and not damage the conductive layer. Jigs are usually made in-house but you can visit electroplating suppliers for ideas. In the electroplating industry, the jig together with the part to be electroformed is called the mandrel. It is immersed in the solution and is the cathodic element of the plating system. Some jigs are made to be used only once, and some are reusable.

One of the problems associated with the jig is that plating starts on the jig and works toward the center. This can result in your plate having thick edges and depressions or voids in the center of the plate. To help minimize this, your jig should be at least 25% larger than the desired finished embossed hologram.

Plating The First Generation Master: The plating process is like any normal plating process in that you have to constantly monitor temperature, current density, pH, and concentration of solution. The solution used in the tank is nickel sulfamate. The anode is a bar of pure nickel and the cathode is your mandrel. Be careful of long immersion because the thicker the plated deposit and the longer the plating cycle the greater the risk of defects. A perfectly flat metal master without flaws is your goal.[16]

Once your shim has been plated, you remove it from the jig and separate the metal shim from the photoresist. Sometimes you can pry the two apart with a razor blade and sometimes it takes a sturdier tool. Great care should be taken at this step because you are essentially working with a piece of metal foil which may be as thin as .05 mm (.002 inches) and it can be easily damaged. After separation, rinse the metal shim in a solvent to remove any resist that might be stuck to it. Inspect your shim to see if it is good or if you have to go back and make another photoresist. The smart thing to do is to make more than one photoresist to begin with, just in case you fail on your first try.

Duplicating Metal Shims: Now that you have your first generation master, you can use your first master to make the next generation of masters. Here are some things to keep in mind:

- Each metal master can make approximately ten "next generation" masters.

• Each time you make a next generation master, your image is reversed left to right.

• Each time you make a next generation master, you lose about 1 cm (.25-.50 inches) from each side.

• The production run from each metal master varies widely with the operator and machine, but by a conservative estimate you should get at least 2,000 holograms per master.

• You should not go beyond three generations of metal shims because there is a little degradation with each step.

• The image on the final shim that you use for stamping should look just like the image you want the customer to see.

This means that you have to plan ahead. First, you have to decide how many holograms you are going to have. If it is 20,000 or less, then you make second generation shims and use them as stamping shims. If you plan for a very large production run, you use the third generation shims as your stamping shims. You may have to reverse the image when exposing the original H-1.

After your first generation shim has been washed with solvent to remove residual photoresist, it is necessary to passivate the first generation shim so that the second generation shim does not permanently bond to the first shim. To do this, the first shim is immersed in a 2.5% potassium dichromate solution for 1-10. minutes and then is rinsed in distilled water.17 It is then mounted in a jig and you proceed to plate the second generation · shim as you did the first. The third generation shims are made in the same manner.

5) Embossing

Now we have our stamping plate and we are ready to emboss. There are three methods of producing embossed holograms.

• Flat bed embossing
• Roll embossing
• Hot stamping (also called hot foil blocking)

The most commonly used material in both flat-bed embossing and roll embossing is thermoplastic, typically PVC (polyvinylchloride), and it is usually .075 mm to .20 mm (.003 to.OOB inches) thick. Polyester is also used but is more difficult to control. The unembossed vinyl base is the viewing side because the embossed hologram surface is only a few thousand angstroms deep and very vulnerable to scratches. This is the reason the stamped hologram has its image reversed left to right.

Flat Bed Embossing: There are three main reasons this method is used.

• The machines are inexpensive (approximately US$3,000.00+).

• Some companies that do a lot of embossing want to have a proof press.

• You are able to do large format holograms that cannot fit on roll or embossing machines.

In flat bed embossing, two flat metal platens are each internally heated, and pressure is applied to the shim and plastic. The platens are then cooled until the temperature is below the plastic flow point. Then pressure is released. The entire cycle is measured in seconds to minutes depending on the type of plastic and machine. This process is obviously slower than roll embossing but it can yield better quality holograms.

Production rates can be increased by stacking the shim and plastic so you have a plastic-shim-plastic-shim-plastic-shim sandwich. If you have a large platen, you can also spread the shims out and make several holograms at a time. Roll plastic can be advanced through the machine between impressions.

The metal shims for platen work have to be rigid and very flat. It is also desirable to have the shims a little thicker since they will be subjected to continual heating and cooling and they sometimes have to be pried apart from the plastic. The shim's metal thickness varies with hologram size from about .25 mm (.01 inches) for a 5 x 5 cm (2 x 2 inches) hologram to 1 mm (.04 inches) for a 30 x 30 cm (12 x 12 inches) hologram or larger.

Roll Embossing: Here are a few main points about the machinery.

• The machinery is expensive (costs are around US$25,000.00+).

• Very fast production rates (30 meterslminute+). Millions of holograms are made using this method.

• Because of the relatively low heat and pressure subjected to the shim at anyone time, the shims last considerably longer than with platen presses.

• This is a good method for any quantity of holographic stickers like the ones you see on cereal boxes, etc.

These machines can emboss masters that are as large as 30 x 60 cm (12 x 24 inches). Much larger thermoplastic roll embossing machinery exists but it has not yet been used for embossing holograms.

Roll embossing machines use a multiple pressure-roller system. The metal master is attached to an internally heated roller. A roll of plastic is passed between this roller and an adjacent pressure roller. The metal master should be about .05 mm (.002 inches) thick and should be flexible enough to conform exactly to the shape of the roller. A new roller is used for each job because the metal masters are permanently attached to the roller. The deviation between masters should be kept to a minimum (±.005 mm or .0002 inches). Variations create dark areas or image voids.

In some systems, a roll of plastic is mounted on the machine. In other systems, plastic is extruded from pellets, calendered, and embossed in one continuous operation. The typical embossed hologram is .076 mm,.003 inches) to .20 mm (.008 inches).

The embossed roll is then aluminized on the embossed side to create a mirror backing. This protects the hologram from scratching and creates the mirror-like backdrop we are all familiar with in embossed holograms. This aluminized backing is also subject to degradation and it is usually covered with a self-adhesive backing and a plastic or paper peel-off called a liner if your desired final roll is a sticker.

Hot Stamping (or Hot Foil Backing): A few of the major points of hot stamping are:

• This is just about the only method used for security applications such as banknotes, etc. because the hologram cannot be removed without destroying the product.

• The final hologram is very attractive since it is almost flush with the product and appears to be a part of the product.

• This is cost-effective for the bindery of a printing company because they can use their hot-foil equipment, with some adjustments, to apply the hologram directly to the printed product.

• Once committed to the hot stamping process, you cannot use the hologram as a sticker, so you should consider your objective carefully.

Hot foil stamping has been around for a long time. If we had a roll of hot stamping foil in our hands and we started at the top and peeled each layer off we would go through the following layers.

• Plastic carrier sheet that comes off when hot stamped (Approximately 25 μm meters thick).

• Wax release layer that melts and releases the plastic carrier sheet when heated (Approximately .05 μm meters thick)

• Lacquer layer (Approximately 1.6 μm meters thick)

• Aluminum foil (Approximately .03 μm meters thick)

• Hot metal adhesive to bond aluminum to printed product. (Approximately 1.0 μm meters thick)

The lacquer layer, although very thin, is three times thicker than the indentation made by our embossing master. The lacquer layer is embossed for holograms before the aluminum foil and hot metal adhesive layers are put on. You obviously have to coordinate this with your foil provider.

In hot stamping, a heated platen presses on top of the above sandwich. The wax release layer and the hot metal adhesive both soften and the plastic carrier comes off while the hot metal adhesive bonds to the product. Our hologram is safe since it is under the lacquer layer. It is also bonded to the product and cannot be removed without destroying it.

6) Bindery (applying the hologram)

In a great many of the cases the company that does the embossing is not the company that affixes the hologram to the product. In both roll embossing and hot stamping the end product is a roll of holograms ready to go to your printer's bindery. The bindery then applies the holograms to your product either by a label machine, by hand, or by hot stamping the hologram onto the product. This last step should not be taken for granted. You have a considerable investment in your product at this stage and there are numerous cases where projects with very large investments have met defeat at the bindery.

There is also a little confusion of terminology in the industry because printers that do hot stamping usually do embossing of printed products as well, if their machines are designed to do both. Therefore, hot stamping and embossing, in the printing world, are closely-related terms. So when you say "Company X is doing the embossing", people can get confused and think that the printer is actually embossing the holograms.

Careful thought should be given to which product is best for your purpose. Labels can be applied in the bindery by hand or by label applicators. Any extra labels can be hand-applied to other short run advertising products. Hot stamping's benefit is that for long runs it is very cost-effective. The hot stamping machine also registers your product extremely accurately.

There are a number of hot stamping machines on the market. Bosc and Kluge are among the more famous and widely used.

Potential Problems At The Bindery: One of the most important areas to supervise your project is at the bindery. Some common problems occur with:

• *Stickers:* Hand application always leads to register problems and needs to be watched carefully. There do exist automatic labeling machines, such as Label-Aire, that can apply 250 to 1,000 stickers per minute. Machines like this help reduce the possibility of human error.

• *Hot stamp and stickers:* If the product you are putting your hologram on is uneven, the hologram does not play back as the ripples in the surface cause ripples in the hologram, thus causing the image to interfere with itself. This is especially critical for hot stamping since the hologram is so thin.

• *Hot stamp and stickers:* If the product on which you apply your hologram is not receptive to your adhesive, the hologram simply does not adhere to the product. Watch out for printed products that have been coated before applying your hologram.

• *Hot stamping:* The heat from the hot stamping machine, when overdone, can cause ripples in the surface of the product and destroy your hologram.

• *Post-application packaging problems:* An example of this would be making a hardbound book that has no dust jacket or a printed ad that will be affixed to a box. Your hologram, in this case, is hot stamped onto smooth plastic or paper that will cover the product. The hot stamping may look fine but later, when the bindery applies the stock (with hologram hot stamped on it) to the product, it may be subjected to a heating process that could warp the stock or product. The bindery people will feel the small amount of warping is well within their normal quality control limits but the heating and warping may be enough to destroy your hologram.

Footnotes:

1. Schell, K.J. (1985). White Light Holograms for Credit Cards, SPIE Proceedings, 523, 331·5

2. Light Impressions, Inc.

3. Burns, J. R (1985). Large Format Embossed Holograms, Proceedings of the SPIE, 523, 7·14.

4. Saxby, G. (1988). Practical Holography, Prentice Hall International.

Bartolini, R.A., Feldstein, N. & Ryan, RJ., (Oct. 1973). Replication of Relief·Phase Holograms for Prerecorded Video, Journ Electro·chemical Society:Solid.State Science & Technology, 120, 1408·13.

Clay, B.R & Gore, D.A. (1974). SPIE Proceedings 45,149·55.

5. Towne Laboratories, Sommerville, NJ.

6. Norman, S.L. & Singh, M.P., (1975). Spectral Sensitivity and Linearity of Shipley AZ1350J Photoresist, Applied Optics, 14,818·20.

7. Peacock, Inc., Philadelphia, PA.

8. Graham, K.A.,Electroplating Engineering Handbook, Van Nostrand Reinhold Company, Third Edition,(1971) p. 479.

9. Electroplating Engineering, op cit. pp. 482·3.

10. Electroplating Engineering, op cit. pp. 212·3.

11. Elnic, Inc., Nashville, TN.

12. Plating Products, Inc., Roselle, NJ.

13. M&T Technical Data Sheet #P·N1·Sn, M&T Chemicals, Inc., Rahway, NJ, Nov. 1979.

14. DiBari, G.A., "Electroforming", Metal Finishing. Guidebook & Directory Issue 1982, Metals and Plastics Publications, Inc. pp. 523·24.

15. Electroplating Engineering, op cit. pp. 212·3.

16. Harstan Technical Bulliten, "Nickel Sulfamate", Harstan Chemical Corp. Brooklyn, NY, May 1973.

17. Fraunhofer, J.A., Basic Metal Finishing, Chemical Publishing Co., NY, 1976, pp. 149·50.

10

Holography Market Place & The Year In Review

The global holography market is composed of thousands of companies, universities, and other organizations involved in developing different aspects of this technology. As the listings in this publication make clear, it is also composed of hundreds of associated equipment and service suppliers, users and investors. This chapter focuses principally on the "core" industry: the group of companies and other organizations that are working with the technology of holography for commercial or internal applications.

That industry is made up of a wide range of players. There are a few publicly-traded holographic companies. Applied Holographics PLC and International Bank Note are two of the most notable. There are a substantial number of major corporations with significant holographic efforts. Polaroid, DuPont, 3M, Pilkington, Dai Nippon, Hoechst Celanese, and ICI all fall into this category. Likewise, a number of medium-sized corporations are actively involved. Newport Research Corp., Ilford, Agfa-Gevaert, Crown Roll Leaf, and Ealing Electro Optics are examples of this tier of the industry. Of course, much of the research and some of the commercialization work takes place within the university or nonprofit

area. MIT in the United States, and Loughborough University and others in Europe are a few examples of these.

In typical pyramid fashion, there are a large number of small companies that make up the bulk of the holography industry. Although the size and focus of these companies ranges widely, there is a tendency for them to group according to applications area. For example, there is a developing worldwide network of artistic hologram producers, distributors and retailers. Often, even relatively small companies tend to be globally oriented within their applications segment. It is not unusual for relatively small embossed hologram producers to have affiliated companies on several continents.

Distribution

Holography is unquestionably a business for the Nineties. Still in its infancy, it has an annual growth rate that is better than many long-established enterprises. To be honest, any business like holography that is on the cutting edge of a new technology poses a lot of risk for those venturing into it. On the other hand, there is much to be made for those who succeed. The positive aspect of this growing field is that there is not a competitor waiting on every corner. Another advantage of this market is that while many mature businesses cringe at the thought of advancing technology destroying their markets, holography thrives on advancing technology.

We can divide the holography market into two broad categories, artistic holography and industrial holography. Artistic holograms can be defined as holograms created for the purpose of being seen and whose value derives, at least in part, from the image presented. Artistic holograms, therefore, can be anything from commercial art like bank card holograms to limited edition fine art holograms. Industrial holography encompasses items like supermarket scanners and holograms used for non-destructive testing of machine parts. Although still developing, there is a network of distributors, wholesalers and shops that sell artistic holograms. We will try to outline how this market works from the top down.

Copyright Holder

The distribution process starts with the copyright holder. As its name implies, copyright means that no one has the right to copy a unique work of art made by someone else without that person's permission. All works of art can be copyrighted and for any unique work of art there exists only one copyright holder. The copyright holder may be a group of people such as a business. This means that every hologram, if properly copyrighted, has only one owner, the copyright holder. Holograms that are not copyrighted can be copied by whoever has the master. This person can make as many copies as he wishes.

Sometimes the copyright holder is a business that pays a manufacturer to make a master hologram and multiple copies. On the other hand, sometimes the manufacturer is the copyright holder and can then make masters and copies itself. In order to keep our discussion as clear as possible, we describe each of the businesses in our distribution chain as separate entities. Therefore, the copyright holder is at the top of the chain and controls all distribution.

Whoever creates a unique work of art that later becomes a hologram is considered the copyright holder of the image unless there is a prior written agreement stating that the art, upon completion, is owned by someone else. This arrangement is common in a

"work for hire" situation. It is also possible to copyright a completed hologram as a work of art provided the objects in the hologram are not already copyrighted by someone else. There are statutory limits stating that after a number of years, a piece of art is held in the public domain and anyone may use the image.

The Manufacturer

The manufacturer is only a manufacturer and gives the buyer the best price for creating multiple copies of the hologram in question. The manufacturer's price is important because the copyright holder estimates the retail price of the product based on the unit cost from the manufacturer. To get the unit cost, take the total bill from the manufacturer and divide it by the number of copies made. Everything depends on volume. The more units you manufacture, the lower the unit manufacturing price. The lower the unit manufacturing price, the lower the retail price you can offer the product for. It is a broad generality, but many copyright holders take their unit manufacturing cost and mark it up approximately five times to get the retail unit price of their product.

The Chain Of Distribution

Let's assume the copyright holder decides to make a hologram. What does the distribution channel look

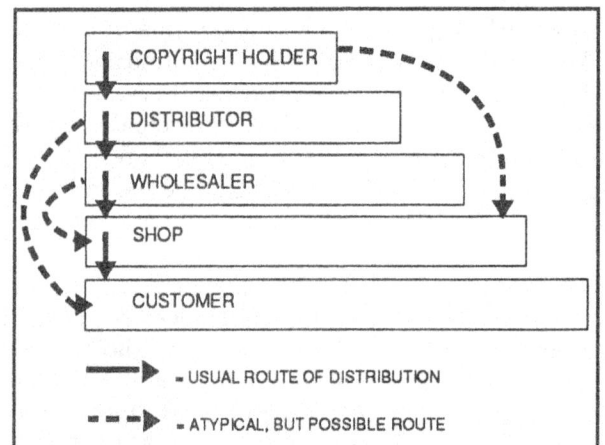

like from this point on?

The levels shown above exist in the holography market and there are some rough numbers that can show, approximately, the relationship between the different levels. It should be pointed out, however, that this is a young industry and frequently a busi-

THE HOLOGRAPHY DEVELOPMENT GROUP

....is
the name of our company.
Our team experience is next to none.
We are a network of professionals dedicated
to providing the business and graphics community
with all the resources it needs to identify and develop
commercially viable holographic products and applications.

Our
project developments
are always very demanding
which is why we only deliver unique solutions.
Over the past ten years we have been 'painting with light' for a
number of clients including Coca Cola, Dupont, Madame Tussauds,
Wellington Crypt Chapel, Midland Bank International, Harp Lager,
Prudential, McDonalds, The Royal Opera, London Zoo.

We will
continue to deliver
'State of the Art' productions by improving the standards
in the industry through better communications.

Allow
us to introduce you
to our new language in multi-dimensional design....

'H' SPACE ™

creative
design tools
design & art direction
modeling & animation
illumination
holographic images
producers
networking
marketing & sales
packaging
distribution
investments

for further information about 'H' Space, or to arrange
an appointment to visit the design showroom,
please call or write to
Andrew Laczynski
at
23 Neville Park Boulevard,
Toronto, Ontario, Canada, M4E 3P5,
telephone (416) 691-9381 • facsimile (416) 691-0407

ness does several of the above functions. There exists one business, for example, that owns copyrights to some holograms that they manufacture; they import and act as a distributor for certain products; they sell products to other wholesalers; they wholesale to stores and have a storefront themselves. They are everything but the customer! There are other businesses, however, that engage in only one level like a shop or distributor. We will discuss each of the above businesses as separate entities in the distribution chain.

We feel it is our duty to give some idea of how discounts work in this industry. It is an impossible task since there is no established order of business. After some research, however, we have come up with what is a rough idea of how a item is discounted. The chart below is for an item costing US$50.00 to the final customer.

BUSINESS	DISCOUNT OFF RETAIL PRICE	GROSS AS % OF RETAIL PRICE	COST TO BUSINESS (UNIT ITEM)	GROSS PER UNIT ITEM
(RETAIL PRICE = US $50.00)				
COPYRIGHT HOLDER	80%	30%	$10.00	$15.00
DISTRIBUTOR	70%	10%	$15.00	$5.00
WHOLESALER	60%	10%	$20.00	$5.00
RETAILER	50%	50%	$25.00	$25.00
CUSTOMER	00%	00%	$50.00	$00.00
		100%		$50.00

Fig. 10.2   Typical costs, margins, and profits for holograms

A rough description of each of the businesses follows.

The Distributor

The distributor buys the product in large volumes and frequently has a contract from the copyright holder that gives him some kind of exclusive territorial right for a product. The copyright holder, on the other hand, sometimes demands that the distributor not sell competing products. The distributor generally handles all import problems such as customs and sometimes handles translation of written material that accompanies the product. Because there are two more discounts that have to be taken before the product reaches the customer, the distributor gets a very large discount from the copyright holder and has the obligation to order and sell large quantities of the merchandise.

The Wholesaler

The wholesaler is the link to the retailers. The essential function of a wholesaler is to get the product into shops. Wholesalers hire sales representatives or commission agents who visit shops and persuade them to buy the product. A good sales representative or agent should help the retailer set up the display, keep inventory current and generally help to ensure that the holograms look good and sell to the consumer.

It is important that the wholesaler's representative visit the retailer regularly. A wholesaler normally carries stock from a number of different suppliers, offering a full range of holographic items to retailers. This is a convenience to the retailer, who does not have to deal with many different hologram suppliers. Frequently the wholesaler has two catalogues. One is used for shops and one is used for direct sales to customers. Wholesalers do not generally have exclusive rights to a product.

The Retailer

Retailers are the point of contact with the public, the place where the holograms are displayed, seen, and bought. Retailers of holograms include specialist hologram shops and galleries, gift shops, department stores, chain stores, stationers, jewelers and toy shops. Some are independently managed by the proprietor, and some are national chains with a central buyer.

Only the buyer in the specialist hologram shop, usually the proprietor, is predisposed to buy holograms. In all other stores the person making the buying decision has to be persuaded that the saleability and profitability of holograms (or products featuring holograms) is preferable to the other goods that are on offer to him. In sales, of course, the personal touch counts, so regular visits and product updates are an important feature of the wholesaler's function.

Discounts

The exact percentage discounts between businesses varies. In some businesses, like publishing, the final retail price of the product is established by the copyright holder and discounts follow a fairly established procedure. Holograms are sold much like gifts in the gift trade. A retailer, for example, gets a catalogue that tells him what it costs to buy an item. The price at which the item is sold is for the retailer to decide. This is called the retail price. The wholesaler sometimes suggests that the retailer mark the item up 100% (which means that the retailer buys at an average of 50% discount off the retail price).

Wholesale Confusion

There is some necessary confusion in the industry about the word wholesaler. Retailers frequently find that the catalogues quote "wholesale prices". To the uninitiated, this implies that the prices being quoted are the prices that a distributor gives the wholesaler (wholesale prices) when actually the prices being quoted are prices at which the retailer buys the merchandise from the wholesaler.

From the distributor or wholesaler's point of view, this is an unfortunate but necessary confusion. Distributors and wholesalers point out that the retail price of a product is established by the retailer. Since wholesalers have different catalogues for retailers and end customers, they need to designate whether the prices being quoted are for the public or not. What do you call the price being quoted to retailers? You can't call it the retail price! Thus, you have to call it the wholesale price, even though it might cause some confusion.

Prices that are quoted to a distributor or wholesaler from the copyright holder are usually given as a percentage discount from the wholesale price (the price quoted to retailers).

Repackaging And Customizing

Holograms can frequently be bought by any dealer in the distribution chain in an unfinished state for a reduced price. There is a small savings that can be made if you are willing to put in the extra work, but the big appeal is that it allows you to repackage or customize the holograms for sale.

Fine Art Holograms

The distribution chain for a fine art hologram, in theory at least, is similar to that of artistic holograms although more straightforward. The day may come when holograms are a fully accepted art medium and part of the established art market; regrettably this is not true today. This means that the traditional gallery distribution system of the art world is not open to most holographic artists, nor do they have the benefit of the guiding hand of an agent, manager or dealer.

The fine art holographer's point of contact with the public is generally through exhibitions, either in specialist or public art galleries. There are several specialist galleries around the world which hold exhibitions and sell work, and many not-for-profit galleries. The specialist galleries are generally commercial operations which put a large mark-up on the artworks (or demand a large discount from the artist), whereas the not-for-profit galleries put a 15-25% mark-up on sales.

While the non-profit organizations "cost" less on any individual sale, they do not usually have continuing displays and representation for the artist. Specialists, on the other hand, do keep a pennanent exhibition and often buy some of the artist's work themselves. Again, because their continued commercial existence depends on sales, the for-profit galleries tend to work hard marke.ting fine art holograms.

Because the market for fine art holograms is not yet fully developed, artists often sell directly to the public. Most of them know the few serious hologram collectors, and are pleased to deal with other buyers who contact them. In "dealing direct" there is often pressure on the artist to extend discounts to collectors "because there is no middle man involved." Extending such discounts, however, undermines the growth of the distribution system and interferes with efforts to build a growing awareness of holographic art.

Custom-Designed Holograms

The markets for security and advertising holograms are similar to each other. They differ from the giftware and art markets in that the end buyer is not a consumer buying one hologram, but a company or group of companies usually buying in large quantities. The exception is the company buying only a few holograms, usually large-format, for use in promotional events such as trade shows or point-of-purchase advertising.

As discussed previously, the end user can deal directly with the producer. Often, however, intermediary agencies bring the producer and client together. These intennediaries fill different functions from those in the giftware business. Instead of buying and selling as principals like a distributor or wholesaler does, they tend to be agents being paid either by the purchaser or the producer. They usually work on a commission in the range of 15-25%.

The intennediary ensures that the client gets a

suitable hologram, well-designed, delivered on time and with the required quality. The intermediary works with client and producer to design the hologram and supervise all stages of its production, from model-making or artwork, through master and proof, to installation in its final environment. This may be hot-stamping onto a credit card, insertion in a brochure or installation in a trade-show exhibition. Everyone wants the hologram to look good - the specialist agency has to ensure that it is well integrated into its display environment and, if it is a promotional piece, that it is designed and installed to fill its potential. This involves liaison with producers, designers, printers, print finishers and above all, the client.

The hologram producer may handle all this inhouse and many have large sales teams. As with any industry, producers find it valuable to deal directly with clients, especially large clients who bring repeat business. However, the purchaser often needs the support and advice of an independent agency. The agency can recommend the best producer for a particular job. It can encourage a client to make the most of holograms by using large format film in a point-of-purchase application to tie in with a brochure using embossed holograms. The agency should also be aware of the many pitfalls that can delay the delivery of the order and work with the suppliers on the purchaser's behalf to ensure that problems are minimized.

If the custom hologram is for a security or anticounterfeiting application then the purchaser will want to ascertain that all stages of production are in secure facilities. Credit card companies and banks especially cannot risk holograms finding their way into the wrong hands. Several embossed hologram producers are now secure, but some have higher levels of security than others.

Distribution - Sign Of A Maturing Industry

Ten years ago the few companies producing holograms dealt directly with their buyers. Today the industry has grown and diversified. It is a sign of the maturing of the industry that these distribution mechanisms are now in place. The distribution chain smooths the path for the customer who wishes to buy holograms, whether that customer is a large corporation or an individual consumer.

Holography 1990-The Year in Review

The holography field saw a number of significant developments take place during 1990. New materials were introduced; major new applications were undertaken; companies started, merged and closed; key litigation took place and the field grew toward maturity. The following collection of abridged articles gives an overview of several of the most significant events of 1990.

Dupont Dives Into Holography Marketplace:

DuPont introduced its new line of OmniDex holographic recording materials at OE-LASE in January 1990. The company mounted a highly visible launch involving an exhibit booth and several conference papers, with many staff attending the symposium.

Twenty years ago, DuPont developed and patented a photopolymer designed for use in holographic recording but abandoned the project when there appeared to be an inadequate market. As interest in holography has picked up over the past five years, the company reviewed its decision and some three years ago created a "new business initiative" group within its Imaging System division. The group was organized as an entrepreneurial venture by a few employees who believed they had identified a business opportunity. Currently this group has grown in size to nearly 30 individuals with an internal "milestone" budget and access to the vast technical and production resources of the chemical giant. The effort has led to the introduction of a family of recording materials suited for a variety of holographic applications.

The company sees two markets for its new material, artistic and industrial uses. The introduction at OE-LASE was aimed at stimulating the artistic market by encouraging holographers to begin using the film commercially. DuPont believes industrial and commercial optical elements will eventually be the larger market. In time, the company hopes to find significant applications for its materials in channel wave guides, sophisticated integrated optical circuits, matched filters for inspection systems, and automobile center-high-mounted stop lamps.

Contact: Krishna Doraiswamy, DuPont Imaging Systems, Optical Elements Venture, PO Box 80352, Wilmington, DE 19880-0352, USA; phone (301)695-1258.

Award For Dennison's Embossed Hologram Paper:

CPS, a Tennessee film and paper converter, has won the non-food award in the AIMCAL (Association of Industrial Metallisers, Coaters and Laminators, USA) Awards for its Dazzle line of holographically embossed gift wrap paper. The paper, produced by the Dennison Manufacturing Company of Massachusetts, is overprinted, cut and packaged as gift wrap by CPS, which introduced it to the market last autumn for the Christmas season. Following its success, Dennison is now introducing the holographic paper to the European market.

Dennison had been working on the product for several years. The company, a diversified paper and stationery manufacturer with world-wide capacity, produces various metallized films, foils and papers through its Metallized Papers Division. It's UK subsidiary, Metallized Films and Papers Ltd. (MFP), undertakes metallization for UK holographic embossers so the company is aware ofthe material tolerances and requirements for holographic products.

In addition to gift wrap, it sees packaging as a prime target market. At the launch of the product Dennison estimated sales potential at $3M in 1990, but according to Harold Isherwood, Managing Director of MFP, it is well ahead of these projections. MFP is now launching the product in Europe. Apparently there is considerable and specific market interest and MFP has already shipped some material. Isherwood pointed out that although the material is currently supplied from the USA, MFP has the capability to produce it when demand warrants it.

Contacts: Marc Wey, Dennison Manufacturing Company, Metallized Paper Division, 275 Wyman St., Waltham, MA 02254-9139, USA; Phone (+1 617) 890 6350.

Harold Isherwood, Metallized Films and Paper Ltd., Thomas Gilchrist Industrial Estate, Blaenavon, Gwent, NP4 9RI, Wales, Phone (+44 495) 790214.

Holographically Secure CA Drivers License:

The State of California plans to issue new drivers licenses with an embossed holographic security overlay produced by American Bank Note. The holograms feature the Seal of the State of palifornia and the Department of Motor Vehicles logo in a repeat pattern. It will be produced in a transparent material to cover the surface of the license. In normal conditions it will not obstruct the license information, but when viewed from the reconstruction angle under appropriate illumination the pattern will be clear, serving as a verification of the license's authenticity.

The State expects to issue approximately 7.5 million licenses annually over a five year period, with a projected total of 30 million. Tony Walter, project leader for the State of California, explained that the hologram is only one of several security features in the new design. Several companies bid for the job, which in June was awarded to National Business Systems Imaging Inc. (NBS), on a bid of $28.5 million for the complete package, including hardware and software. NBS sub-contracted the hologram production to ABNH.

Contact: ABNH, 4 Westchester Plaza, Elmsford, NY 10523, USA; Phone (914)592-2355.

Australia To Issue Notes With OVDs:

The Reserve Bank of Australia has announced that it will be replacing the existing series of currency notes with a new series of plastic currency notes featuring optically variable devices (OVDs). This follows the efficacy and public acceptance testing of the commemorative $10 note issued in 1988, the first with an OVD.

The new series being introduced in early 1991 will be produced by Note Printing Australia, a division of the Reserve Bank, using technology developed by the division with the Commonwealth Scientific and Industrial Research Organization (CSIRO). Plastic is used as the substrate as it is more suitable as a carrier for the OVD, while having greater longevity than paper notes. The plastic notes required the development of special inks and production processes so that standard security techniques as well as new ones could be incorporated.

The Reserve Bank has proceeded cautiously in deciding to introduce these new-style notes. Some 17 million of the commemorative $10.00 notes were introduced in 1988, primarily in the Newcastle district of New South Wales. This limited issue simulated normal penetration and usage of a limited production run and allowed the Bank to monitor reaction.

$50 Plastic Note For Singapore:

The Reserve Bank has also produced a Singapore $50.00 plastic note with OVD for the 25th Anniversary of the State. The OVD is a portrait of Encik Yusof bin-Ishak, Singapore's first President. The note was introduced on August 9 1990, when 300,000 "special edition" commemorative notes were issued, with a another 4.8 million notes for circulation.

Contact: RG. Pearson, Marketing Manager, Note Printing Australia, PO Box 21, Craigieburn 3064, Australia; Phone (+61 3) 303 0444, fax 3030491.

Banknote Focuses On
Marketing/global Strategy:

Holography will grow as a major priority for the U.S. Banknote Co. L.P. (USBC, a Limited Partnership) and International Banknote Co., Inc. (lBK) following their merger according to Morris Weissman, Chairman ofUSBC.

Weissman outlined the companies' plans for leveraging their current strengths in order to increase the volume of holographic sales and to establish a dominant position in the industry as well as certain unresolved issues facing the companies as they complete their merger. The combined sales of the two companies makes the merged entity the second-largest security printer in the world, after De la Rue. Over the next few years, the company plans to make significant changes in its facilities. It will expand its existing Singapore plant to achieve a stronger Asian presence. In the U.S., the company will consolidate its holographic operations in a single new facility. Weissman expects European expansion in the form of a production plant through acquisition or new construction. The company also plans to make major efforts at cross-selling its holographic capabilities to its current security customer base. Weissman pointed out the large number of currency, security documents and identification system customers the companies currently have and indicated that the development of holographic applications for these customers would be a high priority for the new company. Weissman criticized current holographic companies for failing to take an organized, long term approach to developing other nonsecurity, application areas.

When asked about the his financial expectations, Weissman indicated that he expected holographic sales of approximately $30 million for 1990 and would be looking for this figure to double to $60 million within two years. He emphasized that some of his optimism was based on technological progress nearing the commercialization point within the company. High among these is a technique for embossing holograms directly on paper, eliminating the need for pre-embossing foil and hot stamping. He also alluded to machine readable holographic systems the company is already beginning to install through one of its subsidiaries, Laser Vision.

Contact: Mr. Morris Weissman, Chairman, United States Banknote Co., L. P.; phone (212)741-8500.

A.H. Prismatic Acquires Holos Gallery:

Effective September 1, 1990, A H. Prismatic, Inc., the US subsidiary of A H. Prismatic Ltd., acquired all of Holos Gallery's operations, assets and liabilities. A H. Prismatic, Inc. will now have its national sales office in San Francisco and its distribution and administration in New York. The Holos Gallery will continue to operate under the Holos name, while all other operations will now operate under the name A H. Prismatic, Inc. Alan Rhody, previously with Holos, will become National Sales Manager for the combined operation. Gary Zellerbach will continue as a consultant to AH for a few months but will begin devoting more time to the DZ Company which he owns jointly with Dan Cifelli.

Thompson and Zellerbach described the key assets of the sale as the Holos name and the company's customer base, staff, line of products, gallery and traveling exhibit.

Contact: A H Prismatic Ltd., New England House, New England Street, Brighton, Sussex, BN1 4GH, England; Phone (+44 273) 686966, fax 676692.

AH. Prismatic, Sales Office, 1792 Haight Street, San Francisco, CA 94117, USA; Phone (+1 415) 221 4717,8005373631, fax: (+1415) 2214815.

Third Dimension Ltd. Closes:

Third Dimension Limited, previously the leading producer of silver halide film holograms for the gift ware market, closed its doors in 1990. Receivers were appointed at on September 21 by National Westminster Bank. This followed from the company defaulting on a £75,000 loan. The company's assets were purchased in November, 1990 by Light Impressions, Europe, of Surrey, England.

According to the receiver, S. Swaden of Leonard Curtis, his estimates put the company's bank and trade creditors at some £380,000 ($684,000). To offset this, the company has laser and optical equipment (including a Lumonics ruby pulse laser) with a balance sheet value of some £120,000 ($192,000), plus finished stocks.

Contact: Steven Swaden, Leonard Curtis & Co, 30 Eastbourne Terrace, London W2 6LF, England; Phone (+44 71) 262 7700, fax 723 6059 or Light Impressions, Europe, of Surrey, England.

Holography Commercialized
In USSR - Export Trade Sought:

With the political and economic changes in the Soviet Union, many organizations which were 100% State funded are now required to earn up to 60% from other sources. Several of the country's holographic facilities are therefore re-organizing on commercial lines to exploit their products, processes and expertise, with a focus on Western market opportunities.

Leningrad's Joffa Institute- Prof Y. Denysiuk has now become chief of the holography group after many years behind closed doors at the Vavalov State Optical Institute. The Joffa Institute has set up Technoexan, a Soviet-Austrian joint venture, to exploit its resources commercially. Technoexan has ten full-time holographers and access to the Institute's 200 staff. One of the company's products is the "Regina" interferometry camera, a compact stand-alone camera which uses standard plates or photothermoplastic. Technoexan plans to market holograms and PFG-03, the highly respected Russian silver halide emulsion; prices are available from the Austrian office.

At the Leningrad Nuclear Physics Institute of the Academy of Sciences, Dr. Boris Turukhano's main focus is on producing holographic memory disks, linear and radial holographic gratings. The 30 cm diameter memory disks are used for archival holographic storage, holding up to 10,000 pages of graphic text information using Fourier image recording. The holograms are recorded using a HeNe laser on silverhalide emulsion. Viewing of the pages is effected using an argon laser to reconstruct the individual holograms which are displayed on a diffuser screen.

Specified Design Bureau Now Open To West: The Specified Technological Design Bureau has 50 people working in holography. The Bureau was set up three years ago to design high-quality optical and electronic components and because of the nature of its work it received permission to open its doors to the west and to investigate trade possibilities only two months ago. It is setting up a commercial arm to exploit this permission. The Bureau's products link its electronic, software and optical design resources. The Bureau is also growing Bizmuth Silver Oxide (BiSiO) photorefractive crystals which it is keen to export. Its production crystal is 40 x 40 mm, but they can also produce up to 80 mm sq for research purposes. These crystals are sensitive to HeCd lasers.

Holographix Announces Laser Printer Unit:

Holographix Inc. of Burlington, Mass, has announced an innovative scanning unit for laser printers featuring patented holographic systems to direct the beam to the print drum. The core of the HX 310GS Laser Scanning Unit (LSU) is a patented holographic optical element which replaces the existing polygon and lens train. It is produced from a photoresist master using embossing technology to produce a CD-like plastic disk. This disk diffracts and focuses the incoming laser beam.

The LSU also uses a holographic achromatisation strip to correct for dot displacement caused by shifts in operating wavelength, so that the system is effective across a wide spectrum of laser sources. Holographix claims performance superior to existing technologies with a cost advantage of 25-50%. A further advantage of the design is that at minimal incremental cost it can readily be upgraded to 1000 dpi, a 300% improvement over the current laser printer resolution of 300 dpi. This is envisaged as a major future market because current technologies are nearing the limit of their capabilities at 300 dpi. Holographix will be marketing the LSU to OEM manufacturers and already has licensed the design to a Korean manufacturer of printer engines.

Contact: Kenneth K Liu, Holographix Inc., 87 SecondAve., Burlington, MA01803, USA; Phone (+1617) 229 8840, Fax 229 7920.

Appeals Court Upholds District Court In Patent Litigation:

On October 1, 1990, the three judge Federal Appeals Court panel entered a brief decision upholding the District Court's rulings in the litigation between American Bank Note (ABN) and Light Impressions (LII). Judge Markey, writing the unpublished opinion, found that District Court had neither abused its discretion nor committed errors of law in its decisions.

Substantively the ruling left standing the orders entered by the District Court earlier in 1990(no preliminary injunction against LII on claims 11 and 12 of patent No. 3,894,787; and continuation of a preliminary injunction against LII on patent No. 3,838,903). The Court did this without commenting upon the substantive patent law issues presented by the two sides. Effectively, therefore, the ruling reduces the immediacy of the litigation to the industry. The most authoritative determination at present, Judge Aguilar's rulings, holds that the '787 (hologram as product) claims are void for double patenting, and that, while certain past LII techniques infringed the '903 (H1IH2 claims), its current techniques do not. The next step in the proceeding and ultimate trial of all the issues will probably be delayed by Judge Aguilar's conviction on unrelated criminal counts. It is possible, therefore, that the patents will have all but expired prior to a full review of the scope and strength of the claims at issue.

ABN Files Suit Against Chroma Concepts:

American Bank Note Holographics, Inc. (ABNH) has filed a patent infringement suit against Chroma Concepts, Inc., a California hot-stamper which has

recently expanded to cope with actual and anticipated increases in demand for holographic hot stamping. The suit seeks injunctive relief and trebled damages for the alleged "wilful and deliberate infringement."

Chroma could not be reached for comment. It is possible that the company may cross-claim against one or more of its hologram suppliers. Morris Weissman, Chairman of U. S. Banknote, said that the company had decided to file the law suit because Chroma's business competition was becoming significant and because it was on-going in the face of warnings from ABNH. He indicated that no other suits were currently being filed but that others would be if other producers "infringe in a significant way."

Contact: Morris Weissman, Chairman, U. S. Banknote Co., L. P., 345 Hudson St., New York, N.Y. 10014; phone (212) 741-8500.
Gary Zuercher, President, Chroma Concepts, Inc., 17742-B Mitchell, Irvine, CA 92714; phone (714)250-1416; fax (714) 250-4026.

This year-in-review piece was contributed by the editors of Holography News, which is published ten times a year by Reconnaissance, Ltd.

11

Holography Businesses

A

ABBOTT LABORATORIES. Department 93F, (Building AP-9), Routes 43 and 137, Abbott Park, IL 60064 USA. Telephone: (708) 937 4117. Contact: Dr. Gerald Cohn. Company description: Holographic Non-Destructive testing.

ACME HOLOGRAPHY. Established in 1988. 2 Employees at this address. 12 Sunset Road, West Somerville, MA, 02144. USA. Telephone: (617) 6230578. Contact: Betsy Connors. Managing Partners: Betsy Connors, David Chen. Company description: Acme Holography is Boston's first private holography lab. We offer full service in reflection transmission and computer generated holography, including design consultation and H1 mastering for Polaroid Mirage photopolymer holograms.

ACTIVE IMAGE. Established in 1989. 1 Employee at this address. P.O. Box 97, Boulder Creek, CA 95006 USA. Telephone: (408) 338 2405. FAX: (408) 338 9833. Contact: Nelson Poe. Subsidary of Steve Provence Holography. Company description: Exclusive wholesale distributor of packaged embossed holograms by Steve Provence Holography.

ADEL ROOTSTEIN, INC. 205 West 19th Street, New York, NY 10011 USA. Telephone: (212) 645 2020. Company description: Manufacture mannequins for displays andlor holography models.

ADLAS G.m.b.H. & COMPANY KG. Established 1986. 27 Employees at this address. Seeland Strasse 67, D-2400 Luebeck 14, Federal Republic of Germany. Telephone: o 11+(49)-451-390-9300. FAX: 011+(49)-451-390-9399. Contact: Gerhard Marcinkowski, Product Manager. President: Dr. Bernhard Steyer. Sales: Juergen Reingruber. Subsidiary: ADLAs/A-B Lasers, Inc. 4 Craig Road, Acton, MA 01720 USA. Company description: Manufacturer of diode laser-pumped solid state lasers which operate in CW and pulsed mode with wavelengths in IR, visible and UV.

ADVANCED ENVIRONMENTAL RESEARCH GROUP. Route 2, Box 2948, Davis, CA 95616 USA. Telephone: (916) 757-2567. Contact: Richard Ian-Frese, Project Director. Description: Holographic research and development applications include architectural engineering, day lighting, PV enhancement, energy conservation, environmental research, and wavelength selectivity.

ADVANCED HOLOGRAPHICS CORP. Established in 1987. 16 Employees at this address. Sales and Administration: 2469 East Fort Union Blvd. Suite 108, Salt Lake City, UT 84121 USA. Telephone: (801) 943 1809. FAX: (801) 942 7006. Contact: Scott Garrett. General Manager: Mike Janko. Manufacturing and Shipping: 750 West, 200 North, Logan, UT 84321USA. Contact: Keith Gulbranson. Company description: Manufacturer of mass-produced, low cost dichromate holograms in glass, flexible and rigid plastics. Over 50 stock images incorporated into a line of novelty items. Full service ASI supplier; custom work.

ADVANCED HOLOGRAPHICS, LTD. 243 Lower Mortlake Rd.,Unit 11, Richmond, Surrey TW9 2LL, England, United Kingdom. Contact: John Andrews! Don Tomkins. Company description: Artistic hologra phy; presentation support

ADVANCED IMAGING TECHNOLOGIES. "Normanhurst", 7 Broomfield Lane, Hale, Altrincham, Cheshire WA15 9 AP England, UK. TelephoneIFAX: (44) (61) 941-6367. Contact: Glenn P. Wood, Director.

ADVANCED OPTICS, INC. Established 1966.4 Employees at this address. 5 East Old Shakopee Road, Minneapolis, MN 55420, USA. Telephone: (612) 888-0868. Contact: Sharilyn Loushin, Director of Marketing and Sales. President: H.T. Sherman. Company description: Manufacturers of custom and precision optics (first-surface mirrors, windows, beamsplitters, etc) for use in holographic equipment, lasers, and other optical devices in industrial, research, OEM and prototype development applications.

ADVANCE PHOTONICS, A-147 Ghatkopar Industrial Estate, Ghatkopar, Bombay 400 086, India. Telephone: (91)22 582 204. FAX.: (91)22 2024202. Branch office of Newport Corporation, Fountain Valley, CA USA.

AD 2000.946 State Street, New Haven, CT 06511 USA. Telephone: (800) 334 4633. Contact: Jeffrey Levine. Company description: Gallery; wholesale.

AEROSPATIALE-DMSION HELICOPTERES, 2 avenue MarchelCachin, F-93126 La Courneuve Cedex, France. Telephone: (33X48) 389 178. Contact: M. Guignard. Description: Scientific research; industrial research; holographic non-destructive testing; interferometry.

AEROSPATIALE. Established 1970. 1700 Employees at this address. Ets D'Aquitaine, Saint-Medard-en.Jalles, F-33165 Bordeaux, France. Telephone: (33)(56) 57 34 80. FAX: (33)..(56) 57 35 48. Contact: C. LeFloc'h, Engineer. Description: Scientific & industrial research; holographic non-destructive testing; interferometry.

AEROTECH INC., (MAIN HEADQUARTERS), 130 Employees at this address, Electro Optical Division, 101 Zeta Drive, Pittsburgh, PA 15238 USA, Telephone: (412) 963-7470, FAX.: (412) 963-7459, Contact: Steve A. Botos, Marketing Manager. President, Stephen A. Botos; Vice-President, David Kinoel; Sales Manager, Frank Armstrong; Customer Service, Wes Taylor. Subsidiary Companies: Aerotech Ltd- England, Aerotech GMBH.Germany, Aerotech AustraliaAustralia.
Company description: Aerotech manufactures helium neon tubes, power supplies and complete systems for OEM and end users. Other product lines include optical table positioners and precision rotary and linear positioning systems.

AG ELECTRO-OPTICS LTD. 29 Forest Road, Tarporley, Cheshire CW6 OHX, England, United Kingdom. Telephone: (44X829) (73) 3305. Contact: Dr. J .A. Gibson. Description: Distributor of lasers, optics, lab equipment.

AGFA CORPORATION. (US Headquarters). 100 Challenger Road, Ridgefield Park, NJ 07660 USA. Telephone: (201) 440 2500 ext. 4226. FAX.: (201) 440 5733. Contact: Mark Redzikowski, Product Manager. President & CEO: Helge H. Wehmeier. Parent Company: Agfa Gevaert Ltd., England, United Kingdom. Main Headquarters: AGFA GEVAERT, Mortsel, Belgium. Branch offices: Atlanta GA USA; Des Plaines, IL USA; Burbank, CA USA; Brisbane, CA USA; Irving, TX USA.
Company description: Manufacturer of fIlm, plates, emulsions and recording material

AGFA-GEVAERT LTD. (Subsidiary of AGFA GEVAERT N.Y.). 27 Great West Road, Brentford, Middlesex, TW8 9AX. England, United Kingdom. Main Headquarters: AGFA GEVAERT N.V., Mortsel, Belgium. Company description: Manufacturer of film, plates, emulsions and recording material

AGFA GEVAERT N.V. (MAIN HEADQUARTERS). Established 1894. 8000 Employees at this address. Holography Film Dept, Septestraat 27, B2510, Mortsel, Belgium. Telephone: (03) 444 2111. FAX: (03) 444 7094. Contact Person: R. De Winne, Product Manager-Holography.President, E. De Wolf; Sales Manager, H. Deschaumes; Customer Servioe: R. De Winne. Branch offices: AgfaGevaert LTD, England, United Kingdom. Subsidiary Company: Agfa Corporation, USA.
Company description: Manufacturer of film, plates, emulsions and recording material

AG PRISMATIC, Bimbolegge & Bimbogioca SRL, Via Borfuro 12, 1- 24100 Bergamo, Italy. Telephone: (39X35) 213 015. Description: Artistic holography.

A.H. PRISMATIC INC. Established 1986. 8 Employees at this address. 285 West Broadway, New York, NY 10013 USA. Telephone: (212) 219 0440. FAX: (212) 219 0443. Contact: Brian Szpakowski, Sales Manager. Company description: Distributors of exclusive ranges of holographic gifts, toys and jewelry, and film holograms. Point of purchase displays available for all items. Hologram store in Museum of Holography, New York.

A.H. PRISMATIC, INC. (Incorporating Holos Gallery.) Sales Office. 1792 Haight Street, San Francisco, CA 94117. Telephone: (415) 221-4717. (toll free) (800) 537 3631. FAX: (415) 221-4815. Contact: Alan Rhody, National Sales Manager. Customer Service: Catlin Adami. Description: Hologram store in Holos Gallery. A. H. Prismatic also has a traveling exhibition, "Laser Light Frontiers', which has been displayed throughout the U.S. at colleges and museums.

A.H. PRISMATIC, LTD. Established 1982. 15 Employees at this address. New England House, New England Street, Brighton, BN1 4GH England, United Kingdom. Telephone: (44) 0273 686 966. FAX.: (44) 0273 676 692. Contact: Ian Dayus, Sales Manager. President, Barc Thompson; Sales Manager, Ian Dayus; Customer Service, Sheila Bagley. Company description: Manufacturers and distributors of holographic gifts, toys, and jewellery, together with exclusive range of fIlm holograms. Point of purchase displays compliment all products

AIMS OPTRONICS SA/NV, Rue Ferd Kinnenstraat 30, B-1950 Kraainem, Belgium. Telephone: (32X027) 310 440. FAX: (32X027) 318918. Branch office of Newport Corporation, Fountain Valley CA,USA.

AITES LIGHTWORKS. 2148 North 86th Street, Seattle, WA 98103 USA. Telephone: (206) 526 5752. Contact: Edward Aites. Description: Artistic holographer.

AKS HOLOGRAPHIE-GALERIE GmbH. Established in 1985. 3 Employees at this address. Potsdamer Stra13e 10, 4300 Essen 1, Federal Republic of Germany. Telephone: (49X0201) 704 562. Contact person: Gudrun Sott, Administrative Director. President, Detlev Abendroth; Vice-President, Peter Kremer; Sales Manager, Gudrun Sott. Company description: Embossed hologram manufacturer, artistic hologram maker; Buying & selling holograms

ALPHA FOILS, INC. P.O. Box 152, Bernardsville, NJ 07924 USA. Telephone: (908) 766-1500. FAX: (908) 766-1501.Contact: Henry

Ruschman. Company description: Manufacturers of embossed holography foil; company embosses foil and has other embossing services - consulting work on technique and machinery. Main headquarters: Munsterhalden 10, D7816 Munstertal, Germany.german sp. (umlaut)

ALPHA PHOTO PRODUCTS, INC. Established in 1947. 50 Employees at this address. 985 Third Street, P.O. Box 23955, Oakland, CA 94623 USA. Telephone: (415) 8931436 (ext. 1). FAX: (415) 834-8107. Company decription: In addition to photographic equipment and supplies, Alpha distributes holographic plates, mms and processing chemistries by AGFA, llford, Kodak and Photographer's Supply as well as photographic equipment and supplies.

AMAZING WORLD OF HOLOGRAMS. Established in 1984. 6 Employees at this address. Corrigan's Arcade, Foreshore Road, South Bay, Scarborough, North Yorkshire Y011 1PB, England, United Kingdom. Telephone: (44X0723) 354 090. FAX: (44) (0482) 492 286 ref holo. Contact: Carl Racey. Sales Manager, Carl Racey. Parent company: Laser Light Image, Hull, England. Company description: Exhibitors and retailers of mm, glass, embossed, dichromate and related products. Permanent display of 200 holograms updated and changed regularly. Main season May-October. Distributors of mm & glass.

AMAZON. do Ruggero Maggi, C. So Sempione 67, 20149-Milano, Italy. Telephone: (39X02) 349 1947. Company description: Fine art holograms, Stereogram maker; Gallery sells fine art holograms.

AMBLEHURST LTD., Established in 1978. 40 Employees at this address. 52 Invincible Road, Farnborough, Hants GU14 7QU, England, United Kingdom. Telephone: (44) 0252520052. FAX: (44) 0252 373871. Contact: P.M.G. Hudson, Managing Director. Sales Manager, C.L. Hope, Customer Service A. Ackroyd. Company description: Laminates, hot stamp foil and pressure sensitive labelling manufactured in-house in integrated secure plant.

AMERICAN BANK NOTE HOLOGRAPHICS, INC. (MAIN HEADQUARTERS). Established in 1983. 50 Employees at this address. 500 Executive Boulevard, Elmsford, New York 10523 USA. Telephone: (914) 592-2355. FAX: (914) 592 3248. Contact: Russell R. LaCoste, Vice-President, Sales and Marketing. President, Salvatore D'Amato; Vice-President, Russell LaCoste; Parent company: International Bank Note Company. Branch offices: 400 Montgomery Street, Suite 810, San Francisco, CA 94104 USA; 999 Plaza Drive, Suite 400, Schaumburg, IL 60173 USA.
Company description: American Bank Note Holographies is the producer of the world's best known holograms; those seen on Visa, MasterCard and the three holographic covers of National Geographic Magazine.

AMERICAN BANK NOTE HOLOGRAPHICS, INC., (Branch office) 999 Plaza Drive, Suite 400, Schaumburg, IL 60173 USA.

AMERICAN BANK NOTE HOLOGRAPHICS, INC. (Branch office). 400 Montgomery Street, Suite 810, San Francisco, CA 94104 USA.

AMERICAN HOLOGRAPHIC INC. Established in 1976. 30 Employees at this address. P.O. Box 1310, 521 Great Road, Littleton, MA 01460 USA. Telephone: (508) 486 9621. FAX: (508) 486 9080. Contact: Rick Dishman. Description: Design, develop & manufacture optical components & instruments for use in industrial & medical measurements. Using holographic diffraction grating design & maufacturing capability to produce components for unique measurement instruments.

AMERICAN HOLOGRAPHY SOCIETY. 2018 R Street, N.W., Washington, D.C. 20009. Telephone: (202) 667-6322. Contact: Office of the Secretary. Description: Created to participate in the development of the art, science and technology of holography, and to stimulate and preserve education, industry and research .

AMERICAN LASER CORPORATION. Established 1970. 100 Employees at this address. 1832 South 3850 West, Salt Lake City, Utah 84104 USA. Telephone: (801) 972-1311. FAX: (801) 972-5251. Contact: Jean Bonnell, Marketing/Laser Sales. President: Larry Levy. Vice President: Mark Ballard. Branch Office: American Laser GmbH, Hans Pinsel Strasse 9A, 8013 Haar, West Germany. Company description: manufacturer of Argon, Krypton and Mixed gas laser systems from 3 mW to 25 W in air or water configuration .

AMHERST MEDIA. 418 Homecrest Drive, Amherst, NY 14226 USA. Telephone: (716) 874 4450. Description: Fine art hologram maker.

AMITY PHOTONICS CO. Established 1985. 3 Employees at this address. 26 Gibbs Road, Amity Harbor, NY 11701 USA. Telephone: (516) 789-1099. FAX: (516) 226-8701. Contact: Paul Westphal, General Manager. Company description: Manufacturers and distributors of optical lenses, prisms, filters, reticles, mirror and optical sub-assemblies for all purposes. Scientific, medical and industrial. Consulting services available.

ANGSTROM INDUSTRIES, INC. Established in 1988. 2 Employees at this address. 3202 Argonne, Houston, TX. 77098 USA. Telephone: (713) 526 0006. Contact: Frank Davis, President. President, Frank Davis; Vice-President, Vikki Fruit. Description: Artistic holography; embossed holograms:computer-generated to resist.

ANOTHER DIMENSION. 946 State Street, New Haven, CT 06511 USA. Telephone: (800) 334 4633. Contact: Peter Scheier.

A.N. SEVCHENKO RESEARCH INSTITUTE OF APPLIED PHYSICAL PROBLEMS. 220106 Minsk, USSR. Contact: V.P. Mikhaylov. Description: Materials researrch; artistic holography.

ApA OPTICS, INC. 2950 Northeast 84th Lane, Blaine, MN 55434 USA. Telephone: (612) 784 4995. FAX: (612) 784 2038. Contact: Ani! Jain. Company description: Design and manufacture of head-up displays, HOEs.

API Group plc (formerly Associated Paper Industries) Parent Corp. of Whiley Foils Ltd., Silk House, Park Green, Macclesfield, Cheshire SKll 7NU, England, United Kingdom.

APOLLO LASERS, INC. 12351 Research Parkway, Orlando, FL 32826 USA. Telephone: (407) 281-4103. Contact: Steve Qualls. Description: Manufacturer of carbon dioxide & ruby lasers; equipment & supplies for holography.

APPLIED HOLOGRAPHICS CORPORATION, Established in 1989. 15 Employees at this location. 1721 Fiske Place, Oxnard, CA 93033 USA. Telephone: (805) 385-5670. FAX: (805) 385-5671. Contact: Mr. Chris Outwater, Executive Vice-President. Sales Rep: Ms. Fran Lantz. VP Marketing: Mr. D. Buell. Project Manager: Francis 'fuffy. Parent Company: Applied Holographics, PLC., England. Branch office: Riehmond, VA USA. Company description: From embossed credit cards up to meter square holograms, using computer generated or real life graphics, Applied Holographics Corporation (formerly ADD) is the recognized leader in full color holographic stereography.

APPLIED HOLOGRAPHICS CORPORATION. (Branch office).

Eastern Sales Office, 8508 Kenwin Road, Richmond, VA 23235 USA. Telephone: (804) 323-4402. FAX: (804) 320-5547.

APPLIED HOLOGRAPHICS, PLC., (MAIN HEADQUARTERS), Braxted Park, Great Braxted, Malden, Essex CM8 3XB, England, United Kingdom. Subsidiary Companies: Applied Holographics Corporation, Oxnard, CA USA. Company description: Holography, stereograms, narrow and wide web embossing, hot-stamp foil, pressure sensitive, polyester, OPP, acetate, large format holograms, holodisc.

ARBEITSKREIS HOLOGRAFIE B.V. Herman-Josef Bianchi, Boeckelter WEG 47, 4170 Geldern. Federal Republic of Germany. Telephone: (49)(2831) 3034. Contact: Christian Liegeois. Description: Artistic holography.

ARCHEOZOIC INCORPORATED. Established 1989. 5 Employees at this address. 777 Gravel Hill Road, Southhampton, Pennsylvania 18966 USA. Telephone: (215) 322-4915. Contact: Paul E. Yannuzzi, President. Vice President: Greg Sholette. Company description: From product development to custom work. We present the most startling images of earth's past, present and future through a unique fusion of model-making and the highest quality holography.

ARCHITECTURAL GLASS & HOLOGRAPHY. 1 South Street, Great Waltham, Chelmsford, Essex CM3 1 DF, England, United Kingdom. Contact: Jean Bailey. Description: Application ofholography to decorative architectural glass.

ARMSTRONG WORLD INDUSTRIES. P.O. Box 3511. Lancaster, PA 17604 USA. Telephone: (717) 397 0611. Contact: Larrimore B. Emmons. Description: Artistic holography.

ART FREUND HOLOGRAPHY. Established in 1982. 1 Employee at this address. 124 Brookwood Drive, Santa Cruz, CA 95065 USA. Telephone: (408) 426 4436. Description: Artistic holographer.

ARTIGLIOGRAPHY CO. Established in 1987. 4 Employees at this address. 7130 Mohawk West Drive. Indianapolis, IN 46236 USA. Telephone: (317) 823 0069. Contact: Kerry J. Brown. Description: Broker for embossed holography & artistic holography.

ARTIGLIOGRAPHY GALLERY: (Branch of Artigliography Co.) 415 Massachusetts Ave, Indianapolis, IN 46204 USA. Telephone: (317) 684-9855. Contact: Kerry Brown. Description: ,Buying & Selling holograms and holography novelties (I.e. jewelry); consulting; holography education, gallery.

ART INSTITUTE OF CHICAGO (See School of The Art Institute of Chicago)

ARTKITEK. Established in 1986. 1 Employee at this address. 122 Myrtle Avenue, Cotati, CA 94931 USA. Telephone: (707) 664 2330. Contact: Steve Anderson, Proprietor. Company description: Artkitek provides holographic services to the design community: holograms of architectural models, laser speckle interferometry, laser sculpture for lobby displays and special events, HI mastering and purchasing consultation.

ARTPLAY HOLOGRAPHIC STUDIO. (MAIN HEADQUARTERS). Established in 1986. 12 Employees at this address. H-1191 Budapest, Ady Endre lit. 8, Hungary. TelephoneIFAX: (36-1) 127 0412. Contact: Tibor Balogh. President, Tibor Balogh; Vice-President, Zsuzsa Dobranyi; Sales Manager, Zsolt Sz~kely ; Customer Service, Szilvia Bagdan. Subsidiary companies: Galvanoart, Budapest, Hungary. Company description: Besides usual holographic services

the ARTPLAY is the only Eastern European studio offering complex design work applying holograms, shim production and own galvanic services also for other studios and printers.

ART, SCIENCE AND TECHNOLOGY INSTITUTE (ASTI)-HOLOGRAPHY COLLECTION. Established in 1983. 6 Employees at this address. 2018 R Street, N.W.,Washington D.C. 20009 USA, Telephone: (202) 667 6322. FAX: (202) 861 0621. Contact: L. Bussaut, Research Director. Company description: Research and educational organization for the advancement of the art, science and technology of holographyl Research focus: holographic motion picture, imagery/2 galleries for the artworks' promotion/travelling exhibit/Training

ASCOT LASER PICTURE STUDIO, 27 Upper Village Road, Sunninghill, Ascot, Berkshire SL5 7AJ, England, United Kingdom. Telephone: (202) 667 6322. FAX: (202) 861 0621. Contact: Mr. Brodel. Description: Artistic holography; holography education; workshops

ASHLAND ELECTRIC PRODUCTS-VISUAL ELECTRONICS CORP., 80-39th Street, Brooklyn, NY 11232 USA. Telephone: (718) 8553319. Contact: J . Conlan. Description: Manufacturer of optics.

ASOCIACION ESPAnOLA DE HOLOGRAFIA. Avda. Filipinas 38-1A, Madrid 3, Spain. Description: Artistic holography.

ATELIER HOLOGRAPHIQUE DE PARIS. 13, Passage Courtois, F-75011 Paris, France. Telephone: (33X1) 43796918. President, Pascal Gauchet; Vice-President, Jonathan Collins; Customer Service, Dominique Seuray. Company description: Artistic holography; Buying & Selling; Consulting

ATLANTA GALLERY OF HOLOGRAPHY. Established 1989. 75 Bennett Street, Suite E-2, Atlanta, GA 30309 USA. Telephone: (404) 352-3412. Contact: Lisa Murray, DireCtor. Description: Only fine art holography gallery in Southeast US. Specializes in showing internationally recognized holographic artists. One-man, group shows exhibited on every-other-month system; permanent gallery also.

ATOMIKA TECHNISCHE PHYSIK GmbH. D-8000 Munchen 19, Kuglmuellerstrasse 6, Federal RepUblic of Germany. Telephone: (49X89) 152 031. Description: Artistic holography.

AT&T BELL LABS-MATERIALS GROUP. 600 Mountain Avenue, Murray Hill, NJ 07974-2070 USA. Telephone: (201) 582 3000 extension 3086. Contact: Dixon Dudderar. Description: Scientific & industrial research; holographic non-destructive testing.

AUSTRALIAN HOLOGRAPHICS PTY LTD. P.O. BOX 160 Kangarilla, South Australia 5157. Telephone (61) (08) 383 "1255. FAX: (61) (08) 383 7244. Contact: David Ratcliffe. Company description: Manufacturers of white light transmission and reflection holograms, laser transmission holograms, and holographic illumination sources. Specialise in very large format continuous wave holography. Consultancy, model construction, research and courses available.

B

BARR & STROUD, LTD. Caxton Street, Anniesland. Glasgow G13 1HZ, Scotland, United Kingdom. Telephone: (44X41) 954 9601. Contact: George Brown. Company description: Artistic holography eriksberg, Denmark. Telephone: (45XOl)198 208. FAX: (45XOl) 198 747. Branch office of Newport Corporation, Fountain Valley, CA USA. .

BEDDIS KENLEY ENGINEERING LTD. Unit 3, EIland Terrace, Holbeck,Leeds, West Yorkshire LS11 9NW, England, United Kingdom. Telephone; (44X0532) 465 979. Contact: Steve Smith. Company description: Artistic holography; embossed holography; equipment & supplies.

BELJING INSTITUTE OF POSTS AND TELECOMMUNICATIONS. Department of Applied Physics, Holography Laboratory, Beijing 10080, People's Republic of China. Telephone: (86Xl) 668 1255. Contact: Hsu Da-Hsiung. Description: college courses in holography.

BELJING NORMAL UNIVERSITY. Analysis and Testing Centre, Beijing 100875, Peoples Republic of China. Contact: Huang Wanyun. Description: Industrial and Scientific research, NonDestructive testing

BIAS (BREMER INSTITUTE FOR APPLIED BEAM TECHNOLOGY). Established 1977. 40 Employees at this address. 2800 Bremen 33, Klagenfurter Str. 2, Germany. Telephone: 0421-218-01 or 0421-218-5002 (direct line for Prof. Juptner) FAX: 0421-218-5063. Contact: Prof. Dr. Ing, or Werner Juptner. President: Prof. Dr. Ing. Gerd Sepold. Description: Industrial research; holographic nondestructive testing.

BOB MADER PHOTOGRAPHY. P.O.Box 796728, Dallas, TX 75379 USA. Telephone: (214) 690 5511. Contact: Hope Hickman. Description: Artistic holography; pulsed laser portraits; marketing consultant.

BOBST GROUP. 146-T Harrison Avenue, Roseland, NJ 07068 USA. Telephone: (201) 226 8000. Contact: John Torchia. Company description: Manufacturer of hologram applicator machinery.

BRANDTJEN & KLUGE, INC, (MAIN HEADQUARTERS), Established in 1919.80 Employees at this address. 539 Blanding Woods Road, Box 736, St. Croix Falls, WI 54024 USA. Telephone: (715) 483 3265. FAX: (715) 483-1640. Contact: Hank A. Brandtjen III, Vice-President. President, Henry A. Brandtjen, Jr.; Vice-President, Hank A. Brandtjen III, H.M. Williams.Company description: Manufacturer of Hot Stamp Machine

BRIDGESTONE GRAPHIC TECHNOLOGIES, INC. Established 1989. 15 Employees at this address. 375 Howard Street, Bridgeport, CT 06605, USA. Telephone: (203) 366-1595. FAX: (203) 366- 1667. Contact: Rich ZuckerlMike ZubretskylDoug Miller. Company description: Fully integrated manufacturer for embossed, dichromate, and photopolymer holograms. In-house metallizing and converting. Manufacturer of Maximage (tm) holograms. Full product development services.

BRIGHTON IMAGE CRAFT. 7 Bath Street, Brighton, East Sussex BNl 3TB England, United Kingdom. Telephone: (44X0273) 202 069. Contact: Jeff Blyth. Description: Specialist in producing holographic recording material: red-sensitive DCG and photopolymer; photochromic plate for HeNe lasers.

BRITISH AEROSPACE PLC. Established in 1908. 8000 Employees at this address. Sowerby Research Centre, FPC: 267, P.O. Box 5, Filton, Bristol BS12 7QW, England, United Kingdom. Telephone: (44X0272) 366 842. FAX: (44X0272) 36 3733. Contact: Mr. S.C.J. Parker. Chairman, Prof. Smith; Vice-President, John Evans. Subsidiary company: Austin Rover, Royal Ordnance. Branch offices: Filton, Bristol; Stevenage; Warton; Lostock; Bruff; Preston; Plymouth; Weybridge; Chadderton; Hatfield. Company description: Holographic Non-destructive testing.

BRITISH AEROSPACE PLC. (MAIN HEADQUARTERS). 11 Strand, London, England, United Kingdom.

BROOKHAVEN NATIONAL LABORATORY. Building Side, Upton, NY 11973 USA. Telephone: (516) 282 3758. Description: Industrial research.

BURLEIGH INSTRUMENTS, INC. Established 1972. 65 Employees at this address. Burleigh Park, Fishers, NY 14453, USA. Telephone: (716) 924-9355. FAX: (716) 924-9072. Contact: Ann Kost, Marketing Administration Manager. President: David Farrell. Sales Manager: Timothy Van Slambrouck. Company description: Burleigh Instruments, Inc. is a leading manufacturer of Wavemeters and Fabry Perot Interferometers and Etalon Systems for laser diagnostics; Piezoelectric-based micro positioning equipment; Color Center Lasers.

BURNS HOLOGRAPHICS LTD. Established in 1989. 3 Employees at this address. P.O. Box 377, Locust Valley, NY 11560 USA. TelephonelFAX: (516) 674 3130. Contact: Joseph Burns. Company description: Since 1972, Holograms/Stereograms/Editions; Silver halide, Photoresist, Nickel, Embossed with Agam, Dali, Cossette, Nunez, Dieter Jung, Sam Moree, Others; 1979 - Injection-Molded Holograms; 1987 - DesignlDevelopment NY Telephone Hologram CreditCard.

BURTON HOLMES INTERNATIONAL. 1004 Larrabee Street, West Hollywood, CA 90069 USA. Telephone: (213) 652 0970. Contact: Burton Holmes. Company description: Artistic holography; multiplex holograms.

C

CAMBRIDGE LASERS, INC. Established 1989. 4 Employees at this address. 43216 Christy Street, Fremont, CA 94538 USA. Telephone: (415) 651-0110. FAX: (415) 651-1690. Contact: Brian Bohan. Sales Manager: Grahame Rogers. Parent Company: Cambridge Lasers Ltd., Brookfield Business Centre, Cottenham, Cambridge, U.K. CB4 4PS. Branch Offices: Pittenweem, Scotland; Eppertshausen, West Germany. Company description: Tube processing service and repairs from re-gassing to complete rebuilding of broken or damaged tubes.

CAMBRIDGE STEREO GRAPHICS GROUP. Established 1976. 4 Employees at this address. P.O. Box 159, Kendall Square Station, Cambridge, MA 02142-0002 USA. President: Stephen A. Benton, Vice-President, S. Lee Anthony. Company description: consultants in holographic optics and imaging, and producers of prototype holographic items.

CASDIN-SILVER HOLOGRAPHY. (Studio). Established in 1968. 51 Melcher Street, #501, Boston, MA 02210 USA. Telephone: (617) 423 4717. Contact: Harriet Casdin-Silver, Owner. Company description: I have been creating holographic art and interactive hologrsphic environments since 1968. Our company specializes in original holograms for advertising, architectural and theatre settings, expositions. We are consultants, exhibition organizers/designers.

CASDIN-SILVER HOLOGRAPHY. (Office) 99 Pond Avenue, Suite 0403, Brookline, MA 02146 USA. Telephone: (617) 739-6869. Contact: Harriet Casdin-Silver.

CENTER FOR ADVANCED VISUAL STUDIES. Massachusetts Institute of Technology, 40 Massachusetts Avenue, Cambridge, MA 02139 USA. Telephone: (617) 2534415. Description: Scientific, academic research.

CENTER FOR APPLIED RESEARCH IN ART AND TECHNOLOGY. 72 Lange Boomgaardstraat, B 9000 Gent, Belgium. Parent: University Gent, Workshop Holography, Gent, Belgium. Telephone: (32) (91) 626384. FAX: (32) (91) 237326. Contact: Pierre M. Boone. President: J. van der Haute. Description: Organization created for positive interaction between science and art, bringing together people interested in visual communication by holography and related techniques.

CENTRAL MICHIGAN UNIVERSITY, Art Department. Mt Pleasant, MI48859 USA. Telephone: (517) 774 3025. Contact: Richard Kline. Description: Artistic holography; holography education; courses/workshops.

CHERNOVTSY STATE UNIVERSITY, 2 Kotsyubinsky Str., 274012 Chernovtsy, USSR. Telephone: (700) Chernovtsy 44730. Contact: Oleg V. Angelsky, Correlation Optics Department head. Description: High-speed holographic and interference methods for non-destructive measurements of surfaces with surface roughness heights 0.001 - 0.5μm. Measurements of particle density, size and velocity distribution in disperse media.

CHERRY OPTICAL HOLOGRAPHY. 2047 Blucher Valley Road, Sebastopol, CA 95472 USA. Telephone: (707) 8237171. FAX: (707) 823 8073. Contact: Greg Cherry. Company description: Proprietors Greg Cherry and Nancy Gorglione produce art and display editions of silver halide reflection and transmission holograms in all size formats. Services include custom holograms and innovative exhibition installations.

CHIBA UNIVERSITY. Faculty of Engineering, 1-33 Yayoi-cho, Chiba 260, Japan. Telephone: (81X0472) 511 111 ext. 2874. Contact: Jumpei Tsujiuchi. Description: Scientific research; holographic non-destructive testing.

CHIMERIC IMAGES, INC. Established in 1988. 1 Employee at this address. 713 1/2 Main Street, Lafayette, IN 47901 USA. Telephone: (317) 742 0586. Contact: Ellen M. Shew, President. Company description: Chimeric Images, Inc., is a design and marketing firm specializing in Special Eventll'rade Show & Promotional Item holography. Dichromate; Embossed; Photo Polymer Mediums.

CHROMAGEM INC. (Branch office). Established in 1981. 4 Employees at this address. 573 South Schenley, Youngstown, OH 44509 USA. Telephone: (216) 799 0323. FAX: (216) 747 9371. Contact: Steve Lev. President, Thomas J. Cvetkovich; Vice-President, Steve Lev. Company description: A diversified holographic company with multi-service labs specializing in embossed mastering & full color display holography and is currently expanding into retail and distribution. 30 years combined staff experience in holography.

CHROMAGEM INC. (MAIN HEADQUARTERS). 25 Market Street, International Towers, Youngstown, OH 44503 USA.

CINEMA & PHOTO RESEARCH INSTITUTE. NIKFI, Leningradsky, Prospect 47, Moscow USSR. Telephone: (7X1570) 2923. Contact: 1. Nalimov. Description: Scientific research.

CISE SPA TECHNOLOGIE INNOVATIVE. Established in 1946. 600 Employees at this address. via Reggio Emilia, 39, 20090 Segrate, Milano, Italy. Mail address: P.O.Box 12081, 1-20134 Milano, Italy. Telephone: (39)(2) 2167 2634. FAX: (39X2) 2167 2620. Contact:

Mrs. M. Luciana Rizzi, Eng. President, F. Velona; Vice-President, S. Villani; Sales Manager, F. Banfanti; Customer Service, F. Bonfanti. Parent company: ENEL (Italian National Electric Power Authority). Subsidiary company: Conphoebus, Siet, Co. Tim, Tim Tecnopolis CSATA, CNRSM. Company description: CISE is a research company developing innovative technologies and transferring them to industry. In addition to research, CISE performs technological services and produces equipment, instrumentation, and new technological items.

C.ITOH & COMPANY. Central P.O. Box 136, Tokyo 100-91, Japan. Telephone: (81X3) 639 2946. Description: Holography lab equipment.

CITRoEN INDUSTRIE. 35, rue Grange Dame Rose, F -92360 Meudon-la-Foret, France. Contact: Thierry Manderscheid. Company description: Industrial research; holographic non-destructive testing.

CITY CHEMICAL. 132 West 22nd Street, Dept. H, New York, NY 10011 USA. Telephone: (212) 929 2723. FAX: (212) 463-9679. Company description: Equipment & Supplies including photochemistry & emulsions.

COBURN CORPORATION. Established 1973. 100 + employees at this address. 1650 Corporate Road West, Lakewood, NJ 08701 USA. Telephone: (908) 367-5511. FAX: (908) 367-2908. Contact: John White. President: Joseph W. Coburn II. Subsidiary companies: Coburn Europe GmbH, Coburn Japan Corporation, Coburn! Prisma Chile. Company description: Embossing & shim making; training programs ..

COHERENT, INC. LASER GROUP. (MAIN HEADQUARTERS). Established in 1966. 350 Employees at this address. 3210 Porter Drive, Palo Alto, CA 94304. USA. Telephone: (415) 493-2111. FAX: (415) 858-7631. Contact: Christine Brown, Sales Associate. President, Henry Gauthier; CEO, James Hobart; Sales and Marketing Director, Paul Crosby; Customer Service, Herbert Pummer. Branch offices: Coherent (U.K.) Ltd., Cambridge, England; Coherent GmbH, Ober-Roden, Federal Republic of Germany ..

COHERENT (U.K.) LTD. Cambridge Science Park, Milton Road, Cambridge CB4 4RF, United Kingdom. Telephone: (44X223) 420 501. FAX: (44X223) 420 073.

COHERENT GMBH. SenefelderstraBe 10, 6074 Rodermark, Ober-Roden, Federal RepUblic of Germany .. Telephone: (49X6074) 914/0. FAX: (49X6074) 95654.

Company description: Manufactures Innova (tm) 200 and Innova 300 Ion Lasers for holographic applications. Features include Modetrack tm, which eliminates mode-hops, and PowerTrack tm, which automatically and continuously keeps the laser aligned for hands-off performance.

COLLEGE OF MANUFACTURING. Cranfield Institute of Technology, Cranfield, Bedford MK43 OAL, England,United Kingdom. Contact: JM. Burch. Description: Scientific, industrial engineering; holographic non-destructive testing.

COLUMBIA UNIVERSITY. Department of Otolaryngology, 630 West 168th Street, New York, NY 10032 USA. Telephone: (212) 305 3993. Contact: Shyam Khanna. Company description: Medical holography.

CONTINENTAL OPTICAL. 15 Power Drive, Hauppauge, NY 11788 USA. Telephone: (516) 582 3388. Description: Optics and custom orders.

CONTROL OPTICS. 13111 Brooks Drive, Unit J. Baldwin Park, CA 91706. Phone: (818) 813-1990. FAX: (818) 813-1993. Contact: W. Liu, President. Description: Maker of optics and accessories.

CORION CORP. Established 1967. 65 Employees at this address. 73 Jeffrey Ave., Holliston, MA 01746 USA, Telephone: (508) 429-5065. FAX: (508) 429-8983. Contact: Walter J. Lekki. President, Frank Mascis; Vice-President, Walter Lekki; Sales Manager, Michael Mascis. Company description: Corion Corp. manufactures; volume and One-of-a-Kind/Custom and Stock, optical components including coatings, filters, optics and optical assemblies for use in the UV-Visible-IR spectrum

COULTER OPTICAL COMPANY. P.O. Box K , 54140 Pinecrest Mad, Idyllwild, CA 92349 USA. Telephone: (714) 659 2991. Contact: Mary Braginton. Company description: Make telescope mirrors, parabolic mirrors and more. Send for free list of products.

CREATIVE LABEL. 2450 Estes Drive, Elk Grove Village, IL 60007 SA. Telephone: (708) 956 6960. Contact: Jerry Koril. Company description: Bindery application on Kluge (2 stream) and Bobst (4 stream) machines. Call for more information.

CROWN ROLL LEAF, INC., (MAIN HEADQUARTERS). Established in 1970.95 Employees at this address. 91 Illinois Ave., Paterson, NJ 07503 USA. Telephone: (201) 742-4000. FAX: (201) 742-0219. Contact: James Waitts, Holographic Mgr. President, Robert Waitts; Vice-President, Manuel Cueli; Sales Manager, Pamela Herforth; Customer Service, Maggie Carola. Branch offices: Crown Roll Leaf Inc .. , CA; Crown Roll Leaf Inc., IL. Company description: Crown Roll Leaf has been supplying embossing material internationally for 5 years. We have been manufacturing shims and embossed products for 4 years. Please call Jim Waitts with any questions.
CROWN ROLL LEAF, INC., 20705 South Western Avenue, Torrance, CA 90501 USA. Branch office.
CROWN ROLL LEAF, INC., 2456 Elmhurst Road, Elk Grove Village, IL 60007 USA. Branch Office.

CSI. 7 Meadowfield Park South, Stocksfield, Northumberland, NE43 7QA, England, United Kingdom. Telephone: (44X661) 842 741. Company description: Manufacture mirrors; optics.

CVI LASER CORPORATION, CVI WEST. Established in 1986.470 Lindbergh Avenue, Livermore, CA 94550 USA. Telephone: (415) 449-1064. FAX: On request. Contact: Dr. Alex Jacobson. President, Dr. Yu H. Hahn; Vice-President, F.J. Brandiger. Parent company: CVI Laser Corporation. Branch offices: Livermore, CA; Putnam, CT. Company description: Manufactures holographic quality single and multiple element lenses, mirrors, windows, and beamsplitters for all standard holographic laser sources. Free 104-page catalog available.

CVI LASER CORPORATION. (MAIN HEADQUARTERS). 200 Dorado Place SE, P.O. Box 11308, Albuquerque, NM 87192 USA.

D

DAIMLER BENZ AG. D-7000 Stuttgart 60. Federal Republic of Germany. Contact: H.G.Leis. Description: Industrial Research; holographic non-destructive testing.

DAI NIPPON PRINTING CO. Ltd.Central Research Institute 12, 1-Chome Ichigaya-Kagacho, Shinjuku-ku, Tokyo 162, Japan. Contact:
Tokio Kodera. Telephone: (81X03) 266 2310. Company description: Artistic holography; embossed holography; printing applications.

DARKROOM EIGHT LTD. Unit 8 - Impress house, Vale Grove, Acton, London W3 7QH, United Kingdom. Telephone: (44X81) 749 2218. Company description: Artistic holography.

DATASIGHTS LTD. Alma Road, Ponders End, Enfield, Middlesex, EN3 7BB, England, United Kingdom. Telephone: (44X81) 805 4157. Company description: Manufacture mirrors.

DAVID SCHMIDT HOLOGRAPHY. Established in 1985. 3 Employees at this address. 23962 Craftsman Road, Calabasas, CA 91302 USA. Telephone: (818) 992 1541. FAX: (818) 703 1182. Contact: David Schmidt, Owner. Company description: Holography courses offered. Company description: David Schmidt Holography is a full service mass production laboratory specializing in stereograms both cylindrical and image plane formats. We also mass produce reflection and transmission holograms for the trade.

DAVIN OPTICAL LTD. Reliant House, Oakmere Mews, Potters Bar, Hertfordshire, EN6 9xx, England, United Kingdom. Telephone: (44X707) 44445. Company description: Manufacture mirrors; optics.

DAZZLE ENTERPRISES. 425 Southlake Blvd, Richmond, VA 23236 USA. Phone: (804) 379-5500. FAX: (804) 379-6328. Description: bossing specialists; electroplating, silvering and application.

DB ELECTRONIC INSTRUMENTS S.R.L., Via Teano 2, 1-20161 Milano, Italy. Telephone: (39)(02) 646 9341. FAX: (39X02) 645 6632. Branch office of Newport Corporation, Fountain Valley, CA USA.

DEEP SPACE HOLOGRAPHICS. 1328 Dunsterville Avenue, Victoria, British Columbia, Canada V8Z 2X1. Telephone: (lX604) 384-3927. Contact: Karan Wells, Sales Manager. President: Marc de Roes. Company description: Exotic fine art/commercial sculpture and animation, conceptual and industrial design, display merchandising, exhibits and special effects. Since 1980 secured worldwide distribution of our DCG disigns via "Holocrafts". Quality guaranteed!

DELL OPTICS COMPANY, INC. Established 1950. 15 Employees at this address. 25 Bergen Blvd., Fairview, NJ 07022, USA. Telephone: (201) 941-1010. FAX: (201) 941-9524. Contact: Belle Steinfeld, Sales Manager. President: Frank Gervolino. Company description: Custom working of precision optical components.

DENNISON MANUFACTURING COMPANY. Metallizing Division. 300 Howard Street, Framingham, MA 01701, USA. Telephone: (508) 879-0511. Contact: Marc Wey. Company description: Manufacturer of holographic paper.

DEUTSCHE GESELLSCHAFT FUR HOLOGRAFIE E.V. (GERMAN HOLOGRAPHIC SOCIETY). Established in 1989. Lerchenstr. 142 a, D-4500 Osnabrock, Federal Republic of Germany. Telephone: (49X0541) 7102 199. FAX: (49) (0541) 74297. Contact: Dr. Peter Zec, President. President, Dr. Peter Zec; Vice-President, Brigitte Burgmer. Company description: The society was founded to promote awareness of holography, and its members are mainly holographers and artists. To this end, the group intends to organise exhibitions.

DEUTSCHER HOLOGRAPHIE-VERTRIEB. Kolner Strasse 49-51, D-4047 Dormagen I, Federal Republic of Germany. Contact: Thomas Rost. Company description: Artistic holography.

D.G.A. MFG. CO. A SUBSIDIARY OF DIVERSIFIED GRAPHICS, Ltd., 3719 Joy Road, Columbus, GA 31906 USA.

DIALECTICA AB. Skanegatan 87, 6tr, S-116 37 Stockholm, Sweden. Contact: Ambjorn Naeve. Company description: Artistic holography.

DIAURES S.A. HOLOGRAPHIC DIVISION, Via 1 Maggio 2621A, 1-41019 Soliera (Modena) Italy. Telephone: (39X059) 567 274. Company description: Artistic holography; embossed holography; equipment & supplies.

DIE DRITTE DIMENSION. Established in 1987.2 Employees at this address. Frankfurter StraBe 132-134, D-6078 Neu Isenburg, Federal Republic of Germany .. Telephone: (49X06102) 33367. FAX: (49X06102) 36709. Contact: Mr. Carlo Westphal. President, Mr. Carlo Westphal; Sales Manager, Mrs. Elke Hein; Customer Service, Mrs. Elke Hein. Company description: Biggest special shop for holography in Federal Republic of Germany .. Always over 600 different holograms in stock. Comprehensive fine art section. Branch office: Nordwest-Zentrum, Tituscorso, D-6000 Frankfurt/M. 50, Germany. Telephone: (49) (069) 5890349.

THE DIFFRACTION COMPANY, INC. Established in 1957. 15 Employees at this address. P.O. Box 151, Riderwood, MD 21152 USA. Telephone: (301) 666 1144. FAX: (301) 472 4911. Contact: Hugh C. Wynd, President. Vice-President, Christopher W. Wynd; Customer Service, Kim Price. Company description: We offer: A. 58 patterns available in 16 colors with a variety of adhesives. B. Color explosion graphicslmicro-etching an alternative to 3D. C. Custom embossing of holograms. D. Dazzlers-Stickers.

DIRECT HOLOGRAPHICS. P.O. Box 295, Strasburg,USA. Telephone: (717) 687-9422. FAX: (717) 687-9423. Contact: Jacque Phillips, President. Description: The exclusive distributor of products and services for Third Dimension Ltd. Largest and most popular range of images for the specialist gallery and gift market.

DORIS VILA HOLOGRAPHICS. 157 East 33rd Street, NY, NY 10016. Telephone or FAX: (212) 686-5387. Description: Quality laboratory for creative holography with emphasis on large-scale holograms, color-mixing and architectural applications. Available for commissioned work, prototyping, workshops and purchase of existing artwork

DOVECOTE STUDIO. Witham Friary, Frome, Somerset, England, United Kingdom. Telephone: (44X74) 985 691. Contact: Angela Coombes. Company description: Artistic holography; oversize format

DREAM IMAGES. Established 1987. 3 Employees at this address. Postfach 1602, Vermeerweg 15,. D-5047 Wesseling, Federal Republic of Germany. Telephone: (49) 02236 43138 (gallery: 02644-3147). FAX: (49) 02236-82369. Contact: Klaus Thielker, Thomas Ramunke. Company description: Artistic holography; gallery in Castle of Lim-JRhein since 9/89; marketing consultant.

DUPONT COMPANY. (See E.I. DUPONT DE NEMOURS & CO.) DUSTON HOLOGRAPHIC SERVICES INC. Established 1988. 2 Employees at this address. 115 Shannon Street, Ottawa Ontario Canada K1Z 6Y6. Telephone: (613) 722-9004. Contact: Deborah A: Duston, President. Company description: Duston Holographic Svcs consults corporate and government clients on HOEs, remotely sensed Holographic Stereograms and the educational and curatorial aspects of Holography. Deborah Duston is also a well known artist-holographer.

DUTCH HOLOGRAPHIC LABORATORY. Established in 1983. 6 Employees at this address. Kanaal d)ik Noord 61, 5642 JA Eindhoven, The Netherlands. Telephone: (31X40) 817 250. FAX: (31X40) 814 865. Contact: Walter Spierings, Director. President, W. Spierings; Sales Manager, R. Van Oorschot. Company description: Production of photoresist masters, embossed holograms and computer-generated images. Silver halide reflection holograms on glass 50 x 60

cm) and fIlm (1 m square); high volume film copying.

THE DZ COMPANY. Established in 1985. 10 Employees at this address. P.O. Box 5047-R, 181 Mayhew Way, Suite E, Walnut Creek, CA 94596 USA. Telephone: (415) 935 4656. FAX: (415) 935 4660. Contact: Gary Allison, Sales Manager. President, Dan Cifelli; Vice-President, Gary Zellerbach. Company description: Manufacturers of fast selling holographic products for retail sales and promotions. Many of the items can be imprinted for ad specialty and incentives.

E

EALING ELECTRO-OPTICS INC. New Englander Industrial Park, Holliston, MA 01746 USA. Telephone: (508) 429 8370. Contact: Pauline Lebine. Company description: Manufacturer of mirrors & optics.

EALING ELECTRO-OPTICS. Greycaine Road, Watford, Hertfordshire, England WD2 4PW United Kingdom.

EALING SCIENTIFIC LTD. P.O. Box 238, Pointe Claire-Dorval, Quebec H9R 4N9, Canada. Telephone: (514) 631 1B07. Company description: manufactures lasers.

EASTMAN KODAK COMPANY. Scientific Imaging, Department 841-S, 343 State Street, Rochester, NY 14650-0811. USA. Telephone: (1) (800) 242-2424, ext. 12. Canada: (1) (BOO) 387-8773. Description: Manufacturer of holographic equipment & fIlm.

ECOLE NATIONALE SUPERIEURE D'INGENIEURS. clo Groupe de Laboratoires, C.M.R.S., 5, Av. D'Edimbourg, BLD MarechalJuin, F-14032 Caen-Cedex, France. Contact: Jean-Charles Vie not. Description: Scientific & industrial research; holographic nondestructive testing.

E.C. SCHULTZ & COMPANY, Established in 1895. 9 Employees at this address. 333 Crossen, Elk Grove Village, IL 60007 USA. Telephone: (312) 640 1190. FAX: (312) 640 1198. Contact: Bob Schultz President. Company description: Our company makes stamping: embossing debossing and applique dies for the graphic industries. Quality craftmanship and 94 years experience joining for innovative, distinctive and exciting effects in todays demanding market.

EDMUND' SCIENTIFIC COMPANY. 101 East Gloucester Pike Barrington, NJ 08007 USA. Telephone: (609) 547 3488. Compan; description: Mailorder, wholesale, and retail holography products for schools, science fairs, etc.

ED WESLY HOLOGRAPHY. Est. 1978.5331 N. Kenmore Ave Chicago, IL 60648 USA. Telephone (312) 784-1669. Contact: Ed Wesly, Head Poobah. Description: I am an artist making candy for the eyes using Holographic Optical Elements and junk (found objects).

ElF PRODUCTIONS. EIZYKMAN. 19 Rue Jean Jacques Rousseau, F- 75001 Paris, France. Telephone: (33X1) 4236 0631. Contact: Claudine Eizykman. Company description: Artistic holography; multiplex.

E.I. DUPONT DE NEMOURS & CO., INC. Established in 1802. Over 5000 Employees at this address. Optical Element Venture, Experimental Station, P.O.Box 80352, Wilmington, DE 19880-0352 USA. Telephone: (302) 695 4893. FAX: (302) 695 9631. Contact: Paula Bobeck, Customer Service Representative. Director, A. Wasy

D'Cruz; Venture Manager, Lory Galloway; Marketing Manager, Krishna C. Doraiswamy; Senior Product Specialist, Evan D. Laganis. Parent Company: Du Pont.
Company description: Du Pont / Optical Element Venture's advanced photo polymer products feature reliable holographic performance, easy dry processing and high environmental stability. Call us to see why our materials are the best choice for your appliation.

E.I. DUPONT DE NEMOURS & CO., INC., (MAIN HEADQUARTERS),1007 Market Street, Wilmington, DE 19898 USA. Subsidaries: E.I. Dupont de Nemours & Co,. Optical Element Venture, Wilmington, DE USA.

ELAN BIO-MEDICAL. HOLOGRAPHY LABORATORY, 411 Lewis Road, #417, S"an Jose, CA 95111 USA. Contact: Michael Gersonde. Company description: Medical holography.

ELECTECH DISTRIDUTION SYSTEMS, PTE. Ltd. Established 1982.35 employees at this address. 605-A, MacPherson Road, #03- 03, Citimac Industrial Complex, Singapore 1336. Telephone: (65)286 9933. FAX: (65) 284 3256. Contact: Ching-Wat Chia, Division Manager. President: Teck-Mong Sia. Parent Company: Waler Holdings Corporation Main Headquarters: No. 7 Ave. 8A, Eunos Industrial Estate, Singapore 1440. Tel. (65) 7463388. FAX: (65)7433244. Branch offices: Malaysia, Thailand, Indonesia and Brunei. Description: Supplier of Coherent lasers and Newport holographic equipment.

ELECTRO OPTIC CONSULTING SERVICES. (EOCS)Established 1989. 18198 Aztec Court, Fountain Valley, CA, 92708, USA. Telephone: (714) 964-0324. FAX: (714) 536-7729. Contact: Dr. Colleen Fitzpatrick, President. Company description: PhD Physics: HNDT, laster/optical system design; high-density information storage. Silver halide, thermoplastic holography; laser broker. CW, pulsed laser work. Lab facilities for lease. Very successful proposal writing/submission.

ELECTRO OPTICAL INDUSTRIES, INC. Established in 1964. 100 employees at this address. 859 Ward Drive, Santa Barbara, CA 93111 USA. Thlephone: (805) 964 6701, sales ext. 280. FAX: (805) 967 8590. Contact: Joseph Lansing, Applications Engineer. President, Arthur J. Cussen. Company description: Manufacturer of infrared test and calibration instrumentation including: collimators, choppers, blackbody sources, differential temperature sources, FLIR test equipment, radiometers and LLL-TV target simulators.

ELECTRO OPTICS DEVELOPMENTS LTD. Howards Chase, Pipps Hill Industrial Estate, Basildon, Essex, England, SS14 3BE, United Kingdom. Telephone: (44X268) 531344. Contact: Mr. Chris Plumb, Sales Manager. Company description: Equipment & supplies; optics.

ELUSIVE IMAGE. 603 Munger Street # 316, Dallas, TX 75202 USA. Telephone: (214) 720 6060. Contact: Fred Wilbur. Company description: Holography gallery.

ELUSIVE IMAGE. 135 West Palace Avenue, Suite 102, Santa Fe, NM 87501 USA. Telephone: (505) 986 0221. Contact: Fred Wilbur. Company description: Holography gallery.

EMPAQUES Y EVOLTURAS HOLOGRAFICAS SA DE C.V. Established 1990. Pino 343, Local 41. Col. Sta. Ma. La Ribera. 06400 Mexico, D.F. Thlephone: (525) 547-2033. (525) 547-1893. Contact: Dan Lieberman. Description: Manufactures wide-web embossing for packaging, giftwrap, and labeling applications. Manufactures embossed holograms onto paper, polypropylene, polyester, acetate, and pvc materials. A free joint process with

perfect registration.

ENVIRONMENTAL RESEARCH INSTITUTE OF MICHIGAN. Optical Science Lab., ACD, P.O.Box 8618, Ann Arbor, MI 48107 USA. Telephone: (313) 994 1220. Contact: Juris Upatnieks. Company description: Industrial & academic research.

ENVISION ENTERPRISES. Box 2003, Sausalito, CA 94965. Contact: Jeff Allen. Telephone: (415) 459-8232. Description: Produces holography and high-tech shows for entertainment and education. All inquiries for acquisition of new holograms welcome. Marketing "Magic Windows: Pocket Holography" an inexpensive basic reference source on holography.

ERBA. 12, rue Libergier, F-51100 Reims, France. Telephone: (33X26) 884 452. Contact: Bernard Ollier. Company description: Artistic holography; holography education.

EUROLAUNCH LTD. 213 Salisbury Court, Fleet Street, London, EC4Y, England, United Kingdom. Company description: Holographic non-destructive testing.

EVE RITSCHER ASSOCIATES LTD., 73 Allfarthing Lane, Wandsworth, London SW18, England,. United Kingdom. Contact: Eve Ritscher. Company description: Artistic holography education.

EVERGREEN LASER CORPORATION. Established 1985. 10 Employees at this address. Unit B, Commerce Circle, Durham, CT 06422 USA. Telephone: (203) 349-1797 (413) 731-8813. Contact: Phil Fostini, President, or Preston Macy, Sales. Branch Office: 3640 Main St., Springfield, MA 01107 USA. Description: Refurbishing and replacement of Argon, Krypton, HeCd and HeNe laser tubes. Discontinued laser models supported, 6 month warranty, prompt service.

EXCITEK INC. Established in 1984. 7 Employees at this address. 277 Coit Street, Irvington, NJ 07111 USA. Telephone: (201) 372 1669. FAX: (201) 372 8551. Contact: Brian 'furner. President, Brian 'furner; Vice-President, Andrew Dietz; Sales Manager, Al Dietz; Customer Service, Donna McGann. Company description: Excitek Inc. buys, sells, repairs and remanufactures Spectra Physics Ion laser systems and plasma tubes with unmatched attention to best quality, service and cost efficiency.

F

F.A.S.T. ELECTRONIC BULLETIN BOARD. P.O. Box '421704, San Francisco, CA 94142-1704 USA. Telephone: (415) 845 8306. FAX: (415) 841 6311. Contact: Elizabeth Crumley. Company description: Educational materials

FISHER SCIENTIFIC. E .M.D. Division, 4901 West Lemoyne Avenue, Chicago, IL 60651 USA. Telephone: (312) 378 7770. Contact: Sales and Marketing Dept. Company description: Supply science lab equipment, holography kits, lab manuals, lasers and laser related equipment.

FLEXcon COMPANY, INC. Flexcon Industrial Park, Spencer, MA 01562. Telephone: (508) 885-3973. FAX: (508) 885-7000. Contact: Joseph P. Morgan, Jr., Market Development. ext. 379. Company description: Manufacturer of holographic and prismatic materials for packaging, gift wrap and graphic mm markets. Wide-web embossing in excess of 60' width. Range of pressure sensitive films for electronics, medical etc ...

FOCAL IMAGE LTD. P.O.Box 1916, London W11 3QR, England, United Kingdom. Telephone: (44) (071) 229 0107. FAX: (44X071) 727 3438. Contact: Kaveh Bazargan. Company description: Color holography; medical holography

FORD RESEARCH STAFF, SCIENTIFIC RESEARCH LABS. 20000 Rotunda Drive, Room S-1023, SRL, Dearborn, MI 48121 USA. Telephone: (313) 3231539. Contact: Gordon Brown. Company description: Industrial research; Holographic non-destructive testing. Research and development in computer-aided holographic interferometry.

THE FOREIGN DIMENSION. Established in 1989. 5 Employees at this address. Suite B, 81Fl., Central Mansion, 270-276 Queen's Road Central, Hong Kong, Hong Kong. Telephone: (852)(542-0282. FAX.: (852) 541-6011. Contact: Frederic Schvartzman, General Manager. President, Frederic Schvartzman; Vice-President, Nathalie Aboucar; Sales Manager, F. Schvartzman; Customer Service, Stephane Denizot. Company description: Specialists in manufacturing all kinds of holographic products (Watches, keyrings, pendants), we are a French managed company offering Hong Kong competitive prices with high quality standards. Contact us for details.

FOUNDATION IDEE CENTRUM. P.O. Box 222, 5600 MK, Eindhoven, The Netherlands. Company description: Gallery.

FRED UNTERSEHER & ASSOCIATES HOLOGRAPHY. 3463 State Street, Suite 304, Santa Barbara, CA 93105 USA. Telephone: (805) 5686997 FAX.: (818) 549-0534. Contact: Fred Unterseher.

FREE UNIVERSITY OF BRUSSELS. Department of Applied Physics (ALNA), Faculty of Applied Sciences, Brussels, Belgium. Contact: Stephan Roose. Description: Academic and Scientific research.

FRESNEL TECHNOLOGIES INC. 101 West Morningside Drive, Fort Worth, TX 76110 USA. Telephone: (817) 926 7474. FAX.: (817) 926 7146. Contact: Linda H. Claytor. Company description: Manufactures plastic Fresnel lenses & lens arrays from its POLY IR® plastics for use into the infrared; also other optical products for use into the ultraviolet from acrylic & other plastics.

FRINGE RESEARCH HOLOGRAPHICS INC.1179A King Street West, Suite 008, Toronto, Ont~rio M6K 3C5, Canada. Telephone: (416) 535 2323. Contact: Michael Sowdon, Director. Company description: artistic holography; silver halide holograms; pulse portraits; gallery; workshops; travelling exhibit.

FTI JOFFE, Politechnicheskaya 26, Academy of Sciences of the USSR, 194021 Leningrad, USSR. Contact: G.A. Sobolev. Description: Artistic holography.

FUJI PHOTO OPTICAL CO. , Ltd. No. 324, 1-Chome, UetakeMachi, Omiya, Japan. Telephone: (81X04) 866 30111. FAX.: (81X04) 86510 521. Contact: Takayuki Saito. Company description: Industrial research, optics.

FUJI ELECTRIC CO. LTD., Mecatronics Division, 1-12-1 Yurakucho, Chiyoda-ku, Tokyo 100, Japan. Telephone: (81X3) 211 7111. Company description: Manufactures C02 lasers and related equipment.

G

GALLERIE ILLUSORIA, Schwarztorstrasse 70, CH-3007 Bern, Switzerland. Contact: Sandro Del-Priete. Company description: Gallery.

GALVANOART. (Subsidiary of Artplay Holographic Studio) H-1191 Budapest, Ady Endre lit. 8, Hungary.

GALVOPTICS LTD. Harvey Road, Burnt Mills Industrial Estate,Basildon, Essex, SS13 lES, England, United Kingdom. Telephone: (44X0268) 728 077. FAX.: (44X0268) 590 445. Contact: R. D. Wale. President: B.A. J. Wale. Company description: Optics; mirrors, lenses.

GARDENER PROMOTION MARKETING. Established in 1980. 6 Employees at this address. 4165 Apalogen Road, Philadelphia, PA 19144 USA. Telephone: (215) 849 4049. Contact: John Gardener. President, John Gardener; Vice-President, Roy Gunther. Parent Company: Gardener Promotion Products Corp. Company description: As the exclusive package goods marketing representative for Holographic Design, we can show you how holography can be used for problem solving or enhancing opportunities compatible with your objectives.

GENERAL HOLOGRAPHICS, INC. Established in 1978. 6 Employees at this address. P.O. Box 82247, Burnaby B.C., V5C 5P7 Canada. Telephone: (604) 435 6654. FAX: (604) 432 7326. Contact: Paula Simson, Managing Director. President, Bernd Simson; Sales Manager, Paula Simson; Customer Service, Darlene Lafgren. Company description: Distributor of dichromate gift and jewelry items (Holocrafts), silver halide wall and desk decor, and photo polymer (Polaroid, Holos Gallery) for the Canadian market. Custom and stock.

GERALD MARKS STUDIO. 29 West 26th Street, New York, NY 10010 USA. Telephone: (212) 889 5994. Description: Artistic holography; stereo grams; consulting; instruction.

GLOBAL IMAGES, INC. 509 Madison Avenue, Suite 1400, New York, NY 10022 USA. 2556 West 2nd Street, Vancouver, British Columbia, V6K IJ8 Canada. FAX.: (604) 734 2842.Telephone: (212) 759 8606. FAX.: (604) 734 2842. Contact: Walter Clarke. Company description: Manufacturer of holographic embossing machines; equipment for embossing.

G.M. VACUUM COATING LAB, INC. 882 Production Place, Newport Beach, CA 92663 USA. Telephone: (714) 642 5446. Company description: Plate coating.

GRAY SCALE STUDIOS LTD. Established in 1985. 2 Employees at this address. 4500 19th Street, #294, Boulder, CO 80304 USA. Telephone: (303) 4425889. FAX.: (303) 4425889. Contact: George Sivy, President. Company description: Gray Scale Studios, Ltd. specializes in the design and creation of models and sculptures for Holographic Imaging. Consultant services offered, six years experience, samples of work available upon request.

H

HALO POWER-TRACK -- LIGHTING DMSION. McGraw-Edison Corporation, 6 West 20th Street, New York, NY 10011 USA. Telephone: (212) 645 4580. Company description: lighting fixtures.

HICKMOTT & AUSTIN HOLOGRAMS. 11 Castelnau, London SW13 9RP, England,United Kingdom. Telephone: (44X01) 486 5811. Contact: M. Austin. Company description: Artistic holography.

HIGH TECH NETWORK. Skeppsbron 2, S-211 20 Malmo, Sweden. Telephone: (460(040) 350 750. FAX.: (46X040) 237 667. Contact: Christer Agehall. Company description: Artistic holography; security applications.

Holographic T-Shirts and Sweatshirts

FRONT OF SHIRT

BACK OF SHIRT

516

Golden Gate San Francisco

THE BOOKSHELF

Below is a selection of unique or hard-to-find holography items. Some of the books are not found in your local bookshop and can be ordered directly from us, using the order form.

Sweatshirts: The sweat- and T-shirts pictured on the opposite page are from Textile Graphics and can be ordered using the form below. The sweatshirts on the opposite page are soft washable fabric, 50% cotton/poly and are imprinted with non-lead, permanent plastisol ink. Fully machine washable. Specify small, medium or large size.

CABLE CAR T-SHIRT	CABLE CAR SWEATSHIRT	EAGLE T-SHIRT	EAGLE SWEATSHIRT
$19.00(US) or	*$35.00(US) or*	*$17.00(US) or*	*$30.00(US) or*
£12.00 (UK) or	*£22.00 (UK) or*	*£11.00 (UK) or*	*£19.00 (UK) or*
DM 40.00	*DM 70.00*	*DM 34.00*	*DM 60.00*

Holography Handbook by Unterseher, et al. 407pp, $23.00(US), £14.00 (UK), 46.00 (DM)
The world's best selling book on holography. Designed as a do-it-yourself manual of practical holography, this book is laid out in an easy-to-read manner and gives the reader absolutely all the information he needs to buy and construct his own home holography studio. Complete step-by-step instructions with hundreds of illustrations are given on making beginning and advanced holograms. Holography Handbook is currently used by several universities as a lab manual.

Optical Holography by P. Hariharan. 320pp, $28.00(US), £18.00 (UK), 56.00 (DM)
The leading book on the theory of optical holography. Haraharan covers in considerable depth all the major procedures used in holography. Thorough footnotes and references list over 700 works on holography. If you need more depth of understanding and research material, this is unquestionably the book for you.

Practical Holography by Graham Saxby. 488pp, $66.00(US), £38.00 (UK), 132.00 (DM)
For the "technologist or professional". This book covers the practical aspect of holography at all levels. Saxby gives wide coverage of all aspects of the field. This book gives a detailed technical understanding of the basic techniques used in holography and comes with an impressive silver halide hologram in each copy.

TO ORDER *(3 or more items = 10% discount!)*:

Credit card customers can order by phone. Call: (800) 367 0930 or (415) 841 2474
OR FAX your order to: (415) 841 2695.

For orders by mail, fill out the information below and enclose check or credit card information.

NAME: _____

TITLE: _____

COMPANY: _____

ADDRESS: _____

CITY/STATE (PROVINCE): _____

COUNTRY/ZIP CODE: _____

PHONE/FAX: _____

Item description	Qty.	Price	Total
_____			
_____			
_____			
_____			

Subtract 10% if 3 or more items ordered_____
Shipping if outside USA
add $9.00 or £6.00 or DM18.00 _____
Total Cost of order _____

FOR CREDIT CARD ORDER:

CHECK ONE: ☐ MASTER CARD - or ☐ VISA - or ☐ AMERICAN EXPRESS

CREDIT CARD NUMBER: _____

CREDIT CARD EXPIRATION DATE: _____

SINGATURE: _____

ADDRESS ALL CORRESPONDENCE TO: HMP ORDER DEPT.
                                        ROSS BOOKS
                                        P.O. BOX 4340
                                        BERKELEY, CA LIFORNIA 94704
                                        USA

HOECHST CELANESE CORPORATION. 86 Morris Avenue, Summit, NJ 07901 USA. Telephone: (201) 522 7816. Contact: Gunilla Gilberg. Company description: Embossed & artistic holography.

HOL 3 GALERJE FüR HOLOGRAPHIE. Kurfurstendamm 103, 1000 Berlin 31, Federal Republic of Germany. Company description: Gallery & retail shop.

HOLAGE, Established in 1981.1881 Eighth Avenue, San Francisco, CA 94122 USA. Telephone: (415) 564 1840. Contact: Brad D. Cantos. Company description: Fine art holograms; silver halide holograms.

HOLAR SEELE KG. Wasserwerksweg 10-14, D-2960 Aurich 1, Federal Republic of Germany. Telephone: (49) (41) 10005. Company description: Fine art holograms.

HOLAXIS CORPORATION, Established 1984. 499 Farmington Ave, Hartford, CT 06105 USA. Telephone: (203) 232 2030. FAX: (203) 236-3767. Contsct: Martin A. Berson. President: Martin Berson, Vice-President: Gary Haber. Company description: Holaxis' specialties include: Large format holograms and wide-web embossing of pressure sensitive and hot stamp foils. High quality mastering, photoresist transfers, high speed labeling, and die-cutting are also included.

HOLICON CORPORATION. Established in 1987. 906 University Place, Evanston, IL 60201 USA. Telephone: (708) 491 4310. FAX: (708) 491 7955. Contact: Dr. Hans Bjelkhagen. President, Dr. Max Epstein; Vice-President, Dr. Michel Marhic; Sales Manager, Dr. Michel Marhic; Customer Service, Dr. Hans Bjelkhagen. Subsidiary company: Holographic Industries, Inc. Company description: Holicon Corporation specializes in silver halide holograms, pulse or CW, in particular, portraits. Large-format reflection or transmission holograms are made as well as mass production of film holograms.

HOLO 3 . Established in 1986. 7 Employees at this address. roe de l'Industrie, 68300 Saint-Louis, France. Telephone: (33)(89) 69 82 08. Contact: Mrs. J. Striebig, Deputy Director. President, Prof. P. Smigielski; Vice-President, Mr. A. Weber; Sales Manager, Ms. J. Striebig. Company description: Non Profit National Organization depending on French Ministry of Research and Technology and transferring optical technologie$ from the Research Institute of Saint-Louis France, towards industrial applications.

HOLO ARP. KAMAKURA INC. 7-10-8 Ginza Chuo-Ku, Tokyo, Japan. Telephone: (81)(03) 574 8307. FAX: (81)(03) 574 8377. Contact: Yumiko Shiozaki. Company description: Artistic holography production; gallery; wholesaler; equipment & supplies for holography; holography education.

HOLOCOM. Lange Strasse 51, D-2117 Kakenstorf, Federal Republic of Germany. Telephone: (49)(04186) 8510. Contact: Johannes Matthiesen. Company description: Artistic and embossed holography.

HOLOCOM HOLOGRAPHIE. 13, rue Charles V., Faculte des Sciences et des Techniques, F- Paris, France. Telephone: (33)(1) 948) 04 0058. Contact: Alan Baraton. Company description: Artistic holography.

HOLOCOR I.B.F. PRINTING INC. Established in 1986. 4 Employees at this address. 95 des Sulpiciens, L'Epiphanie, Qu~bec JOK lJO, Canada. Telephone: (514) 588 6801. FAX: (514) 5884898. Contact: Jean-Robert Bernier, President. President, Jean-Robert Bernier; Sales Manager, Jean-Robert Bernier; Customer Service, Jean-Bertrand Neron. Parent company: I.B.F. Printing Inc. Company description: We focus our knowledge in what you want to see: "Holographic micro-engraving". We devote our energy in what you need: "Performance". Holocor® from electroming (shims) to final embossed hologram.

HOLOCRAFT INTERNATIONAL. Established 1982. P.O. Box 152, Lake Forest, IL, 60045-0152 USA. Telephone: (708) 234-7625. FAX: (708) 615-0835. Contact: William Crist, Jr. Company description: Artistic holography marketing.

HOLOCRAFTS: DIVISION OF CANADIAN HOLOGRAPHIC DEVELOPMENTS LTD. Established in 1979. 20 Employees at this address. Box 1035, Delta, B.C." V4M 3T2 Canada, Telephone: (604) 946-1926. FAX: (604) 946-1648. Contact: Karoline Cullen, Managing Director. President, Gary Cullen; Vice-President, Barry Michelitsch. Company description: Holocrafts specializes in the manufacture of dichromate reflection holograms. We oITer prompt delivery of stock and custom production in a variety of shapes, sizes and products.

HOLOCRAFTS EUROPE LIMITED. Established 1990. 5 Employees at this address. Barton Mill House. Barton Mill Road. Canterbury, Kent, CT1 lBY England. Telephone: 0227 463223 FAX: 0227 450399. Contact: Chris Luton, President. Company description: Specialists in the manufacture and supply of dichromate products. Also altering a complete range of holographic gift products including stock foils and holographic jewelry.

HOLODESIGN. 1, Boulevard de la Republique, F-95600 Eaubonne, France. Telephone: (33)91) 39 593 954. Contact: Thierry Garcon. Company description: artistic holography.

HOLODESIGN STUDJES. Rebenstrasse 20, CH-4125 Riehen, Switzerland. Telephone: (41)(61) 672 342. Company description: Marketing consulting.

HOLO-DIMENSIONS INC. 3577 Rue de Bullion, Montreal, Quebec, Canada H2X 3Al. Telephone: (514) 845 4419. Company description: Artistic holography.

HOLOFAR LAB (SRL). Piazza Acilia No.3, Int. 3,Rome, Italy 00199. Telephone: (39)(6) 8395 517. Company description: Artistic holography.

HOLOFAX LIMITED. Netherwood Road, Rotherwas Industrial Estate, Hereford HR2 6JZ, England, United Kingdom. Telephone: (44) (432) 278 400. Company description: Silver halide reflection; Photochemistry; Vibration isolation tables.

HOLOFLEX COMPANY. RR 3, Box 381, Urbana, IL 61801 USA. Telephone: (217) 684 2321. Contact: Donald Barnhart. Company description: artistic holography; silver halide holograms.

HOLO GMBH HOLOGRAFIELABOR OSNABRüCK. Minderner-Str. 205, D-4500 Osnabrock, Federal Republic of Germany. Telephone: (49)(0)(541) 7102 173. FAX: (49)(541) 7102 176. Contact: Vito Orazem. Company description: Holograms up to 1 x 1 m (master + first copy =$8,000.); embossed holography. HOLO-design: floor-tiles, door signs, lamps.

HOLOGOS LTD. Established 1985. 18198 Aztec Ct., Fountain Valley, CA, 92708, USA. Telephone: (714) 964-0324. FAX: (714) 536-7729. Contact: Dr. Colleen Fitzpatrick, President. Company description: PhD Physics, educational consultant. Experience on elementary, high school, college, grad school levels. Lectures, demonstrations, workshops, educational materials. Optical laboratory design and construction. Lab facilities for lease. Artwork/laser broker.

HOLOGRAFIA GALLERIA. Jaakonkatu 3, 2nd floor, SF-00100 Helsinki, Finland. Telephone: (358) (06) 941909. Company descripion: Gallery, retail shop.

HOLOGRAFIA GALLERIA. (Branch of Starcke KY). c/o Science Center, Dulu, Finland. Telephone: (358X39) 360 700. FAX: (358X39) 07 905. Company description: gallery & retail shop

HOLOGRAFICA. 8 Hylda Court, St. Albans Road, NW5, London, England, United Kingdom. Description: Artistic holography.

HOLOGRAFIE - HOFMANN LABOR. Carl-Hermann-Gaiserstrasse 20, 7320 Groppingen, Federal Republic of Germany. Contact: Martin Hofmann. Company description: Full service artistic holography; buying & selling; holography education; equipment & supplies.

THE HOLOGRAM. P.O. Box 9035, Allentown, PA 18105 USA. Telephone: (215) 434 8236. Contact: Frank DeFreitas, Publisher. Description: Free Newsletter on Holography. Contact for more details.

HOLOGRAMA LABORATOIRE HOLOGRAPHIQUE. 41 rue Marinzano, CH-1227 Geneva, Switzerland. Telephone: (41)(022) 422 144. Contact: Yves Rossignol. Company description: Embossed; pulsed portraits.

HOLOGRAMAS DE MEXICO. Established in 1984. 50 Employees at this address. PINO 343, Local 3, Col. Sta. Ma La Ribera, 06400 Mexico, D.F. Mexico. Thlephone: (525) 5411791 ot (525) 547-9046. FAX: (525) 5474084. Contact: Dan Lieberman. Company description: Hologramas de Mexico has a vertically integrated factory where we do our own photoresist plates, embossing, and electroforming; a hot-melt unit where we do adhesives, die-cutting, hotstamping and labeling.

HOLOGRAM EUROPE SPRL. Avenue Voltaire 137, 1030 Brussels, Belgium.Telephone: (32)(2) 242 7284. Contact: J . B. Boulton. Company description: Retail shop

HOLOGRAM. INDUSTRIES, Established in 1984. 12 employees at this address. 42-44, rue de Trucy, 94120 Fontenay sous bois, France. Telephone: 1 4394 1919. FAX: 1 439 40032. Contact: Hugues Souparis, President. Sales Manager: Brigette Degaire. Company description: Hologram. Industries produces high quality display and embossed holograms. We have an integrated production line, from lab to embossing. Hologram. Industries initializes graphic holograms and 3D stereograms for security and advertisements.

HOLOGRAM LAND. Established 1989. 3 Employees at this address. 43 Main Street SE, #131 Riverplace, Menneapolis, MN, 55414, USA. Telephone: (612) 378-9618. Contact: Susan Rickert, owner. Mailing address: 1117 Marquette Ave, Suite 105, Minneapolis, MN 55403. Branch office: Halos: 3001 Hennepin Ave #E202, Minneapolis, MN 55408. Description: Hologram Land is the Twin Cities' ftrst hologram gallery and shop, and offers a wide range of holographic products, diffraction-based merchandise, and optical novelties and toys. At Riverplace, along the historic Mississippi Mile.

HOLOGRAMM WERKSTATI & GALERIE, GALLERIE FUR HOLOGRAMME, Established in 1984. 2 Employees at this address. Via Principale 30,CH - 7649 Castesegna, Switzerland. Telephone: 08241718. FAX: 082 412 68. Contact: Horst Gutekunst,

Director. Company description: Creative workshop, developments, looking for new and attractive ways for hologram making

HOLOGRAM ROADSHOW. Longlear House, Warminister, Wiltshire, 12 Queen Square, Bath, Avon BA1 1WU, England, United Kingdom. Telephone: (44X225) 339 333. Company description: retail shop; travelling exhibit.

HOLOGRAMS AND OTHER STRANGE THINGS. Established in 1987. 2 Employees at this address. 3200 West Oakland Park Boulevard, Lauderdale Lakes, FL 33311 USA. Telephone: (305) 739 9634. Contact: Dennis Drucker, President. Vice-President, Harriet Drucker. Company description: A retail store specializing in holographic products, and other three-dimensional and illusory-type items.

THE HOLOGRAM SCHOPPE. P.O. Box 318, 591 Tonawanda, Buffalo, NY 14202 USA. Contact: Maureen McNamara. Company description: retail sales of holography.

HOLOGRAM WORLD. Established 1987. 4 Employees at this address. 1212 lf2 Dixon Boulevard, Cocoa, FL 32922 USA. Telephone: (407) 631 3615 and 1 (800) 8824656. Contact: Susan K. Harrison. President: Joel Todd Hays. Company description: Retail shop, mailorder, wholesale. Finest selection of 3D artwork, jewelry, stationery, toys, stickers, rainbow makers and optical illusions. Ask about our success guarantee. Call for wholesale or retail catalogs.

HOLOGRAPHIC APPLICATIONS, Established in 1985. 2 Employees at this address. 21 Woodland Way, Greenbelt, MD 20770 USA. Telephone: (301) 345 4652. FAX: (301) 345 4653. Contact: Suzanne St. Cyr, President. Company description: Technical and Marketing Services for manufactures of holographic products. Education, Design Consultation, Product Development, Vendor Selection, Project Management, and General Contracting for endusers of holography.

HOLOGRAPHIC ART. Established in 1988. 4 Employees at this address. Werderstraße #73, Bremen 2800, BR Deutschland, Federal Republic of Germany. Telephone: (49X421) 555 690. FAX: (49X421) 556 202. Contact: Hartmut Finé. Barbel Rathje, Contact. Company description: Manufacturing & distribution of ftne art holographic jewelry. Wholesale of a wide variety of holographic products. Agency offers competent service for any private & commercial need: concept design, model making, manufacutring process.

HOLOGRAPHIC CONCEPTS. 14 Cove Road, Forestdale, MA 02644 USA. Telephone: (508) 477 2488. Contact: George Willenborg. Company description: Artistic holography marketing; consulting.

HOLOGRAPHIC CONCEPTS, INC. 1711 St. Clair Avenue, St. Paul, MN 55115. Telephone: (612) 698-6893. Contact: Stephen Sugarman. Description: Consulting for educational and industrial needs; consulting for artists; classes in holography for individuals; display design; silver halide mastering; mm or glass plate copies; commercial product research and design.

HOLOGRAPHIC CREATIONS. 26- Rue Daniel Stern, Paris 75015, France. Telephone: (33X1) 45 78 8742. Contact: J-C Raverat de Boisheu . Company description: artistic holography, custom work.

HOLOGRAPHIC DESIGN, INC. Established in 1979. 6 Employees at this address. 1084 North Delaware Avenue, Philadelphia, PA 19125 USA. 400 West Erie Street, Chicago, IL 60610 USA. Thlephone: (215) 425 9220. FAX: (215) 425 9221. Contact: D. Miller.

Branch office: Robert Sherwood Holographic Design, Inc., Chicago, IL USA. Holograpic Products Inc., Richmond, UT USA. Company description: HDI provides holograms for a variety of display, promotional, advertising, packaging, and architectural applications. We offer a complete range of services to take your project from concept to final product.

HOLOGRAPHIC DESIGN SYSTEMS, INC. Established 1981. 8 Employees at this address. 1134 West Washington Blvd., Chicago, IL 60607 USA. Telephone: (312) 8292292. Contact: Robert Billings. Description: HDS has the capability to produce all types of holograms for display, advertising, marketing, promotion pop, direct mail, packaging and architectural uses. We make reality from ideas.

HOLOGRAPHIC DIMENSIONS, INC., 9235 SW 179 Terrace, Miami, FL 33157 USA. Telephone: (305) 255 4247. Contact: Kevin Brown. Offices: Miami, New York, Buenos Aires, Santiago. We specialize in commercial photoresist origination and replication. Large volume production runs are our specialty. Clients include Coca Cola, Diners' Club, and Warner Brothers.

HOLOGRAPHIC IMAGES INC., Established in 1982. 6 Employees at this address. 1301 Dade Boulevard, Miami Beach, FL 33139, USA. Telephone: (305) 531 5465. FAX: (305) 531-3029. Contact: Larry Lieberman; CEO, Frank Millman; Customer Service, Peg Lieberman. Company description: Produces multi-color reflection holograms on film. Specializing in limited edition art holograms - recorded from artwork created by artists from varied media. Custom images for corporations, commissioned editions.

HOLOGRAPHIC INDUSTRIES, INC. (MAIN HEADQUARTERS). Established in 1988. 5 Employees at this address. P.O. Box 1109, Libertyville, IL 60048 USA. Telephone: (708) 680 1884. FAX: (708) 680-0505. Contact: Robert Pricone, President. President, Robert Pricone; Secretary, Max Epstein; Sales Manager, Robert Pricone; Customer Service, David Epstein. Parent company: Holicon Corp., Evanston, IL USA. Branch offices: Light Wave Gallery, Chicago, IL USA; Light Wave Gallery, Schaumburg, IL, Light Wave Gallery, Dearborn, MI, USA. Company description: Holographic Industries designs and operates retail galleries/gift shops in major shopping centers. We produce our own pulse holographic images, and can obtain nearly any holographic product world-wide.

HOLOGRAPHIC LABEL CONVERTING, INC. Telephone: (612) 941-0922. Contact: Scott LaBelle. Description: Holography finisher specializing in labels, tags, magnets and matchboxes. Guaranteed machine-applied labeling. Serialized 3D holograms. All quantities welcomed.

HOLOGRAPHIC MARKETING, INC. Established 1986. 2 Employees at this address. 9250 S.W. First Street, Plantation, Florida 33324, USA. Telephone: (305) 474-9965. FAX: (305) 472-0101. Contact: Marc R. Rapke, President. Branch Office: Lido Beach, NY. Company description: Consultant to foreign and domestic corporations on applications of embossed holography; broker for fine art holograms; exporter of artistic holograms, jewelry and novelties.

HOLOGRAPHIC PRODUCTS INC. (MAIN HEADQUARTERS). Established in 1975. 18 Employees at this address. 365 North, 600 West, Logan, UT 84321. Telephone: (801) 7524225. FAX: (801) 752-4467. Contact: Dave Rayfield, President. Vice-President, Hollie Rayfield; Customer Service, Marina Heidt. Branch offices: Holographic Design Inc., Philadelphia, PA USA. Company description: At Holographic Products Inc., we manufacture a full line of stock dichromate items. Custom holography is available in sizes up to 14" x 14".

HOLOGRAPHIC RESEARCH PTY LTD. Lot 9, Industry Drive, South Tweed Heads, NSW, 2486 Australia. Phone: (61) 75-54-3988. FAX: (61) 75-24-6625. Contact: David Toyer or Jim Booth. Description: Specializing in volume production of holograms (silver halide and photopolymer). Australian agent for The Lasersmith. Embossed foil holograms, full production facilities from model-making through mastering to mounting and display preparation.

HOLOGRAPHIC SERVICE. 10 via Civerchio, 1-20159 Milan, Italy. Telephone: (39X02) 688 7067. Company description: Marketing consultants

HOLOGRAPHICS INTERNATIONAL. BCM Holographics, London WClN 3xx, England, United Kingdom. Telephone: (44X01) 584 4508. Contact: Sunny Bains. Company descripton: artist's magazine.

HOLOGRAPHICS NORTH INC., Established in 1984. 7 Employees at this address. 444 South Union Street, Burlington, VT 05401 USA. Telephone: (802) 658-2275. FAX: (802) 862-6510. Contact: Dr. John Perry, President. Vice-President, Barbara D. Perry; Research, Jeff Klute. Company description: Designersl Producers of large format holography up to 44" x 72" (1.1 x1.8m.) Known worldwide for the highest quality commercial and fme art display work. Design, model building, production, installation and consulting services.

THE HOLOGRAPHIC STUDIO, LTD. Established in 1987. 3 Employees (includes subcontractors) at this address. 2525 York Avenue, Vancouver, British Columbia V6K 1E4 Canada. Telephone: (604) 734 1614. FAX: (604) 734 2842. Contact: Melissa Crenshaw. Company description: The studio produces quality limited edition multicolor reflection holograms. In addition, we have vast experience in the production of single color and achromatic reflection transfers from ruby pulse masters.

HOLOGRAPHIC STUDIOS. 240 East 26th Street, New York, NY 10010 USA. Telephone: (212) 686 9397. FAX: (212) 4818645. Contact: Jason Sapan. Company description: Artistic holography; marketing & lighting consultants; holography workshops.

HOLOGRAPHICS (UK) LTD. 32 Lexington Street, London WlR 3HR, England, United Kingdom. Telephone: (44X071) 437 8992. FAX: (44X01) 494 0386. Contact: Jon Vogel. Company description: Embossed & artistic holograms; pulsed portraits; DCG; large format.

HOLOGRAPHIE KONZEPT. Korberstrasse 3, D-6000 Frankfurt 50, Federal Republic of Germany. Company description: Artistic holography.

HOLOGRAPHIE LABOR / MIKE MIELKE. Georgenstrasse 61, D·8000 Munich, Federal Republic of Germany. Telephone: (49X89) 271 2989. FAX: (49X89) 271 1375. Company description: Stereograms; silver halide holograms; DCG; retail shop.

THE HOLOGRAPHY DEVELOPMENT GROUP. 23 Neville Park Boulevard, Toronto, Ontario M4E 3P5, Canada (Showroom at 602 King Street West, Toronto, by appointment only). Telephone: (416)-691-9381. FAX: (416)-691-0407. Contact: Andrew Laczynski. Sales Manager: Dale Blue. Customer Service: Lidka Schuch. Branch Office: 1 Moss Hall Crescent, Finchley, London, N12 8NY England. Telephone: (81) 445-5321. FAX: (81) 446-8047. Company description: Research & development, custom packaging.

HOLOGRAPHY INSTITUTE. P.O. Box 446, Petaluma, CA 94953 USA. Telephone: (707) 778 1497. Contact: P. Pink. President, Jeffrey Murray. Company description: Classes: Holography education for teachers, artists, commercial designers; Workshops for hobbyists-all ages, all levels. Group or individual instruction. Commercial: embossing masters; fine artispecial editions; design, consulting, research.

HOLOGRAPHY LTD. Established in 1986. 5 Employees at this address. 21 Hakomemiut Str., Herzlia Pituah, Israel. Contact: David Livneh, Shimon Hameiri. Parent company: The Third Dimension Ltd. Company description: Buying & Selling holograms; educational holography

HOLOGRAPHY NEWS. (MAIN HEADQUARTERS). Established in 1987. 3 Employees at this address. 3932 McKinley Street N.W., Washington, D.C. 20015 USA. Telephone or FAX:(202) 966-3464. Contact: Lewis Kontnik. Publisher, Lewis Kontnik; European Editor, Ian Lancaster. Parent company: Reconnaissance, Ltd. Branch Office: Surrey, England, United Kingdom. Company description: holography News-The International Newsletter of the Holography Industry. Published ten times a year.for the latest news, information, analysis and features on commercial activity, R&D, corporate news, patents and conferences.

HOLOGRAPHY NEWS. (Branch Office).l Erica Court, Wych Hill Place, Woking, Surrey GU22 OJB, England, United Kingdom. Telephone or FAX: (44) 483740689. Contact: Ian Lancaster.

HOLOGRAPHY WORKSHOPS--LAKE FOREST COLLEGE (see Lake Forest College Holography Workshops)

HOLOGRAPHY WORLD. (Branch of Art Science and Technology Institute). 800 K Street N.W., Techworld Plaza, Suite 54, Washington, D.C. 20001 USA. Telephone (202) 408-1833. Description: Exhibition, Education and Research interactive center between professionals (artists and laboratories) and the public. Every 2 years organizes Washington International Exhibit of Holography. Produces large travelling exhibit "Hologram: Image of the Future". Contact: Odile Meulien, Development Director, or Frank van der Kemp, Chairman.

THE HOLOGRAPHY YEARBOOK. Rita Wittig Fachbuchverlag, 10 Chemnitzer Strasse, D-5142 Huckelhoven 1, Federal Republic of Germany. Telephone: (49X2433) 84412. FAX: (49X2433) 86356. Contact: Prof. Dr. Siegmar Wittig. Company description: Rita Wittig Publishing provides the broadest range in holography books worldwide. The Holography Yearbook is a comprehensive annual inventory of holography and its applications. Other publications: textbooks and catalogues.

HOLO-IMAGES, INC. 167 Washburn Road, Briarcliff Manor, NY 10510 USA. Telephone: (914) 941 8811. Contact: David DeBitetto. Company description: Artistic holography.

HOLO-LASER. Established in 1978. 3 Employees at this address. 6, rue de la Mission, Ecole, 25480 Miserey, France. Telephone: (33) (1) 45 315 275. FAX: (33X1) 48 331 702. Contact: Dr. Jean Louis, H., Tribillon. Branch offices: Besanfon, France; Paris, France. Company description: Embossed holography & equipment; artistic holography; buying & selling; education.
HOLO-LASER. (Branch Office). 4, rue du Refuge, 25000 Besanfon, France.
HOLO-LASER. (Branch Office). 12, rue de Vouille, 75015 Paris, France.

HOLO-LASER TECH, LTD. Established 1982. 7 Fraser Avenue, #16, Toronto, Ontario M6K 1Y7 Canada. Telephone: (416) 534-7419. Contact: Glenn Strazds, President. Parent company: Laser Gallery.

HOLOMART-- PREMIUM TECHNOLOGY LTD. 9 Brunswick Centre, London WCIN 1AF, England, United Kingdom. Telephone: (44X01) 353 4212. FAX: (44) (01) 353 0684. Contact: Tanya. Company description: buying & selling holograms.

HOLOMAT. Established 1988. 1 Employee at this address. 741 East Gorham Street, Madison, WI 53703 USA. Telephone: (608) 255-3580. Contact: Matt Hansen. Company description: Artistic holography and holographic engineering consulting.

HOLOMEDIA ABIHOLOGRAM MUSEUM. P.O. Box 45012, Drottninggatan 100, 10430 Stockholm, Sweden. Telephone: (46X08) 105 465. FAX: (46X08) 107 638. Contact: Mona Forsberg. Company description: Broker for embossed and artistic holography; Buying & selling holograms; Holography education; Gallery.

HOLOMEDIA INC., Established in 1977. 7 Employees at this address. 3-15-22, Takaban, Meguro-ku, Tokyo 152 Japan. telephone: (81) (03) 793 2321. FAX: (81X03) 793 2322. Contact: Takao Kawahara, Marketing Director. President, Masato Nakajima; Customer service, Hibiki Tsuge. Company description: Holomedia is a reputable company producing Display Dichromate Holograms. High quality, the world's brightest, and in wide sizes (50Ox500mm).

HOLOMETRIC AB. Bjornasvagen 21, S-113 47 Stockholm, Sweden. Telephone: (46X08) 790 9780. Contact: Ingegard Dirtoft. Company description: Equipment & supplies; holographic nondestructive testing.

HOLOMEX LTD. Established in 1987. 2 Employees at this address. 4 Borrowdale Avenue, Harrow HA3 7PZ, England, United Kingdom. Telephone: (44X01) 427 9685. Contact: Michael Anderson, Managing Director. President, Michael Anderson; Vice-President, Susan Anderson. Company description: The main company product is the Viewcam holographic camera which can make and display silver halide transmission and reflection holograms up to a maximum size of 10in x 10in.

HOLOMORPH VISUALS, INC . P.O. Box 1405, Stn. Desjardins, Montreal, Quebec, Canada H5B 1H3. Telephone: (819) 872-3622. Contact: Mr. Kenneth Chalk. Company description: Artistic holography.

HOLO-OR LTD. Established in 1989. 7 Employees at this address. P.O. Box 1051, Kiryat Weizmann, Rehovot, Israel. Telephone: (972X8) 469687. FAX: (972X8) 466 378 .. President, Grossinger I; Vice-President, Uri Levy; Sales Manager, Nissim Greisas. Company description: Manufactures computer-generated diffractive optical elements by VLSI techniques. Catalog elements and custom designs. Substrates include ZnSe, GaAs and various glasses. Has "DOE" station--dedicated workstation for element design, mask generation.

HOLOPRINT ROSOWSKI. Postfach 1164, Lindenau 23, D-4174 Issum 1, Federal Republic of Germany. Telephone: (49X02835)1684. Description: Workshops, embossed & artistic holography, buying & selling--wholesale.

HOLOPRODUCTION. Established in 1986. 35 rue Abbatucci, 68330 Huningue, France. Telephone: (33X89) 69 82 08. Contact: Mrs. J. Striebig, General Manager. Company description: Embossing consultants; mass-manufacturing; artistic presentation consultants; holography education; medical research; HNDT; Lab installation; equipment and supplies.

HOLOPUBLIC, KLAUS UNBEHAUN. Established in 1985. Hirschstrasse 84, D-5600 Wuppertal-2,Federal Republic of Germany. (FRG). Telephone: 0202 84118. Contact: Klaus Unbehaun. Presi·

dent: Klaus Unbehaun. Company description: Klaus Unbehaun, owner of "Holopublic", is working as a media journalist (especially commercial holography). He is making "Lasergraphics · artwork for exhibitions and publishing the newsletter "Holography and 3-D Software".

HOLOS. Established 1990. 3001 Hennepin Avenue, Space E202, Minneapolis, MN 55408 USA. Telephone: (612) 824-1404. Contact George Robinson. Description: Branch office of Hologram Land.

HOLOS ART GALERIE. 4 Place Grenus, 1201 Geneva, Switzerland. Telephone: (41)(022) 325 191. Contact: Pascal Barre. Company description: Gallery, retail sales.

HOLO-SERVICE. Neuensteinerstrasse 19, CH-4153 Basel, Switzerland. Telephone: (41)(061) 502 287. Contact: Edgar Bar. Company description: Artistic holography.

HOLO-SERVICE.FRIES. Eulerstrasse 55, CH-4051 Basel, Switzerland. Telephone: (41)(061) 22647. Contact: Urs Fries. Company description: Artistic holography.

HOLOS GALLERYJ A.H. PRISMATIC. Established in 1979. 10 Employees at this address. 1792 Haight Street, San Francisco, CA 94117 USA. Telephone: (415) 2214717. FAX: (415) 2214815. Contact: Sales Manager, Alan Rhody. Gallery Manager: Chris French. Company description: Holos Gallery is one of the world's oldest and largest distributors of holographic products. Specializing in holographic novelty products, dichromates, film holograms and excellent new lines of photo polymer holograms.

HOLO/SOURCE CORPORATION. Established in 1985. 5 Employees at this address. 21800 Melrose Avenue, Southfield, MI 48075 USA. Telephone: (313) 355 0412. FAX: (313) 3550437. Contact: Lee Lacey, President. President, Lee Lacey; Vice-President, Robert Levy; Sales Manager, Bill Seydel; Customer Service, Robert Levy. Company description: Holo/Source manufactures fine quality embossed holograms and colorful diffraction grating patterns for catalog and magazine covers, direct mail marketing projects and point-of-purchase displays.

HOLO-SPECTRA. 7742-B Gloria Avenue, Van Nuys, CA 91406 USA. Telephone: (818) 994 9577. FAX: (818) 994 4709. Contact: R. Arkin. Company description: . artistic holography consulting; embossed holography; wholesale buying & selling; silver halide; DCG; lasers, mirrors, lenses; mters, pinholes, isolation tables.

HOLOSYSTEMS INC. P.O.Box 6810, Ithaca, NY 14850 USA. Telephone: (607) 273 1187. Contact: Jonathan Back. Company description: artistic holography.

HOLOTEC CC. P.O. Box 5144, Brackengardens, 1452 Transvaal, South Africa. Telephone: (27)(011) 864 1292. Contact: Mandy Van Der Molen. Company description: Artistic holography.

HOLOTEC PLC. 7 Cameron Road, Seven Kings, Essex, IG1 3DF, England, United Kingdom. Telephone; (44)(01) 597 8004. Contact: Janet Ives, Managing Director. Company description: Artistic holography marketing consultants.

HOLOTEK LTD., 300 East River Road, Rochester, NY 14623 USA. Telephone: (716) 424 4996. FAX: (716) 424 4958. Contact: Charles Kramer. Company description: H.O.E's and scanners.

HOLOTEK, S.A. Established in 1988.4 Employees at this address. Carretera de Santander, Granda 47,33199 Granda-Siero, Asturias, Spain. Telephone: (34)(985) 79 35 26. FAX: (34)(85) 27 18 53. Con-

tact: Julio Ruiz Garcia, President. Company description: Holotek works on distribution of embossing and gifts with holograms through big stores and makes custom holograms. Holotek is open to all. For more information please contact us.

HOLOTRON SRL. 46 via Tolstoi, 1-20146 Milan, Italy. Telephone: (39)(02) 479697. Company description: Marketing consultants.

HOLOVISION. 43 Pall Mall, London SWIY 55G, England, United Kingdom. Telephone: (44)91) 839 5622. Company description: Retail shop.

HOLTRONIC. Melchior-Huber Strasse 25, D-8011 Ottersberg, Post Pliening, Federal Republic of Germany. Telephone: (49)(08121) 81005. Contact: Dieter Basler. Company description: Artistic holography, pulsed portrait, HOEs, embossed holography.

HOWARD SMITH PRECISION OPTICS. 61 Lancaster Road, New Barnett, Hertfordshire, EN4 BA5, England, United Kingdom. Telephone: (44)(1) 441 7878. Company description: Manufacture mirrors, lenses.

HUGHES AIRCRAFT CO.--LASER PRODUCTS. 6155 EI Camino Real, Carlsbad, CA 92008 USA. Telephone: (619) 931 3252. Contact: Marcia Berg. Company description: C02 Lasers for sale.

I

IAN GINN HOLOGRAPHY (See our ad on page .....) Established 1990. 3 Employees at this address. Postbus 116, 5070 AC Udenhout, The Netherlands. Telephone: (31) 4241 4358. FAX: (31) 4241 4368. Contact: Ian Ginn. Customer Service: Riny Alberts or Rolanda Van Drie!. Description: Fourteen years' experience in product development, eight in holography. Consistently produce the best-selling mm holograms on the market. Available through most wholesalers/distributors. Write, call or fax for catalogue.

IAN M. LANCASTER HOLOGRAPHICS CONSULTANCY. Established in 1986. 1 Employee at this address. 1 Erica Court, Wych Hill Park, Woking, Surrey GU22 OJB England, United Kingdom. Telephone: (44)(483) 740 689. FAX: (44)(483) 740 689. Company description: Consultant, Curator; European Editor, Holography News; founder, Third Dimension Limited; former Director, Museum of Holography, NY. Specialising in display holography; business development, market studies, marketing concepts, art and display exhibitions.

IBERO GESTÃO - GESTÃO INTEGRADA E TECNOLOGICA LDA. Established in 1988. 7 Employees at this address. Apartado 1267,4104 Porto Codex, Portugal. Telephone: (351-2).301276. Contact: Filipe Vallada P. Norais, President. Vice-President, Figueroa Gonyalves; Sales Manager, L. Abrunhosa, Customer Service: Fatima. Parent Company: InterEuropeia, Portugal. Company description: Artistic holography; buying & selling; marketing & educational consulting; access to NDT labs.

IBM ALMADEN RESEARCH CENTER. K69/803, 650 Harry Road, San Jose, CA 95120 USA. Telephone: (408) 927 1937. Contact: Glenn Sincerbox. Company description: Manufacturer of HOE's and optical heads; scientific holography research.

IBOU INC. Established in 1984. 5 Employees at this address. CP 214, Cap-de-Ia-Madeleine, Quebec, G8T 7W2 Canada. Telephone: (819) 295 5229. FAX: (819) 295 5229. Contact: Jean-Pierre Marchand. President, Jean-Pierre Marchand; Sales Managaer, Brigitte Gagnon. Parent company: Graphie (Edition); ET (retail & commercial sales). Main Headquarters: Qu~bec, Canada. Company description: Buying & selling holograms; Consulting; Manufacture fme art & silver halide holograms.

IBOU INC. (MAIN HEADQUARTERS) 306 Notre-Dame, Champlain, Quebec, G8T 7W2, Canada.

ICI AMERICAS. Concord Pike, Wilmington, DE 19897 USA. Telephone: (302) 575 3087. Company description: Optics; HOE's; gratings.

ICON HOLOGRAPHIC. 11 Uxbridge Street, London, W8, England, United Kingdom. Company description: artistic holography.

IDHOL. Boite Postale 7, F. 89340 Saint-Agnan, France. Telephone: (33)(16) 8696 1929. Contact: Jacques Bousigue. Company description: Fine art holographics; silver halide holograms; presentation consultant.

ILFORD LIMITED. Established in 1880. Mobberley, Knutsford, Cheshire WA15 7HA, England, United Kingdom. Telephone: (565) 50000. FAX: (44)565 872 734. Parent Company: International Papers Corporation. Company description: Equipment & supplies for holography, fUm, plates, photochemicals.

ILFORD PHOTO CORP. West 70 Century Road, Paramus, New Jersey 07653 USA. Telephone: (201) 265 6000. FAX: (201) 265-0894. Contact: Ek Sachtler, Sales & Inquiries. President: Tony Crupi. Vice President: John Lenhart. Parent Company: International Paper Co.

ILLINOIS INSTITUTE OF TECHNOLOGY, Mechanical & Aerospace Engineering. Engineering Building #1, Room 2460, Chicago, IL 60616 USA. Telephone: (312) 567 3249. Contact: Cesar Sciammarella. Company description: Holographic interferometry; industrial holographic research; Non-destructive testing.

ILLINOIS VALLEY MAGNETIC RESONANCE. 4005 Progress Boulevard, Peru, IL 61354 USA. Telephone: (815) 223 8674. Contact: Dr. John L. Mori. Description: Scientific holography research.

IMAC INTERNATIONAL, INC. 1301 Greenwood, Wilmette, IL 00091 USA. Contact: J. Kauffmann. Company description: Holography marketing consultants.

IMAGEN HOLOGRAPHY, INC. Established 1988. 4 Employees at this address. P.O. Box 10837, Aspen, CO 81612 or 135 Brush Creek Road, Suite B44, Snowmass Village, CO 81615 USA. Telephone: (1) 800-43-PULSE and (303) 923-2905. Contact: Alan Moreterud, President. Marketing Director: Greg Riddell. Subsidiary company: Ski Holographics, Pulse Holographic Enhancements. Company de scripion: Imagen Holography Inc. is a brokeringlconsulting firm specializing in the application of all forms of holography for new products in the sporting goods, apparel and printed marketing material industries.

IMAGES COMPANY. Established in 1982. 14 Employees at this address. P.O. Box 313, Jamaica, NY 11419 USA. Telephone: (718) 706-5003. Contact: Ellen Persch, Customer Service. President, John Panico; Vice-President, David Channer; Sales Manager, Ruth Enivoi. Company description: Images Company sells holographic equipment targeted to educational institutions, students and private holographers. Equipment available: development kits, mounting kits for lenses, beamsplitters, mirrors. Spatial filters, display lights, safe lights, filters

IMAGES COMPANY SOUTH. P.O. Box 2251, Inverness, Flor-

ida, 32651 USA. Telephone (904) 344-8540. Contact: Laura Iovine, owner. Description: Images Company South sells lighting equipment for the display of reflection holograms. We stock surplus holographic equipment such as HeNe lasers and power supplies.

IMAGING & DESIGN. Established in 1987. 7 Employees at this address. 1101 Ransom Road, Grand Island, NY 14072-1459 USA. Telephone: (716) 773 7272. Contact: Keith Allen. Company description: Imaging division distributes fUm, chemicals, darkroom/processing and safety equipment and supplies. Design division direct markets custom or stock embossed and silver images to ad agencies, converters, corporate end-users

IMPERIAL COLLEGE OF SCIENCE. Optics Section, Blackett Laboratory, London SW7 2BZ, England, United Kingdom. Telephone: (44) (71) 589 5111. Contact: J. Dainty. Company description: Courses in holography; scientific holography research; particle measurement.

INFOTECH INTERNATIONAL--Holography Division. Established 1990. 3 Employees at this address. 3607 West Magnolia Blvd., Suite 2, Burbank, CA 91505 USA. Telephone: (818) 845-7997. FAX: (818) 845-0312. Contact: Robert F. Cranford, Director. Main headquarters: 18034 Ventura Blvd., Suite 139, Encino, CA 91316 USA. Description: Infotech International is a distributor of quality holograms and holographic art.

INFRARED OPTICAL PRODUCTS, INC. Established 1975. 10 Employees at this address. 120 Secatogue Ave, P.O. 292, Farmingdale, NY 11735 USA. Telephone: (516) 694-6035. FAX: (516) 694-6049. Contact: Barry or Jim Bassin. Company description: Manufacturer of infrared lenses, windows, reflectors, beam splitters, computer-designed IR lens systems and non-linear optical coatings.

ING.-AGENTUR FUR NEUE TECHNO LOGIE IN OPTIK UND PRECISION ENGINEERING. D-7771 Frickingen 2, Federal Republic of Germany. Contact: P. Langenbeck. Company description: Holographic non-destructive testing; industrial research.

INGENIEUR BuRO GEIGER. Established 1982. 2 Employees at this address. Dieding 7, D-8017 Ebersberg, Federal Republic of Germany. Telephone: (08092) 6583. FAX: (08092) 31658. Contact: Mr. Thomas Geiger. Company description: Embossed; artistic holography

INRAD, INC. Established 1973. 70 Employees at this address. 181 Legrand Avenue, Northvale, NJ 07647 USA. Telephone: (201) 767-1910. FAX: (201) 767-9644. Contact: Maria Murray, Manager, Federal R&D Programs. President: Dr. Warren Ruderman. Sales: Lawrence Kosiba. Subsidiary Company: Inrad Optical Systems. Company description: Manufacturer of nonlinear materials, harmonic generation systems, electro-optic and acousto-optic devices and drivers. Also provides optical components, assemblies and optical coatings for the UV, visible and IR.

INSTITUTE OF ART & DESIGN, UNIVERSITY OF TSUKUBA. Established in 1970. 50 Employees at this address. 1-1, Tennodai, Tsukuba, Japan 305. Telephone: (81X0298) 53 2833. FAX: (81)(0298) 53 6508. Contact: Shunsuke Mitamura, Professor. Description: Artistic holography, Holography education

INSTITUTE OF ELECTRONICS BSSR. Established in 1962. 600 Employees at this address. Academy of Sciences-Minsk, 22 Logoiski Trakt, 220841 Minsk-90, USSR. Telephone: (7) Minsk 65-35-14. Contact: Yuri Morgun. President, VA Pilipovich; Vice-President, A.A. Kovalev. Description: Development & manufacturing of highly coherent monopulse lasers & double-pulse lasers with high spectral radiance based on ruby, YAG, neodymium for applications in holography, holographic interferometry, and holographic systems.

INSTITUTE OF NUCLEAR PHYSICS. Leningradska obI., 188350 Gatchina, USSR. Contact: A.M. Bekker. Description: Scientific research.

INSTITUTE OF OPrHALOMOLOGY. Jud Street, London WC1, England, United Kingdom. Telephone: (44)(071) 387 9621. Contact: John Marshall. Company description: Medical holography.

INSTITUTE OF OPTICAL SCIENCE/CENTRAL UNIVERSITY. Chung-Li 32054, Taiwan, R.O.C. Telephone: (886)(3) 425 7681. FAX: (886)(3) 425 8816. Contact: Mr. Tang Yaw Tzong. Description: HOEs, academic research.

INSTITUTE OF PHYSICS. Ukrainian Academy of Sciences, Prospect Nauki 46, 252 650 Kiev 28, USSR. Telephone: (7) 22 2158. Contact: Vladimir Markov. Description; artistic; reflection holography; research in recording materials.

INSTITUTE OF PLASMA PHYSICS AND LASER MICROFUSION, P.O. Box 49, 00-908 Wroclaw, Poland. Contact: Zbigniew Sikorsky. Description: academic research

INTEGRAF. P.O.Box 586, Lake Forest, IL 60045 USA. Telephone: (708) 234 3756. FAX: (708) 615 0835. Contact: Tung H. Jeong, President. Company description: The main business of Integraf is to distribute holographic film and plates. We also carry pre-packaged processing chemicals, and a variety of stock holograms.

INTERCHANGE STUDIOS. 15 Wilkin Street, London NW5 3NG, England, United Kingdom. Telephone: (44)(071) 267 9421. Company description: workshops.

INTEREUROPEIA, (MAIN HEADQUARTERS), Rua Antonio Rodrigues Rocha 248, Vila Nova Gaia, Portugal. Subsidiary company: Thero Gestao-Gestao Integrada E Tecnologica LDA, Portugal.

INTERFERENCE HOLOGRAM GALLERY. 1179A King Street West, '!bronto, Ontario, Canada M6K 3C5. Telephone: (416) 535 2323. Company description: Gallery, production facility for fringe research.

INTERFERENS HOLOGRAFI D.A. MUSEUM, GALLERI, STUDIO. Established 1989. Exhibitions: Domkirkeodden, N-2300 Hamar, Norway. Studio and office;: Halvor Heels Gt. 6, N-2300 Hamar, Norway. Telephone: (065) 25050 or (065) 30659. Contact: Olav Skipnes, Director or Bodil Skipnes, Art Director. Description: Norway's first and largest permanent holographic exhibition. Separate departments for artwork and museum exhibits. Works of artists like Alexander, Doris Vila, Dan Schweitzer, FrithioIT Johansen and many others shown.

ION LASER TECHNOLOGY INC., Established 1983. 35 Employees at this address. 263 Jimmy Dolittle Road, Salt Lake City, UT 84116 USA. Telephone: (801) 537 1587. FAX: (801) 537 1590. Contact: Richard G. Collier, VP. President, Lynn Barney; Vice-President, Kevin D. Ostler; VP/Sales Manager, Richard G. Collier; Custormer Service, Don Zanelli. Company Description: Manufacturer of air-cooled argon lasers.

ISAST/LEONARDO. (MAIN HEADQUARTERS) Established in 1981. 6 Employees at this address. P.O. Box 75, 1442A Walnut Street, Berkeley, CA 94709 USA. Telephone: (415) 845 8306. FAX: (415) 841 6311. Contact: Kate Sholly. President, Dr. Roger F. Malina; Sales Manager Candace Hansen, Customer Service, Kate Sholly. Branch offices: 8 rue Emile Dunois, 92100 BoulogneiSeine, France;

8000 Westpark Drive, McLean, VA 22102 USA. Company description: Publisher of Journal LEONARDO. Special Issues on Holography as an Art Medium; 1989: $45. Holography Theme Pack: $23. Electronic newsletter, database and Directory: Holography Hotline on MCI, WELL.

ISAST/LEONARDO. (Branch office). 8 rue Emile Dunois, 92100 BoulogneiSeine, France

ISASTILEONARDO. (Branch office). 8000 Westpark Drive, McLean, VA 22102 USA.

ISRAMEX CO. LTD., 25, Arlozorov Street, Tel-Aviv, 62-488, Israel. Telephone: (972)(03) 243 333. FAX: (972)(03) 223 202. Branch office of Newport Corporation, Fountain Valley, CA USA.

J

JAEGER GRAPHIC TECHNOLOGY, J.G.T.--HOLOFOIL S.A., Established 1983. 22 Employees at this address. 20 Avenue des Desirs,B-1140 Brussels, Belgium. Telephone: 00-322-7359551. FAX: 733 1035. Contact: M. Jaeger, President. Vice-President, J. Curci; Sales Manager, R. Doree; Customer Service, H. Majeri. Company description: JGT Brussels is specialized in all kinds of hot stamping technology and runs a separate "HOLOFOIL" department for holographic & diITraction stampings on all graphic and security materials. Worldwide contacts.

JAMES RIVER PRODUCTS. 5420 Distributor Drive, Richmond, VA 23225 USA. Contact: Drurey Baugn. Telephone: (804) 233 9145. FAX: (804) 231 7891. Company description: Manufacturer of holographic embossing machines.

JAYCO HOLOGRAPHICS. Established in 1986. 15 Employees at this address. 29-43 Sydney Road, Watford, Herts, WD1 7PY England, United Kingdom. Telephone: (44) 923 246 760. FAX: (44) 923 247 769. Contact: Rohit Mistry, President. Company description: Complete production service for embossed holograms. Embossing masters thru to fully fmished product. Sixteen years of experience enbles Jayco to offer outstanding quality of product and service at competitive prices.

JODON INC., (MAIN HEADQUARTERS) Established 1963. 15 Employees at this address. 62 Enterprise Drive, Ann Arbor, MI 48103 USA. Telephone: (313) 761 4044. FAX: (313) 761 3322. ContactlPresident: John Gillespie; Vice-President, Mike Gillespie. Company description: Manufacture of Helium Neon Lasers, Laser systems, specialty laser tubes, optical and electro-optical instruments and systems. Holographies films, plates and chemicals. Engineering services.

THE JOHNS HOPKINS UNIVERSITY. Dept of Physics and Astronomy, Baltimore, MD 21218 USA. Telephone: (301) 338 7385. Contact: Homaira Akbari. Description: Scientific holography research; Particle measurement.

JOURNAL OF LASER APPLICATIONS. 4143 Merriweather Road Thledo, OH 43623 USA. Telephone: (419) 885 4803. FAX: (419) 885: 5895. Contact: Jack Dyer, Managing Editor. Company description: Quarterly peer-review technical journal for the laser industry.

JR HOLOGRAPHICS. Suite 1660, 100 Wilshire Blvd., Santa Monica, CA 90401-1135 USA. Telephone (213) 393-2388. FAX: (213) 393-8611. Company description: JR Holographics Inc. acts as a licensee for celebrity images of both company logos and individuals. We expect to place holographic images in major retail markets and product packaging areas.

K

KAISER OPrICAL SYSTEMS, INC. P.O.Box 983, 371 Parkland Plaza, Ann Arbor, MI 48106 USA. Telephone: (313) 665 8083. Contact: B.J. Chang. Marketing Director: Jim McNaughton. Company Description: HOEs; H.U.D.s

KAROLINSKA INSTITUTET, School of Dentistry, Box 4064, S-141 04 HUDDINGE, Sweden. Telephone: (46X08) 774 0080. Contact: Hans Ryden. Description: Medical holography.

K.C. BROWN HOLOGRAPHICS. 22 St. Augustine's Road, Camden Town, London NW1 9RN, England, United Kingdom. Telephone: (44)(1) 4822833. Contact: K.C.Brown. Company description: Pulse portraits; artistic holography.

KENDALL HYDE LTD., Established 1972. 22 Employees at this address. Kingsland Industrial Park,Stroudley Road, Basingstoke, Hants.,RG24 OUG, England, United Kingdom. Telephone: (44) 0256 840 830. FAX: (44) 0256 840 443. Contact: M. D. Kendall, Managing Director. President, M. D. Kendall; Vice-Prseident, D.J. Hyde, A. Edwards; Customer Service, C. Birch. Company description: Thin film optical coating engineers manufacturing windows, mirrors and beam splitters up to 3 metres. Coatings for laser applications, conductive coatings and front surface mirrors.

KEYSTONE SCIENTIFIC CO. Established in 1985. 4 Employees at this address. P.O. Box 22, Thorndale, PA 19372 USA. Telephone: (215) 384-8092, Toll free: (BOO) 4629129. FAX: (215) 384 8093. Contact: Ed Kelly, President, or Claudette Gasbarro. Company description: Manufacturers of automatic film and plate processors, film transports, film and plate holders. Distributors of Agfa, Ilford and Kodak holographic films, plates and chemicals.

KINETIC SYSTEMS. 20 Arboretum Road, P.O. Box K, Roslindale, MA 02131 USA. Telephone (617) 522-8700. FAX: (617) 522-6323. Company description: Manufacturers of Vibraplane standard and special Honeycomb optical tables in three grades up to 5' x 12' x 24". larger sizes available by butt splicing. Also vibration isolation support systems.

KODAK COMPANY. (See EASTMAN KODAK COMPANY).

KOLBE-DRUCK, COLOCO GMBH & CO. KG, (MAIN HEADQUARTERS).Established 1828. 140 Employees at this address. Im. Inndustrigeliinde 50, Postfach 1103, D-4804 Versmold, Federal Republic of Germany. Telephone: (05423 2431 (-5). FAX: 05423 41230. Contact: Sven Deutschmann, Product Manager-Holography. President, Claus-Peter Bohlmann; Sales Manager, Claus-Peter Llohlmann; Customer Service, Jorg Niggebrugge. Subsidiary company: Kolbe-Holografie-Collection, Coloco Printpartner. Company description: Kolbe-Druck is a printing-company known for print specialities on plastic substrates. Complete embossing facilities and application in-house. Kolbe-Holografie-Collection offers a wide-range of standard motifs.

KONING EN HARTMAN, Elektrotechniek B.V., Energieweg 1, NLDelft 2627, Netherlands. Telephone: (31X015) 609906. FAX: (31) (015) 619 194. Branch office of Newport Corporation, Fountain Valley, CA USA.

KRAFTWERK UNION AG. D-4330 MulheimlRuhr, Federal Republic of Germany. Contact: Gerhard Schoenbeck. Company description: Holographic non-destructive testing; industrial research

KREISCHER OPrICS, LTD. Established 1948. 9 Employees at this address. 906 N. Draper Rd., McHenry, IL 60050 USA. Telephone: (S15) 344-4220. FAX: (815) 344-4221. Contact: Cody Kreischer, Pres-

ident. Description: Custom manufacturer of master and production test glasses, optical flats, lenses, condensers, cylinders, windows, filters, prisms, mirrors, beamsplitters, substrates, magnesium fluoride coatings. Consulting services in optical design.

KYOTO TECHNICAL UNIVERSITY. Dept. of Photographic Technology, Matsugasaki, Sakyo-ku, Kyoto 606, Japan. Contact: Toshihiro Kubota. Description: Artistic holography: DCG, Color, Reflection holograms.

L

LABOR DR. STEINBICHLER, Established 19BO. 12 Employees at this address. Am Bauhof 4, D-8201, Neubeuern, Federal Republic of Germany. Telephone: (0049) 8035 1018. FAX: (0049) B035 1010. Contact: Dr. H. Steinbichler, President. Vice-President, J. Engelsberger; Sales Manager: T. Franz. Company description: Holographic investigations, developments on contract basis; application laboratory for: vibration analysis, non destructive testing, deformation measurements, contour measurements, image analysis; pulsed and CW-lasers, motor test bench, computer based evaluation.

LABOR FUR HOLOGRAFIE, Arp Forst 38, Wesel D-4230, Federal Republic of Germany. Telephone: (49) 281) 52837. Contact: A. Fuchtenbusch. Company description: artistic holography; holography education; fine art holograms.

LAKE FOREST COLLEGE CENTER FOR PHOTONIC STUDIES (HOLOGRAPHY WORKSHOPS). Sheridan and College Road, Lake Forest, IL 60045 USA. Telephone: (708) 234 3100 ext 340/343. FAX: (708) 234-6487 and (708) 615-0835.Contact: Tung H. Jeong. Description: Each summer during the second week of July, Lake Forest College offers a 5-day hands-on workshop for participants who have no prior experience in holography. An advanced 5-day workshop follows the third week of July. Write for information.

LAMBDA ANALYTICAL LABORATORIES. 55 Webster Avenue, New Rochelle, NY 10801 USA. Telephone: (914) 654 9117. Company description: Holographic non-destructive testing, optics testing.

LAMBDAITEN OPrICS, Division of Optical Corp. of America.Established 1986. 12 Employees at this address. One Lyberty Way, Westford, MA 01886 USA. Telephone: (508) 692-8140. FAX: (508) 6929416. Contact: George Olmsted. President: DA Johnson, Vice-President: G. Olmsted. Company description: Products: Precision, large aperture (to 36" diam.) aspheric mirrors for holographic production systems.

LAMINEXIHIGH TECH UK LTD. Bromfield Industrial Estates, Mold, Clwyd CH7 lJR, England, United Kingdom. Telephone: (44) (0352) 58444. Contact: Keith Green. Company description: artistic holography; security holograms.

LASART LTD. Established in 1985. 9 Employees at this address. P.O.Box 703, Norwood, CO 81423 USA. Contact: Steven Siegel, Partner. Partner, August Muth. Company description: Lasart, Ltd. specializes in custom DCG work, from modelmaking, mastering and quality finishing. This includes production and limited edition jewelry, Swiss watches, and medium format composite Sculpture.

LASER AFFILIATES. 2047 Blucher Valley Road, Sebastopol, CA 95472 USA. Telephone: (707) 823 7171. FAX: (707) 823 8073. Contact: N. Gorglione. Company description: Laser Affiliates is an award-winning non profit organization that designs innovative holographic and laser theatrical productions, installations and exhibitions. Services include curatorial guidance, videotapes and media lectures

LASER APPLICATIONS, INC. (DIVISION OF LASERMETRICS INC) Established 1965. 38 Employees at this address. 12722 Research Parkway, Orlando, FL 32826 USA. Telephone: (407) 380 3200. FAX: (407) 381 9020. Contact person: Joseph Salg, General Manager.President, Robert Goldstein; Vice-President, J.Salg; Sales Manager, B. Bernard; Customer Service, R. Lusigen. Parent Company: Lasermetrics, Inc. New Jersey, USA. Company description: Holographic non-destructive testing; manufacturer ruby/yag lasers; HOE manufactured; Holography equipment.

LASER ARTS. Established in 1985. 1712 Cathedral Street, Plano, TX 75023 USA. Telephone: (214) 423 0158. Contact: M. Talbott. Company description: Holographic consultants and implementers. Commercial utilization of holography, trade shows, unique promotions and museum exhibits (design, build, rent or sell). Venture capitalists consultants. Professionals in business, art, technology and applications.

L.A.S.E.R. CO. 1900 Grove Drive, Haymarket, VA 22069 USA. Telephone: (703) 754 2526. Contact: Jim Bowman. Company description: Fine art holograms; lighting consultant.

LASER ELECTRONICS PrY., LTD., Established 1967. 20 Employees at this address. P.O. Box 359, Southport, Queensland, 4215, Australia. Telephone: 61 75 53 2066. FAX: 61 75 53 3090. Contact: N. Walden, Managing Director. Sales Manager, R.C. Holberton; Customer Service, L. Darcy. Company description: Laser Electronics designs and manufactures an extensive range of lasers and laser systems across seven industry categories including scientific, educational, and research units. Custom systems can also be developed.

LASER FARE LTD. 15 Industrial Lane, Johnston, RI 02919 USA. Telephone: (401) 231 4400. FAX: (401) 231 4674. Contact: Rich Zucker. Company description: Artistic holography; Equipment; HOEs.

LASERFILM ECKHARD KNUTH - MULTIPLEX-HOLOGRAPHIE, Milchstrasse 12, D-8000 Munich, Federal Republic of Germany. Company description: Artistic holography.

LASER FOCUS WORLD. 1 T~hnology Park Drive, P.O. Box 989,Westford, MA 01886 USA. Telephone: (508) 692 0700. FAX: (508) 692 0525. Company description: Laser trade magazine; annual catalogue.

LASERGRAFICS. 3 Employees at this address. Peris y Valero 130, 46006 Valencia, Spain. Telephone: (34X96) 333 3013. Contact: Santiago Relanzon. Parent Company: Holos-Holos. Company description: Artistic holography.

LASER GRAPHICS. Established in 1988. 5 Employees at this address. 5, Cotta Street, Thessaloniki 55337, Greece. Telephone: (30) (031) 908 087. FAX(30)(031) 234 173. Contact: Yannis Palamas. Xanthippos Vissios. Notis Kaponis. Company description: artistic holography.

LASERGRUPPEN HOLOVISION AB . . Osthammarsgatan 69, S - 115 28 Stockholm, Sweden. Telephone: (46)(08) 663 9908. Contact: Jonny Gustafsson. Company description: Artistic holography.

LASER HOLOGRAPHICS, INC. 1179 King St. West, Unit 111, Toronto, Ontario, Canada M6K 3C5. Telephone: (416) 531 4656. FAX: (416) 229 6724. Contact: Charles Demicher. Company description: Embossed holography broker; Marketing consultants; wholesale

LASER INSTITUTE OF AMERICA. Education Division, 12424 Research Parkway #l30, Orlando, FL 32826 USA. Telephone: (407) 380-1553. FAX: (407) 380-5588. Contact: Beverly Richards. Company description: educational holography

LASER INTERNATIONAL. 19 Normanton Rise, Holbeck Hill, Scarborough, N Yorks Y011 2XE, England, United Kingdom. Telephone: (44)(0723) 366 096. Contact: Keith Dutton. Company description: Gallery.

LASERION HANDELS GMBH. Postfach 110268, 2800 Bremen 11, Federal Republic of Germany. Company description: Artistic holography; commissions.

LASER IONICS INC., (MAIN HEADQUARTERS). Established 1966. 25 Employees at this address. 701 South Kirkman Road, Orlando, FL 32811 USA. Telephone: (407) 2981561. FAX: (407) 297 4167. Contact: Drew Nelson, Business Development Manager. President, Richard Demmer; Vice-President, William Newell. Parent Company: Trimedyne, Inc. Company description: Manufacturer of gas ion lasers including Argon, Krypton and mixed gases. Specializing in high power requirements needing stable power in a compact package.

LASERLABBET. Box 521, SE 581 06 Linkoping, Sweden. Telephone: (46) (13) 123 377. Contact: E.A.Jonsson. Company description: Artistic holography.

LASER LABS, INC. 8000 W.110th Street, Suite #115, Overland Park, KS 66210 USA. Telephone: (913) 383 7639. Contact: Steven Craft. Company description: Medical holography.

LASER LIGHT DESIGNS, 2412 Kennedy Way, Antioch, CA 94509 USA. Telephone: (415) 754 3144. Contact: Michael Mallott. Company description: Gallery; Retail shop; Wholesale

LASER LIGHT EXPRESSIONS PrY. LTD. Established in 1984. 3 employees at this address. 3 Gibbons Street, Telopea, New South Wales, Australia 2117. Telephone: (612) 890 1233. FAX: (612) 890 1243. Contact: Rosemary Sturgess, Marketing Manager. President, John A. Tobin. Subsidiary companies: Optical Security Systems; Optical Control and Display & Holoptics. Company description: Since 1984, LLE has produced commercial holograms and diffractions for security and display applications on time, on budget. We can provide complete artwork, photography, holography and conversion capabilities in-house.

LASER LIGHT IMAGE. 101 Spring Bank, Hull, HU3 1BH, England, United Kingdom. Telephone: (44X0482) 26744 . FAX (44)(0482) 492 286. Contact: Carl Racey. Subsidiary company: Amazing World of Holograms. Company description: Artistic holograms and equipment.

LASER LIGHT LTD. 57 Grand Street, New York, NY 10013 USA. Telephone: (212) 226 7747. Contact: Abe Rezny. Company description: Artistic holography.

LASERMEDIA. 2046 Armacost Ave., Los Angeles, CA 90025 USA. Telephone: (213) 820 3750. FAX: (213) 2079630. Company description: Install laser light show exhibitions.

LASERMET LIMITED, Five Oaks, Sway Road, Brokenhurst, Hants S04 27RX, England, United Kingdom. Telephone: Lyming.
ton (0590) 23075. Contact: Dr William F. Fagan. Company description: Holographic Non-Destructive testing, Instruments, Research Contracting, Consulting, Metrology, Inspection, Safety.

LASERMETRICS, INC. 196 Coolidge Avenue, Englewood, NJ 07631. Telephone: (201) 894-0550. Contact: Robert Goldstein. Subridiary company: Laser Applications, Inc. Company description: Industrial research; laser manufacturing.

LASER OPTICS, INC. Established 1966. 30 employees at this address. 111 Wooster St., Bethel, CT 06801. (mailing address) P.O. Box 127, Danbury, CT 06813 USA. Telephone: (203) 744-4160. FAX: (203) 798-7941. Contact: Henry Louis, Sales Mgr. President: Philip L. Heinrich. Company description: A complete line of laser and optical components for ultraviolet, visible and infrared applications from 250 nm to 16 microns, including focusing lenses, windows, cavity components, prisms, beamsplitters, mirrors and coatings.

LASER RESALE INC. 54 Balcom Road, Sudbury, MA 01776 USA. Telephone: (508) 443 8484. FAX: (508) 443 7620. Contact: Jack Kilpatrick, System Sales. Company description: Laser Resale provides a marketplace for buying and selling pre-owned lasers, laser systems and associated equipment. Currently available holographic lasers are He:Ne, 15-70 mW, and, argon, 100 mW - 20 W.

THE LASERSMITH, INC. 1000 West Monroe Street, Chicago, IL 30607 USA. Telephone: (312) 733 5462. Contact: Steven Smith. Company description: Artistic holography.

LASER TECH INDUSTRIES. 3173 Texas Avenue, Simi Valley, CA 93063. Telephone: (8'18) 583-3406. FAX: (818) 889-5605. Contact: Stanley Cherubin, Customer Service. Company description: Supplies complete support, parts and service on all lasers. Also manufactures mirrors, lenses, rods and other parts for lasers.

LASER TECHNOLOGY, INC. 1055 West Germantown Pike, Norristown, PA 19403 USA. Telephone: (215) 631 5043. Contact: Tom Gleason, Sales Manager. President, John Neuman. Company description: Manufacture equipment for laser-based NDT; Holography and Shearography equipment and inspection services.

LASERWORKS. P.O. Box 2408, Orange, CA 92669 USA. Telephone: (714) 832 2686. Contact: Selwin Lissack. Description: Holographic artist.

LASING S.A., Marques de Pico Velasco, 64, E-28027 Madrid, Spain. Telephone: (34X01) 268 3643. FAX: (34X01) 4073624. Branch office of Newport Corporation, Fountain Valley, CA USA.

LASIRIS INC. (MAIN OFFICE) Established in 1985. 5 Employees at this address. 3549 Ashby, Ville St. Laurent, Que, Canada H4R 2K3. Telephone: (514) 335 1005. FAX: (514) 335 4576. Contact: Alain Beauregard, President. President, Alain Beauregard. Branch office: Quebec. Canada. Company description: Embossed holography; artistic holography; buying & selling; industrial research; HOEs; holographic non-destructive testing.

LASIRIS INC. (Branch office). 840 Ste. Therese, Quebec, Quebec. Canada, GIN 1S7. Telephone: (418) 683 3530. FAX: (418) 6825594. Contact: Alain Beauregard.

LAWRENCE BERKELEY LABORATORY. University of California. Building 80-101, Berkeley, CA 94720 USA. Telephone: (415) 486 4000. Contact: Malcolm Howells. Description: Industrial & academic holography research.

LAZA HOLOGRAMS. (Branch Office) Established in 1983. 4 Employees at this address. 47 Alpine Street, Reading, Berkshire, England RG1 2PY United Kingdom. Telephone: (44) 0734589026. FAX: (44) 0734 571974. Contact: Chris Lambert, Owner. President, Chris Lambert; Customer Service, Carole Lambert. Main Headquarters: 68-72 Katesgrove Lane, Reading, Berkshire, RG 1 2ND England. Company description: Specialist mass-producer of high quality film reflection hologrsms, large or small quantities. Copy service from your master. Full custom service available. Wide range of stock film holograms

LAZA HOLOGRAMS. (MAIN HEADQUARTERS) 68-72 Katesgrove Lane, Reading, Berkshire, RG 1 2ND England.

LAZAP INC. Established 1982. 2614 N. Sweetbriar Drive, Claremont, CA 91711, USA. Telephone: (714) 624-9923. Contact: Richard Cook, President. Description: Manufacturers of and consultants for NDT and HOE applications. Manufacturers of various types of lasers.

LAZART HOLOGRAPHICS. Established in 1985. 2 Employees at this address. 22 Erina Valley Road, Erina, New South Wales 2250, Australia. Telephone: (61X043) 676 245. FAX: (61X043) 652 306. Contact: Brett Wilson, Director. Company description: Artistic holography; buying & selling holograms.

LAZER WIZARDRY. Established in 1987. 2 Employees at this address. 11022 West Oregon Place, Lakewood, CO 80226 USA. Telephone: (303) 987 9438 Contact: Richard M. Osada, Owner. Company description: Wholesale distribution.

LCPC--LAB CENTRAL DES PONTS ET CHAUSSEES. 58 Boulevard Lefebvre, F-75015 Paris, France. Telephone: (33X1) 4532 3179. Contact: Jean-Marie Caussignac. Company description: Industrial research; holographic non-destructive testing.

LENINGRAD SUBSIDIARY IN MACHINERY SCIENCE. Academy of Sciences of the USSR, Bolshoi Av. 61, 199178 Leningrad, USSR. Telephone: (7)(247) 9185. Contact: Juri Ostrovsky. Description: Scientific research.

LENOX LASER. Established 1981. 18 employees at this address. 1 Green Glade Court, Phoenix, MD 21131 USA. Telephone: (301) 592-3106. FAX: (301) 592-3362. Contact: Joseph P. d'Entremont, President. Sales Manager: Don Barnes. Customer Service: Gary Thornton. Company description: Laser-systems laboratory specializing in laser drilling, electron beam welding, Edm machining, and water jet. Offers pre-fabricated aperture kits.

LEONARD KURZ GMBH & CO. Schwabecher Strasse 482, Postfach 1954, D-8510 Firth, Federal Republic of Germany. Telephone: (49)(0911) 71410. Company description: Manufacturer of embossing equipment; broker for hologram embossing.

LES PRODUCTIONS HOLOLAB! 3970, Boulevarde St. Laurent, Montreal, Quebec H2W 1Y3, Canada. Telephone: (514) 849 4325. Contact: Marie-Christiane Mathieu. Company description: artistic holography.

LETTERHEAD PRESS INC. 155 North 120th Street, Dept. HM, Wauwatosa, WI 53226 USA. Telephone: (414) 258 1717. Contact: Mark Mulvaney, President. Company description: Holographic embossing and printing.

LET THERE BE NEON. P.O. Box 337, Canal Street Station, New York, NY 10013 USA. Telephone: (212) 226 4883. Contact: Rudy Stern. Company description: Gallery, retail shop.

LEXEL LASER, INC. Established 1972. 69 Employees at this address. 48503 Milmont Drive, Fremont, CA 94538 USA. Telephone: (415) 770-0800. FAX: (415) 651-6598. Contact: Brian Rossi, Product Manager. President: C. T. Liu. Sales Manager: Len Gold-fme. Company description: Lexel produces the highest quality Argon and Krypton laser systems. In particular, Lexel specializes in production of single frequency systems which are very stable over a variety of environmental situations.

LIGHT-BUCKE-BURO.Bornemannstrasse 10, D-6000 Frankfurt 70,Federal Republic of Germany. Contact: Walter Classen. Company description: Gallery.

LICONiX. Established 1972. 42 Employees at this address. 3281 Scott Boulevard, Santa Clara, CA, 95054 USA. Telephone: (408) 4960300. FAX (408) 492 1303. Contact: Carmen Jordan, Mng. Marketing Services. President, M.W. Dowley; Sales Manager, Randy Kimball; Customer Service, Greg Springer. Company description:

LICONiX, long the recognized leader in Helium Cadmium laser technology, also suplies semiconductor diode laser systems and a recently introduced line of ion lasers.

LIGHT CONSTRUCTION, INC. 2154 Dundas Street West, Suite #503, Toronto, Ontario, M6R 1X3 Canada. Telephone: (416) 533 4692. FAX: (416) 533 0572. Contact: Michael Page. Company description: Fine art holograms; Large format; independent educational facility.

LIGHT ENGINEERING. 12 New St. Johns, St. Helier, Jersey, Channel Islands, England, United Kingdom. Telephone: (44)(534) 30614. Contact Anthony Hopkins. Company description: Gallery, retail shop.

LIGHT FANTASTIC PLC. (MAIN HEADQUARTERS). Established in 1981. 18 Employees at this address. 4EIF Gelders Hall Road, Shepshed, Leicestershire LE12 9NH, England, United Kingdom. Telephone: (44)(509) 600 220. FAX: (44)(509) 508 795. Contact: Managing Director, Peter H.L.Woodd; Sales Manager, Paula Foulkes-Williams. Branch offices: Light Fantastic, Gallery of Holography, London, England. Company description: Light Fantastic PLC (Est. 1981) is a fully integrated holographic business providing the creative and technical services that produce innovative standard and custom-designed holograms of the highest quality. Total service covers embossing and finished product.

LIGHT IMPRESSIONS EUROPE PLC. 5 Mole Business Park 3, Leatherhead, Surrey KT22 7BA, England, United Kingdom. Telephone: (44) 0372 386677. FAX: (~) 0372 386548. Contact: John Brown, Managing Director. Description: Branch Office of Light Impressions Inc., Santa Cruz, CA USA.

LIGHT IMPRESSIONS, INC. Established 1979. 15-20 Employees at this address.149-B Josephine Street, Santa Cruz, CA 95060 USA. Telephone: (408) 458 1991. FAX: (408) 458 3338. Contact: Fred Black, President. President, Fred Black; Sales Manager, Kevin Samson. Branch Office: Light Impressions PLC, Surrey, England, Hologramas de Mexico. Company description: Light Impressions is an integrated, full-service commercial holography company. We produce custom and stock hologram masters and emboss metallized polyester. Diecutting and hot stamping are also offered.

LIGHT WAVE GALLERY. (Branch of Holographic Industries, Inc.) D-208 Woodfield Mall, Schaumburg, IL 60173 USA. Telephone: (312) 240 5344. Contact: Robert Pricone. Company description: Gallery, retail shop.

LIGHT WAVE GALLERY. (Branch of Holographic Industries, Inc.) North Pier, 435 East lllinois Street, Chicago, IL 60611 USA. Telephone: (312) 3211123. Contact: Shana Wills. Company description: Gallery, retail shop.
LIGHT WAVE GALLERY. (Branch of Holographic Industries, Inc.)

N 322 Fairlane Towne Centre, Dearborne, MI 48126 USA. Telephone: (312) 935-7627. Description: Gallery, retail shop.

LINDA LAW HOLOGRAPHICS. 8 Crescent Drive, Huntington, NY 11743 USA. Telephone: (516) 3516056. Company description: Holographic artist; Holography education.

LINE LITE LASER CORPORATION. Established 1982. 12 Employees at this address. 430 Ferguson Drive, Bldg. 4, Mountain View, CA 94043 USA. Telephone: (415) 969-4900. FAX: (415) 969-5480. Contact: John Evans, Marketing Manager. Company description: Manufacturer of low and medium-power CW sealed CO (2) lasers. Also low, medium, high-power, CW and pulsed Nd:YAG lasers; laser gases; power supplies for gas and solid state lasers.

L.I.R.E.R.A. 12, rue Libergier, F-51110 Reims, France. Telephone: (33) 26 884 452. Contact: Michel Grosmann. Company description: Scientific research; HOEs.

LITTON SYSTEMS CANADA LTD. 25 Cityview Drive, Rexdale, Ontario M9W 5A7 Canada. Telephone: (416) 249-1231. Contact: Romuald Pawluczyk. Company description: holographic nondestructive testing; electro optics.

LOS ANGELES SCHOOL OF HOLOGRAPHY. P.O. Box 851, Woodland Hills, CA 91365 USA. Telephone: (818) 703 1111. FAX: (818) 703 1182. Contact: Jerry Fox. Company description: The Los Angeles School of Holography offers a 3 day class. Students learn all phases of holography, and produce both laser viewable transmission and white light viewable holograms in silver halide format

LOUGHBOROUGH UNIVERSITY OF TECHNOLOGY. Dept. of Physics, Loughborough, Leicestershire LE11 3TU England, United Kingdom. Telephone: (44X509) 263 171. Contact: Nick Phillips. Description: Embossing masters/shims; Scientific, industrial research. The University and Markem Systems (UK) are participating in a joint venture, Advanced Holographic Laboratories.

LULEA UNIVERSITY OF TECHNOLOGY. Dept. of Mechanical Engineering, S-951 87 Lulea, Sweden. Contact: Nils-Erik Molin. Description: Industrial research; holographic non-destructive testing.

LUMONICS INC., (MAIN HEADQUARTERS) Established 1971. 180 Employees at this address. 105 Schneider Road,Kanata, Ottawa, Ontario K2K 1Y3, Canada. Telephone: (613) 592 1460. FAX: (613) 592 5703. Contact: Dr. Jim Higgins. President, D. J . James; Vice-President, R.S. Sandwell; Sales Manager, R. Rayman; Customer Service, K. Perkins. Subsidiary Companies: Lumonics Marking Corp. Camarillo, CA USA, Lumonics Materials Processing Corp.,Livonia (Detroit) Michigan USA, Lumonics JK Division, Rugby England. Company description: Lumonics is a manufacturer of high power pulsed ruby lasers for portrait holography and engineering holocameras for NDT. Other products include laser marking and materials processing systems.

LUMONICS LTD., Established 1972. 180 Employees at this address. Cosford Lane, Swift Valley, Rugby, Warwickshire, CV21 1QN, United Kingdom. Telephone: (44) 0 788 570321. FAX: (44) 0 788 579824. Contact: George Synowiec, Sales and Marketing Manager, Scientific Products. Parent Company: Lumonics, Inc. Ottowa,
Canada. Branch Offices: Lumonics, Brussels. Lumonics, Munich. Lumonics, Paris. Company description: Lumonics manufactures pulsed lasers for a range of industrial and scientific applications includirig pulsed ruby lasers for Holography. Single pulse and mul

tiple pulse units available for commercial, research and NDT applications.

LUND INSTITUTE OF TECHNOLOGY. Department of Physics, Box 118, S-221 00 LUND, Sweden. Telephone: (46X046) 107 656. Contact: Sven-Goran Pettersson. Description: Color H-1; holography education; academic research.

LURE. Institut d'Optique, BP 147, F-91403 Orsay, Cedex, France. Telephone: (33)(1)(69) 416 846. Contact: D. Joyeux. Company description: Academic research.

M

MACSHANE HOLOGRAPHYILASER ARTS PROGRAMS. Established in 1985. 2 Employees at this address. 512 West Braeside Drive, Arlington Heights, IL 60004 USA. Telephone: (708) 398 4983. Contact: Jim MacShane, Vice-President, President, Elaine MacShane. Parent company: Laser Arts Educational Programs, SUNBOWS. Company description: Design and manufacturing of SUNBOWS (tm), sculptural, architectural, and gift embossed holographic products, educational programs and artistic holography.

MAGIC LASER. Established in 1985. Quartier de L'horloge, 4 rue BranWme, 75003, Paris France. Telephone: (33)(1) 4274 3578. FAX: (33)(1) 4274 3357. Contact: Anne-Marie Christakis, Manager. Sales Manager, Thierry Gueguen. Company description: Buying & Selling artistic holography.

MAGIC LASER LABORATORY. 6 rue Marie-Stuart, F-75002, Paris, France.

MAGIC LIGHT HOLOGRAFIE - GALLERIE. Bahnhofsplatz 2, D-8000 Munich 2, Federal Republic of Germany. Telephone: (49) (089) 595 981. Company description: Artistic holography; Gallery.

MAN ENVIRONMENT, INC. 2251 Federal Avenue,Los Angeles, CA 90064 USA. Telephone: (213) 477 7922. Contact: Gary Fisher. Company description: Artistic holograms, holography systems, equipment R&D, I-step rainbow holography, multiplex.

MARKEM SYSTEMS LTD. Ladywell Trading Estate, Eccles New Road, Salford, Mnachester M5 2DA England, United Kingdom. Telephone: (44X61) 789 8131. FAX: (44X61) 707 5315,. Contact: Jane Oliver. President, Jeff Lomax; Vice-President, Ken Williamson. Branch offices: Advance Holographies Laboratories, Loughborough University; Marketm offices at High Wycombe, Kent, Glasgow, Rugby, Halifax, Pendleton.
Company description: "One-Stop-Shop"service in embossed hot stamping foil and laminating mm, including everything from origination to foil manufacture. Also large format, silver halide service for exhibitions and permanent installation.

MARTINSSON ELEKTRONIK AB. Instrumentvagen 16, Box 9060, 80126 09 HAGERSTEN, Sweden. Telephone: 946X08) 744 0300. FAX: (46X08) 744 3403. Contact: Per Skande. Company description: Artistic; pulsed portraiture; equipment & supplies.

MARUBUN CORPORATION, 8-1 Nihombashi Odemmacho, ChuoKu, Tokyo, 103, Japan. Telephone: (81)(03) 648 8115. FAX: (81) (03) 648 9398. Branch office of Newport Corporation, Fountain Valley, CA USA.

MARWELL AB. Kyrkbacken 27, S-l71 50 Solna, Sweden. Telephone: (46)(8) 838 261. Company description: Artistic holography; fine art holography.

MASSACHUSETTS INSTITUTE OF TECHNOLOGY. M.I.T. Media Laboratory, Spatial Imaging Group, 20 Ames Street, E15- 421, Cambridge, MA 02139 USA. Telephone: (617) 253 0632. FAX: (617) 258 6264. Contact: Linda Conte. Description: College holography courses; curriculum development

MATT HANNIFIN CO. P.O. Box 4574, Austin, TX 78765 USA. Telephone: (512) 452 7444. Contact: Matt Hannifm. Company description: Holography exhibition installer and spectacle consultant (openings). Previously director of installation-Museum of Holography, NY. Licenced, experienced high-power laser and firework show operator. Manufacturer hand-craft.ed boomerangs with holograms.

MAZDA MOTOR CORP. 3-1, Shinchi, Fuchu-cho, AId-gun, Hiroshima, Japan. Telephone: (81X082)282 1111. FAX: (81) (082) 285 9746. Company description: Industrial holography research; Holographic non-destructive testing; Interferometry.

McCAIN MARKETING. Established in 1974.4 Employees at this address. 10962 North Wauwatosa Road 76W, Mequon, WI 53092, USA. Telephone: (414) 242 4023. Contact: Richard McCain, President. President, Richard McCain; Vice-President, Clare McCain; Sales Manager, Richard McCain; Customer Service, Clare McCain. Company description: Act as lillson between commercial advertisers (including Fortune 500 companies) and holographers. Educate clients to holography, develop advertising promotions, and educational applications using holography. Recommend professional holographic specialists as needed

McMAHAN ELECTRO-OPTICS. Established 1971. 2160 Park Avenue, (Orlando Division), Winter Park, FL 32789 USA. Telephone: (407) 645-1000. FAX: (407) 644-9000. Contact: Robert McMahan, Sr. Branch office: R&D Division, Research Triangle Park, North Carolina, USA. Company description: McMahan Electro-Optics manufactures a laser-based NDT system for testing composite aerospace components and assemblies.

MEDIA INTERFACE, LTD. 167 Garfield Place, Brooklyn, NY 11215 USA. Telephone: (718) 788 4012. Contact: Ronald Erikson.Company description: Artistic holograms, holography education consulting.

MELLES GRIOT, (MAIN HEADQUARTERS) Established in 1969. 100 Employees at this address. 1770 Kettering Street, Irvine, CA 92714 USA. Telephone: (714) 261-5600. FAX: (714) 261-7589. Contact: Lisa Tsufura, Technical Manager. Sales Administrator, Paul Kenrick; Vice-President of Marketing & Sales, Kevin Chittim; Sales Manager, Candice Bauccio; Customer Service, Sigi Hennessey. Parent Company: J. Bibby & Sons. Company description: Melles Griot is a major manufacturer of off-the-shelf and custom tables and isolation equipment, laser, lenses, mounting hardware, positioners, polarizers, coated optics, detectors, collimators and spatial filters.

MEREDITH INSTRUMENTS. Established in 1978. 6 Employees at this address. 6403 North 59th Avenue, Glendale, AZ 85301 USA. Telephone: (602) 934 9387. Contact: Chad Andersen. President, Dennis Meredith; Sales Manager, Mary Moraine. Company description: Specializing in surplus inventories of He-Ne lasers as well as argon and diode lasers, Meredith Instruments is the USA's largest laser discount dealer. Free Catalogue.

MESSERSCHMITT - BOELKOW-BLOHM. ZENTRALE, Entwicklung MBB, Postfach 801109, D-BOOO Munich 2, Federal Republic of Germany. Company description: Scientific research; color holography.

METAMORFOSI OLOGRAFIA ITALIA SRL. Established in 1983.

10 Employees at this address.Via Leeco 6, 20124 Milano, Italy. Telephone: (39)(2) 204 9943. FAX: (39)(2) 204 1625. Contact: Eva Aprile, Manager. President, Eva Aprile; Vice-President, Silvia Aprile; Sales Manager, Manuela Polenta; Customer Service, Francesca Cominelli. Company description: Producer of the flrSt and original

HOLOTIME, the interchageable hologram watch. This year we are producing small size DCG holograms. Consulting to Italian customs marketing.

METAPLAST ELECTROCHEMICALS CORP., Established 1963. 3 Employees at this address. 67 Whitson Street, Hempstead, New York, 11550 USA. Telephone: (516) 4814530. FAX: (516) 4817320. Contact: J . L. Lester, President. Company description: Manufacturer of conductive coatings and silver spray, plating & electroforming equipment are offered for plating on plastic and other nonconductive surfaces. Electroforming consultants to aerospace, electronic, phonographic, computer, holography & toy industries.

METRO LASER. Established 1988. 7 Employees at this address. 18006 Skypark Circle, Suite 108, Irvine, CA 92714-6428 USA. Telephone: (714) 553-0688. FAX: (714) 553-0495. Contact: James D. Trolinger, Cecil F. Hess, Partners. Description: Specialist in laser and optical diagnostics for wind tunnel testing, non-destructive testing, ballistics, combustion research, particle sizing and flow diagnostics. Both R&D and specialized instrumentation provided.

METRO LOGIC INSTRUMENTS, INC. P.O. Box 1458, Coles Road at Route 42, Blackwood, NJ 08012 USA. Telephone: (609) 228 8100. FAX: (609) 228 6673. Contact: Debbie Stecker. Company description: A developer of HeNe lasers, optics and electronic instruments for education since 1968. Manufactures low-cost lasers and accessories for schools and colleges, bar code scanners for commercial applications.

MGD PRODUCTIONS. 5982, Rue Durocher, Outremont, Quebec, Canada. Telephone: (514) 278-4593. FAX: (514) 987-4651. Company description: MGD Productions (Georges Dyens) produces large format holograms to be integrated in architectural or multi-media projects in collaboration with designers or architects.

MICRAUDEL. Established in 1980. 6 Employees at this address. 93, rue Adelshoffen, F -67300 Sehiltigheim, France. Telephone: (33) 8881 3293. Contact: Philippe B).lrger, Directeur Technique. Company description: Electronic and information applications; holographic non-destructive testing.

MINCHIMPROM. 101851 Moscov, USSR.Contact: N.S. Gafurova. Description: Pulsed portraiture, artistic holography.

MIND'S EYE: HOLOGRAPHIC CONSULTANTS. 17329 Zola Street, Granada Hills, CA 91344 USA. Telephone: (818) 360 6023. Contact: Stephen Roth. Company description: Marketing consultant.

MIRAGE HOLOGRAMS LTD. Unit 2 Brook Lane, Business Centre, Brook Lane North, Brentford, Middlesex TW8 OPP England, United Kingdom. Telephone: (44 X 1) 568 2454. Contact: Lynne Heslop.Company description: Silver halide hologram maker. M.I.T. (See Massachusetts Institute of Technology)

MITSUBISHI HEAVY INDUSTRIES LTD., Nagasaki Technical Institute, 1-1 Akunoura-machi, Nagasaki 850-91 Japan. Contact: M. Murata. Description: Holographic non-destructive testing; industrial research.

MITUTOYO MEASURING INSTRUMENTS. 18 Essex Road, Paramus, NJ 07652 USA. Telephone: (201) 368 0525. Contact: Joe Seriff. Company description: Manufacturers of precision measuring

instruments including holographic linear tracking systems.

MODERN OPTICS--A DIVISION OF V-TECH, INC. 120 Employees at this address. 270 East Bonita Avenue, Pomona, CA 91767, USA. Telephone: (714) 596-7741. FAX: (714) 596-9033. Contact: Irvin Miller, Assistant to the President. Parent company: V-Tech, Inc.

MOELLER WEDEL OPTISCHE WERK. Rosengarten 10, 2000 Wedel, Federal Republic of Germany. Company description: Artistic holography.

MOSCOW PHYSICAL ENGINEERING INSTITUTE. Kashirskoe Shosse 1, Moscow, 115409 USSR. Contact: Alexander Larkin. Description: Artistic holography; scientific holography research.

MULTI-PURPOSE HOLOGRAMS (MPH). Established 1986. 5 Employees at this address. 6 rue de l'orme, 75019 Paris France. Telephone: (331) 42009866. FAX: (331) 47434744. Contact: Fani Adam or Edwina Alva. Branch Offices: 13 Place de la Nation, 75011 Paris, France, 3 Rue des Fontenelles, 92310 Sevres, France. Main headquarters: Fedriadon Str, 113.64, Athens, Greece. Company description: Holograms' conception, realization and application is our domain. We undertake all kinds of research in holography, not only scientific (mastering) but also artistic, such as self-illuminated holosculptures.

MUNDAY SPATIAL IMAGING. 39 Pyrcroft Road, Chertsey, Surrey KT16 9HT, England, United Kingdom. Telephone: (44)(0932) 564 899. Contact: Rob Munday. Company description: Specializes in holograms of museum artifacts. Display Reflection and Rainbow holography ofCW, Pulsed and Stereogram images. ComputerMdeo imaging holographic stereogram system. Collection of 200 holograms for exhibition.

MUs LASER WORKS. 1328 Dunsterville Avenue, Victoria, B.C., Canada V8Z 2X1. Telephone: (604) 479-4357. Contact: Ron Meuse, President. Company description: Holographic and photographic services, 3-D photography. Low volume production holographic services. Can provide lab rental and technical assistance. Laser light show and laser effects production and rental.

MUSéE DE L'HOLOGRAPHIE: 15 a 21 Grand Balcon, Forum des Halles, BP 180, 75001 Paris, France. Telephone: (33X!) 42743357. Contact: Anne-Marie Christakis, Manager. Sales Manager, Thierry Gueguen. Company description: Gallery; Mail order; Holography education.

MUSEUM FuR HOLOGRAPHIE & NEUE VISUELLE MEDIEN. Established in 1979. 7 Employees at this address. Pletschmllhelenweg 7, D-5024 Pulheim 1, Federal Republic of Germany. Telephone: (49)(02) 233 385 1053. FAX: (49)(02) 238 52158. Contact: Matthias Lauk, Director. Curator: Hans-Peter Ott. Company description: The first museum of holography in Europe. Permanent showroom including classical holographic artworks. Guided tours. Workshop program. Consultation and organisation of national and international exhibitions. Holography Consulting.

MUSEUM OF HOLOGRAPHY. 11 Mercer Street, New York, NY 10013 USA. Telephone: (212) 925 0581 or 925-0526. FAX: (212) 334-8039. Contact: Martha Tomko, Director. Curator: Sydney Dinsmore. Education: Susan Cowles. Description: Founded in 1976; the only not-for-profit museum in the world devoted to holography. It has the world's largest collection of holography including important archival material. Publishes holosphere journal.

MUSEUM OF HOLOGRAPHY/CHICAGO. 1134 West Washington Boulevard, Chicago, IL 60607 USA. Telephone: (312) 226 1007. Contact: L. Billings. Description: Gallery; Retail shop; Wholesale; Mail order; Holography Education.

MUSEUM OF THE FINE ARTS RESEARCH & HOLOGRAPHIC CENTER. 1134 West Washington Boulevard, Chicago, IL 60607 USA. Telephone: (312) 2261007. Contact: John Hoffmann. Description:Gallery; Hands on Workshops.

MKW INDUSTRIES. 1269 West Pomona St., Suite 110, Corona, CA 31720. Telephone: (714) 278-0563. Contact: Mike Kenny. Description: Distributor of HeNe and Argon lasers.

MYERS PRINTING, INC. 914 East Gier Street, Lansing, MI, 48906, USA. Telephone: (1) (800) 333-3645. FAX: (517) 482-0550. Contact: Jim Dick or Larry Leece. Company description: Four years' experience in applying holograms. We have worked with holograms generators and others applicators across the U.S. We look forward to the opportunity to work with additional holography experts.

N

NASA MARSHALL SPACE FLIGHT CENTER. Space Sciences Laboratory, ES 73, Huntsville, AL 35812 USA. Company description: Scientific holography research; interferometry.

NATIONAL HOLOGRAPHIC STUDIOS, INC. Technology Development Center, Kresson-Gibbsboro Road, Voorhees, NJ 08043. Telephone: (609) 784-3800. FAX: (609) 784-2471. Contact: Mike Chappelle. Company description: Artistic holography; fine art limited editions, holographic stereograms; commercial H-l. Custom presentation support.

NATIONAL PHYSICAL LABORATORY. Queens Road,Teddington, Middle sex, England TW11 OLW United Kingdom. Telephone: (44) (81) 977 3222. Contact: D. Robinson. Company description: Scientific and industrial research; holographic non-destructive testing.

NEOVISION PRODUCTIONS. Established 1976. P.O. Box 74277, Los Angeles, CA 90004 USA. Telephone: (213) 387 0461. Contact: Bill Hilliard. Company description: Fine art originals, producing holograms for home and industry, consulting.

NEWBOLD WELLS COMPANY. 33 Paul Street, London EC2A 4JU, England, United Kingdom. Telephone: (44)(1) 6381471. Company description: Artistic holography: Embossed & silver halide holograms.

NEWCASTLE UPON TYNE POLYTECHNIC. Department of Physics, Ellison Building, Newcastle upon Tyne, NEl 8ST. England, United Kingdom. Telephone: (44)(091) 235 8453. FAX: (44) (091) 235 8017. Contact: Dr. A.E. MacGregor. Department contacts, Graham Rice; Paul Dunnigan. Description: Comprehensive program of short courses in holography for beginners and advanced students alike in new holographic laboratories. Ongoing program of consultancy and research. SpeCialise in CW work; equipped with argon-ion lasers

NEW CLEAR IMPORTS LTD. 27 Burrard Street, St. Helier, Jersey, Channel Islands, England, United Kingdom. Telephone: (44)(534) 30614. Contact Anthony Hopkins. Company description: Gallery; retail shop.

NEW DIMENSION HOLOGRAPHICS. 27 Nurses Walk, The Rocks, Sydney, New South Wales 2000, Australia. Telephone: (61) (2) 247 6063. FAX: (61) 2 419 8670. Contact: '!bny Butteriss. Company description: Artistic holography sales; Educational holography consultant.

NEWPORT CORPORATION. (MAIN HEADQUARTERS). Established in 1969. 400 Employees at this address. 18235 Mt. Baldy Circle, P.O.Box 8020, Fountain Valley, CA 92708-8020 USA. Telephone: (714) 963 9811. FAX: (714) 963 2015. Contact: Anna Amarandos. President, '!bm Galantowicz; Vice-President, Dean Hodges; Sales Manager, Jim Doty; Customer Service, Frank Aranda. Branch offices: Spectra Physics Pty. Ltd., Victoria, Australia; Aims Optronics SAINV, Kraainem, Belgium; Antonio A. Santos, Rio de Janeiro, Brazil; Technical Marketing Associates, Ontario, Canada; SuperbIn Co. Ltd., Taipei, Taiwan; BBT Instrumenter ApS, Frederiksberg, Denmark; Photonetics, SA., Marly Le Roi, France.; Advance Photonics, Bombay, India; Isramex Co. Ltd.,Tel-Aviv, Israel; dB Electronic Instruments S.R.L., Milano, Italy; Marubun Corporation, Tokyo, Japan; Koning en Hartman, Delft, Netherlands; Electech Distribution Systems, Singapore; Lasing S.A. , Madrid, Spain. Company description: Newport Corporation is a designer and manufacturer of laserlholographic systems, E/O components, optics, spatial filters, optical & beamsteering instruments, magnetic bases, fiber optic components, vibration isolation systems, and holographic recording materials. Subsidiary companies: Newport Ltd., Herts., United Kingdom; Newport Instruments AG, Schlieren, Switzerland.

NEWPORT GMBH, (EUROPEAN HEADQUARTERS). Bleichstrasse 26, D-6100 Darmstadt, Federal Republic of Germany. Telephone: (49) 061 5126116. FAX: (49) 061 5122639. Branch office of Newport Corporation, Fountain Valley, CA. USA.

NEWPORT INSTRUMENTS AG, Giessenstrasse 15, CH-8952 Schlieren, Swtzerland. Telephone: (41) (01) 7402283. FAX: (41) (01) 740 2503. Subsidiary of Newport Corporation, Fountain Valley, CA. USA.

NEWPORT: Kyokuto Boeki Kaisha. 7th Floor, New Otemachi Bldg., 2-1, 2-Chome, Otemachi, Chiyoda-ku, Tokyo 100-91, Japan. Branch office of Newport Corp., Fountain Valley CA, USA.

NEWPORT LTD., Pembroke House, Thompsons Close, Harpenden, Herts. AL5 4ES, United Kingdom. Telephone: (44) (058) 276 9995. FAX: (44) (058) 276 2655. Subsidiary of Newport Corporation, Fountain Valley, CA. USA.

NEWPORT HOLOGRAMS. 3412 Via Oporto, Suite 2, Newport Beach, CA 92663 USA. Telephone: (714) 675 1337. Contact: David Schaffner. Company description: Gallery and retail shop; selling reflection and DCG holograms, jewellry and novelties.

NEW YORK HALL OF SCIENCE, Established in 1964. 100 Employees at this address. 47-01 111th Street, Corona, NY 11368 USA. Telephone: (718) 699 0005. Contact: John Driscoll, Manager, Arts and Technologies Program. Company description: The New York Hall of Science is New York's only hands-on science and technology museum. Lasers and optics demonstrated daily. Color hologram depicting quantum atom is on display.

NEW YORK HOLOGRAPHIC LABS. P.O.Box 20391, Tompkins Square Station, 176 East 3rd Street, New York, NY 10009 USA. Telephone: (212) 254 9774. or (212) 674 1007. Contact: Daniel Schweitzer or Samuel Moree. Company description: Private Commissions, Fine Art Holograms, Hands-on Tutorials.

NEW YORK INSTITUTE OF TECHNOLOGY. Center for Optics, 100 Glen Cove Avenue, Glen Cove, NY 11542 USA. Telephone: (516) 686 7863. Contact: Mauric Haliva. Description: Non-Destructive Testing; Industrial research; and Interferometry.

NORGES TEKNISKE HOGSKOLE. Institute for Almen Fysikk, Sem Saelandsv 7, N-7034 Trondheim-NTH, Norway. Telephone: (47)(07) 593 422. FAX: (47)(07) 592 886. Company description: Holographic non-destructive testing; scientific holography research.

NORLAND PRODUCTS, INC. 695 Joyce Kilmer Avenue, P.O. Box 145, North Brunswick, NJ 08902 USA. Telephone: (908) 545 7828. FAX: (908) 545 9542. Contact: Jean Spalding. Company description: Manufacture and distribute holographic photochemistry.

NORTH AMERICAN HOLOGRAPHICS INC. P.O. Box 451, 103 East Scranton Avenue, Lake Bluff, IL 60044 USA. Telephone: (708) 234 4244. Contact: Gary Lawrence. Company description: Holographic portraits; Holography marketing & applications; Trade shows & exhibits.

NORTHERN ILLINOIS UNIVERSITY. Department of Physics, DeKalb, IL 60115-2854 USA. Telephone: (815) 753 1772. Contact: Thomas Rossing. Description: Scientific holography research; interferometry.

NORTHWESTERN UNIVERSITY, FIBEROPTIC LABORATORY. Dept of Electrical Engineering, Technological Institute, Evanston, IL 60208 USA. Telephone: (708) 491 4427. Contact: Rudy Haidle. Description: Optical processing, Forensic holography.

NORTHWESTERN UNIVERSITY, Dept of Biomedical Engineering,Evanston, IL 60208 USA. Telephone: (708) 491 2946. FAX: (708) 491 4133. Contact: Hans Bjelkhagen. Description: Artistic, Commercial, Interferometry, Scientific, Medical, and Industrial Research, Portraits, Fiberoptics, Pulsed Laser, Holographic NonDestructive Testing, Educational Holography.

NOVATOR RESEARCH CENTER. PI. Lesi Ukrainki 1, Kiev-196, 252196 USSR. Telephone: (7X296)8048. Contact: Georgi Dovgalenko. Description: Medical holography.

O

ODHNER HOLOGRAPHICS, Established in 1988. 1 Employee at this address. 833 Laurel Avenue, Orlando, FL 32803 USA. Telephone: (407) 894 7966. Contact: Jeff Odhner, President. Company description: A manufacturer of custom transmission, reflection, and rainbow holograms on silver halide film; standard sizes are 8"x 10" and 4"x 5"; educational and holographic NDT services are also available.

OMNICHROME, Established in 1981. 90 Employees at this address. 13580 Fifth Street, Chino, CA 91710 USA. Telephone: (714) 627-1594. FAX: (714) 591-8340. Contact: Ray Reid, Vice President, Sales. President: Bill Hug. Company description: A manufacturer of Argon and HeCd lasers ranging in power from 3 to 600mW, at wavelengths from 325nm to 514 nm.

ONTARIO COLLEGE OF ART. 100 McCaul Street, Toronto, Ontario M9W 1W1, Canada. Telephone: (416) 977-5311. Contact: Michael Page. Company description: Holography courses

ONTARIO HYDRO. RESEARCH DMSION, 800 Kipling Avenue, KR 128, Toronto, Ontario M8Z 584 Canada. Telephone: (416) 231-4111. Contact: David Mader. Company description: holography research; holographic non-destructive testing.

ONTARIO SCIENCE CENTRE. 770 Don Mills Road, Don Mills, Ontario, M3C 1T3 Canada. Telephone: (416) 429 4100.FAX: (416) 429-5961. Contact: Mr. David Szanto. Company description: Gallery; courses in holography

OP-GRAPHICS (HOLOGRAPHY) LTD. Unit 4: Technorth, 7 Harrogate Road, Leeds LS7 3NB, England, United Kingdom. Telephone: (44)(0532) 628 687. FAX: (44)(532) 374 182,. Contact: Valerie Love, Nick Hardy. Company description: Manufacturer of silver halide film holograms. Carries stock display holograms for retail, including pulsed reduction and mutiple exposures. Supplier of galleries and gift-shops in Europe and U.S.; also takes on commissioned work.

OPTICAL IMAGES. Established in 1974. 3 Employees at this address. 1309 Simpson Way, Suite J, Escondido, CA 92029 USA. Telephone: (619) 746 0976. FAX: (619) 746 6141. Contact: Donald C. Broadbent, Owner. Company description: Optical Images, formerly Broadbent Development Lab, is an independent, privately owned holographic facility producing HOE's and display holograms in various recording materials. Donald Broadbent has 24 years experience in holography.

OPTICAL LABORATORY. 3, Rue de Universite, 67000 Strasbourg, France. Description: Artistic holography.

OPTICAL SURFACES LTD. Godstone Road, Kenley, Surrey, CR2 5AA, England, United Kingdom. Telephone: (44X1) 668 6126. Company description: Artistic holography.

OPTICAL WORKS LTD. Established 1938.17 Employees at this address. Main headquarters: Ealing Science Centre, Treloggar: Lane, Newquay, Cornwall TR7 1HX, England. Contact: E.O. Frisk. Managing Director. Telephone (44) 0637-877222. FAX: (44) 0637 877211. Company description: Artistic holography.

OPTICS PLUS INC. 1369 East Edinger Avenue, Santa Ana, CA 92705 USA. Telephone: (714) 972 1948. FAX: (714) 835-6510. Contact: John Goetten. Company description: Manufacture optics; precision tool mounts (including lens and mechanical mounts).

OPTILAS B.V. Established in 1976. 23 Employees at this address. P.O. Box 222, 2400 AE Alphen, AID Rijn, The Netherlands. Telephone: (31) 01720 31234. FAX: (31) 01720 43414. Contact: A. Kooi. Managing Director. Sales Manager, P. Drok. Company description: Sales/service/engineering of electro-optical and vacuum related products.

OPTISCHE FENOMENEN. Nederlandse Stichting Voor, Waarneming & Holografie, Warenarburg 44, NL 2907 CL Capelle aid Jjssel. The Netherlands. Contact: Jan M. Broeders. Company description: Monthly Newsletter-- Subsidiary of Dutch Foundation of Perception & Holography.

ORIEL CORPORATION. 250 Long Beach Boulevard, Stratford, CT 06497, USA. Telephone: (203) 377-8282. Contact: Scott Heidemann. Advanced Exposure Group.

ORIEL SCIENTIFIC LTD. (Division of Oriel Corporation).P.O. Box 31, Leatherhead, Surrey, KT22 7AU, England, United Kingdom. Telephone: (44) (0372) 378822. Company description: Artistic holography.

OXFORD HOLOGRAPHICS. Established in 1984. 3 Employees at this address. 71 High Street, Oxford OX1 4BA, England, United Kingdom. Telephone: (44)0865 250 505. FAX: (44) 0865 790 353.

Contact: Nick Cooper. Company description: Oxford Holographies has both a very well established retail and an expanding wholesale operation, based on keen pricing. We look forward to hearing from you.

OXFORD UNIVERSITY. Department of Engineering Science, Parks Road, Oxford OX1 3PJ England, United Kingdom. Telephone: (44) (0865) 273 805. Contact: D. J. Cooke. Description: Holography education; Industrial research.

P

PACIFIC HOLOGRAPHICS, INC. 3003 Brandon Circle, Carlsbad, CA 92008 USA. Telephone: (619) 4719044. Contact: Eric Van Hamersveld. Company description: Main area of business is embossing and photoresist masters; stereogramming; small to medium formats.

PALCO MARKETING, INC. 11110 48th Avenue North, Plymouth, MN 55442. Telephone: (612) 559·5539. Contact: Jim Paletz, President. Company description: Selling a complete line of holographic products at wholesale. Distributor of holographic pictures, watches, keychains, stickers and related products. Exhibits at trade shows throughout U.S ..

PENNSYLVANIA STATE UNIVERSITY. Applied Research Laboratory, P.O.Box 30, State College, PA 16804 USA. Contact: C.S. Vikram. Description: Scientific research; particle density testing.

PENNSYLVANIA STATE UNIVERSITY. Dept. of Electrical Engineering, 121 East University Park, Pennsylvania 16802 USA. Contact: Francis Yu. Description: Color holography.

PERCEPTION HOLOGRAPHY. Thornton Marketing Ltd. Aketon Close, Haggs Lane, Follifoot, Harrogate, North Yorks., England 3 G3 1A2 United Kingdom. Telephone: (44X93) 782 323. Contact: Mike Burridge. Description: Fine art holography, holography eduestion.

PHANTASTICA. Hohenweg 14, 5760 Arnsberg 1, Germany. Telephone: (49) 2932-81917. FAX: (49) 2932-29441. Contact: Gerd M. Albrecht. Company description: Makers and distributors of articles related to embossed holograms and diffraction foil, including earrings and other jewelry, badges, pens, mobiles. Main focus is street, crafts and Christmas markets.

PHASE ·R CORPORATION. Box G-2, Old Bay Road, New Durham, NH 03855 USA. Telephone: (603)859 3800. Company description: We manufacture laser equipment. Call for more information.

PHOENIX HOLOGRAMS. Established in 1986. 3 Employees at this address. Elmar Spreer, Traubenstr. 41, D-2913 - Apen, Federal Republic of Germany. Telephone: (49)044 89 15198. President, Elmar Spreer. Company description: Phoenix Holograms specializes in fmding individual solutions for sophisticated problems. We do silver halide holograms up to 1m2 for: architecture, interior decoration, promotion/corporate design, education, exhibitions, portraits.

PHOTOGRAPHERS FORMULARY, INC. Established 1988. P.O. Box 5105, Missoula, MT, 59806-5105 USA. Telephone: (800) 922-5255. FAX: (406) 543 6232. Company description: Holographic photochemistry; holography kits. Contact: Bud or Lynn Wilson.

PHOTONETICS SA, 52 Avenue de l'Europe, F-78160 Marly-LeRoi, France. Telephone: (33)(013) 916 3377. FAX: (33)(013) 916 5606. Branch office of Newport Corporation, Fountain Valley, CA USA.

PHOTONICS DIRECTORY. (Laurin Publishing Co. Inc.) P.O. Box 4949, Berkshire Common, Pittsfield, MA 01202 USA. Telephone: (413) 499 0514. Company description: Publishers of information on Optical components; Optics, Electro-Optics, Lasers, and Imaging technology .

PHOTONICS SYSTEMS LABORATORY. 7 Rue de L'universite. 67000 Strasbourg, France. Telephone: 88 355150. Contact: P. Meyrueis, Director. Description: Education, NDT.

PHOTON LEAGUE OF HOLOGRAPHERS. Established in 1987.2 Employees at this address. 110 Sudbury Street-Unit B, Toronto, Ontario M6J 1A7 Canada. Telephone: (416) 531 7087. FAX: (416) 536-4291 clo Online Graphics. Contact: Meryn Cadell or Heidi von der Gathen, Associate Directors. Company description: Artist-run, nonprofit, holography studio. Photon league is committed to the production of fme art holography, education, dialogue & curation of exhibitions. Technical workshops are ongoing throughout the year.

PHYSICAL OPTICS CORPORATION. Established 1985. 70 Employees at this address. 2545 W 237th Street, Suite B, Torrance, CA 90505 USA. Telephone: (213) 530-1416. FAX: (213) 530-4577. Contact: Kevin Rankin, Application Engineer. President: Dr. Joanna Jannson. Sales Manager: Michael Feeney. Main Headquarters: 20600 Gramercy Place, Torrance, CA 90501. Description: Phototonies- based high tech company involved in research, development and manufacture of holographic optical elements, including reflection and transmission HOE's, diffraction gratings and display holograms.

PHYSICS INSTITUTE. LATVIAN SSR Academy of Sciences, 229021 Riga-Salaspils, USSR. Established 1946. Contact: Dr. Kurt Shvarts. Telephone: 007-0132/947642. Description: Scientific research on recording materials.

PILKINGTON P.E. LTD. Glascoed Road, St. Asaph, Clwyd, England, LL17 OIl, United Kingdom. Telephone: (44X745) 588 000. FAX: (44) 745 584 258. Contact: Andrew Head, Product Manager. President: Mr. Graham Smart. Vice President: Dr. J. Allister McQuoid. Description: Artistic holography; DCG manufacturer; HOEs. Main headquarters: Pilkington PLC, Prescot Rd, St. Helens, Lancashire, England.

POINT OF VIEW DIMENSIONS, LTD. Established in 1982. 2 Employees at this address. 45-2903 River Drive South, Jersey City, NJ 07310 USA. Telephone: (201) 626 8844. Contact: Neal Lubetsky, President. President, Neal Lubetsky; Vice-President, Wendy Barkin. Company description: Point of View Dimensions specializes in the conceptualization, design, and execution of holography (all formats and all sizes) for exhibitions, trade shows, premiums, points of sale, brochures, and annual reports.

POLAROID CORPORATION. Established in 1937. 100 Employees at this address. Holography Business Group,730 Main Street-1A, Cambridge, MA 02139 USA. Telephone: (800) 237 5519 or (617) 577-4307. FAX: (617) 577 5434. Contact: Phillip E. Mestancik, Marketing Manager. Sales Program Manager: Lorraine Dikmak. Company description: Polaroid manufactures high quality photopolymer holograms for high volume applications, both custom & stock holograms are available. Polaroid's high quality & dependability are products of twenty years of experience in holography.

POLYMER HOLOGRAPHICS, INC. Box 20732, V.O.C. Sedona, Arizona 86341 USA. TelephonelFAX: (602) 634-8823. Contact: Robert C. Latterman, President. Description: Manufactures high quality photopolymer holograms using DuPont OmniDex photopolymer. Stock and custom holograms. Labels for printing, packaging, advertising. Replication service. HOE's. Transmission and reflection holograms; design/production of advertising pieces.

PORTSON, INC. (LASER IMAGES). Established in 1986. 14 Employees at this address. 9201 Quivira, Overland Park, KS 66215-3905 USA. Telephone: (913) 4927010. FAX: (913) 4927099. Contact: Steve Larson. President, Steve Larson; Vice-President, Roger Larson; Sales Manager, Anne Larson. Company description: Manufacturer of stock and custom holograms and holographic products; total in-house production capabilities in dichromate, silver halide, and photo-resist. Stock products include jewelry, watches, calculators, and framed art.

Q

QUé SERA SERA. P.O. Box 29, 9700 AA, Groningen, The Netherlands. Telephone: (31)(050) 140417. FAX: (31)(050) 144142. Contact: H.T. Vogel, Joop DeKens. Company description: Artistic holography; DCG holograms; Wholesale; Workshops; Manufacturer of optics (mirrorsllenses).

R

RADIO SHACKII'ANDY CORPORATION. Bilston Road, Wednesbury, West Midlands, WSlo 7JN, England. Description: Equipment & supplies.

RAINBOW SYMPHONY INC. Established in 1975. 6860 Canby Road 11120, Reseda, CA 91335 USA. Telephone: (818) 708-8400. FAX: (818) 708-8470. Contact: Mark S. Margolis, President. Company description: Manufacturer of uniquely designed holographic and diffraction products for the gift, novelty, advertising, specialty, premium incentive, souvenir and museum markets.

RALCON. Established in 1985. 4 Employees at this address. Box 142, 8501 South 400 West, Paradise, UT 84328 USA. Telephone: (801) 245 4623. FAX: (801) 245 6672. Contact: Richard Rallison, CEO. President, Richard Rallison; Vice-President, Scott Schicker. Company description: Design, development and fabrication of volume holographic optical elements, (HOEs) including gratings, scanners, multi focus devices, heads ~p and down displays and notch filters formed in dichromated gelatin or photo polymer

RALPH CULLEN HOLOGRAPHICS / UKOS. Established in 1980. 20 Employees at this address. 84 Wimborne Road West, Wimborne, Dorset, BH212DP, England, United Kingdom. Telephone: (44)(202) 886 831. FAX: (44)(202) 742 236. Contact: Ralph Cullen, Director. Parent company: U.K. Optical Supplies. Branch office: Ralph Cullen Holographics/UKOS, London, England. Company description: A Consultancy-Design Service which in association with UK OPTICAL SUPPLIES (Manufacturing) provide customized holographic optical components. Advice on component selection and laboratory/studios designed to any budget is available.

RANDY JAMESIHOLOGRAPHY. Established in 1976. 1 Employee at this address. P.O. Box 305, Santa Cruz, CA 95061 USA. Telephone: (408) 458 4213. Company description: Commercial and fine art holography since 1974. Extensive background in all forms of display holography: design, mastering, and production. Custom quotes, stock price list available.

RAVEN HOLOGRAPHICS LTD. Established 1987. 15 employees at this address. Old Saw Mills, Nyewood, Near Petersfield, Hampshire,

GU3 15HX, England, United Kingdom. Telephone: 0730-821612. FAX: 0730-821260. Contact: Mr. Terry Woodd, sales director. President: Mr. Stuart Ainslie-Brown. Vice President: Mr. Mike Medora, technical director. Company description: Fine art holograms

RECONNAISSANCE, LTD. (MAIN HEADQUARTERS). Established in 1984. 3 Employees at this address. 3932 McKinley Street N.W., Washington D.C., 20015 USA. Telephone or FAX: (202) 966-3464. Contact: Lewis Kontnik. Publisher, Lewis Kontnik; European Editor, Ian Lancaster. Subsidiary company: Holography News. Branch office: Surrey, England, United Kingdom. Reconnaissance, publisher and corporate consultants on the holography industry. Publisher of Holography News; programme consultant to Holography '90 Conference and PIRA's HoloPack '90. Undertakes strategic and market studies for single or multiple clients.

RECONNAISSANCE, LTD. (Branch Office).l Erica Court, Wych Hill Place, Woking, Surrey AU22 OJB, England, United Kingdom. Telephone or FAX: (44)483 740 689. Contact: Ian Lancaster.

REAL IMAGE HOLOGRAPHICS (ACK ACK DESIGNS)31 Clerkenwell Close, London, EC1 England, United Kingdom. Description: Fine art holograms.

RED BEAM, INC. 9011 Skyline Blvd., Oakland, CA 94611. Telephone: (415) 482-3309. FAX: (415) 482-1214. Contact: Lon Moore, Director. Company description: Specializes in the design and production of master HI's for commercial mass production. Produces its own line of commercial holograms. Clients include Activision, AT&T, llford, NFL (Superbowl ticket), Polaroid.

REEL IMAGE. P.O. Box 566, Pacifica, CA 94044 USA. Telephone: (415) 355 8897. Contact: Roy Bradshaw, Owner. Company description: Fine art holograms

REGAL PRESS INC .. REGAL HOLOGRAPHICS DIVISION, 129 Guild Street, Norwood, MA 02062 USA. Telephone: (617) 769 3900. Contact: William Duffey. Company description: Holographic embossing, application; Artistic holography.

RESIST MASTERS, INC. 816 Minion Drive, Ypsilanti, MI 48198 USA. Telephone: (313) 481-1980. Contact: Don Shelton. Description: Maker of photoresist plates.

REYNOLDS METALS CO. Flexible Packaging Division, 6603 West Broad Street, Richmond, VA 23230 USA. Telephone: (804) 281 3969. Contact: Rich Patterson. Company description: Embossed holography on packaging materials.

RICHARD BRUCK HOLOGRAPHY. Established in 1987. 2 Employees at this address. 3312 West Belle Plaine 112, Chicago, IL 60618 USA. Telephone: (312) 267 9288. Contact: Richard Bruck. Company description: A shop producing quality low-volume runs. Custom and commercial work. Fine art originals. Experienced installation and consultation

RICHMOND HOLOGRAPHIC STUDIOS LTD. Established in 1979.4 Employees at this address. 6 Marlborough Road, Richmond, Surrey, England, TW10 6JR United Kingdom. Telephone: (44)(81) 940 5525. FAX: (44)(81) 948 6214. Contact: Edwina Orr, Director. Co-Director, David Traynor. Company description: Our principle services include mass produced film holograms up to 500 x 700 mm. suitable for horizontal and vertical displays. Pulsed mastering. Achromatic and multi-colour reflection and transmission copies. Technology transfer consultancy.

RICHTER ENTERPRISES Established 1984. 2 employees at this

address. 640 19th Street, Manhattan Beach, CA 90266-2509 USA. Telephone: (213) 546-5107. FAX (213) 545-6757. Contact: Thomas A. Richter, President. Company description: International sales representative and distributor of optical and electronic components, CCD's, CCD cameras, Bragg cells, grating and subsystems. Call for complete information on purchasing or selling your requirements.

RITCHER HOLOGRAMS. Adolf Kolping Strasse 16, D-4050 Monchengladbach, Federal Republic of Germany. Description: Artistic holography.

ROBERT SHERWOOD HOLOGRAPHIC DESIGN, INC. Established in 1985. 4 Employees at this address. 400 West Erie Street, Suite #405, Chicago, IL 60610 USA. Telephone: (312) 944 3200. Contact: Kurt Kellison, Customer Service. President, Robert Sherwood; Vice-President, Greg Hoskins. Branch office of Holographic Design, Inc., Philadelphia, PA. Company description: At Holographic Design, our designers, holographers, and account service Personnel provide you with the highest quality standards this new and exciting technology can offer.

ROBINSON HOLOGRAM LIGHTING SYSTEMS. 5 Hillside Cottages, Owismoor Road, Camberley, Surrey GU15 4SU, England, United Kingdom. Telephone: (44)(0344) 762 739. Contact: Anthony Robinson. Company description: Holography lighting consultant

ROCHESTER INSTITUTE OF TECHNOLOGY. One Lomb Memorial Drive, P.O. Box 9887, Rochester, NY 14623 USA. Telephone: 716) 475 2770. FAX: (716) 475-5804.Contact: Arnold Lungershausen. Description: RIT offers holography instruction as part of its Imaging and Photographic Technology program. Practical applications are stressed. The faculty do industrial consulting and offer occasional workshops.

ROCHESTER PHOTONICS CORPORATION. Established in 1989. 3 Employees at this address. 80 0' Connor Road, Fairport, NY 14450 USA. Telephone: (716) 377-7990. FAX: (716) 377-7913. Contact: GM. Morris, President. Vice-President, Dean Faklis. Company description: Designer and manufacturer of dilTractive optical elements for use in precision electro-optical systems. Supplier of optical system design services for military and commercial uses.

ROLLS-ROYCE PLC--ADVANCED RESEARCH LABORATORY. P.O.Box 31, Derby DE2 8BJ, England, United Kingdom. Contact: Ric Parker, Group Leader, Optical Sciences. Telephone: (332) 242424. FAX: (332) 249936. Company description: This aircraft engine manufacturing company has developed and used holographic techniques over the past twenty years. Now offering consultancy and contract research in holographic NDT, vibration analysis and flow visualization. Branch offices: Derby, Bristol, Leavesden, Newcastle, Coventry, Glasgow, Atlanta, Greenwich. Main headquarters: Rolls-Royce PLC, 65 Buckingham Gate, SWIE 6AT.

ROSS BOOKS. Established in 1977. P.O. Box 4340, Berkeley, CA 94704 USA. Telephone: (415) 841 2474. FAX: (415) 8412695. Contact: Brian Kluepfel, Editor. President, Franz Ross. Editorial Consultant: Elizabeth Yerkes. Accounts Administrator: Edith Katz. Company description: Publisher of holography books and resources; FAX us (number above) or call for our free catalogue (800) 367 0930 (Toll-Free within US).

ROTTENKOLBER HOLO-SYSTEM GMBH. Henschelring 15, D-8000 KirchheimlMunich, Federal Republic of Germany. Telephone: (49)(089) 2014911. Company description: Holographic non-destructive testing; Industrial research; manufacturer of holographic and imaging systems

ROWLAND INSTITUTE FOR SCIENCE. 100 Cambridge Parkway, Cambridge, MA 02142 USA. Telephone: (617) 4974657. Contact: Jean-Marc Fournier. Description: Scientific holography research.

ROYAL COLLEGE OF ART. Holography Unit, Darwin Building, Kensington Gore, London SW7 2EU, England, United Kingdom. Telephone: (44XOl) 584 5020. Course leader: Rod Murray. Description: MA (RCA) Holography; two-year post-graduate course. Explores creative holography in both fine art and design-related fields. Facilities include Pulse, Krypton, HeNe, Stereogram and Computer Graphics.

ROYAL INSTITUTE OF TECHNOLOGY. Department of Industrial Metrology, S-10044 Stockholm, Sweden. Telephone: (46)(08) 790 7823. FAX: (46)(08) 790 8219. Contact: Lennart Svennson. Description: Interferometry; Industrial and scientific holography research; Holography education; Holographic non-destructive testing.

ROYAL PHOTOGRAPHIC SOCIETY. Salisbury College of Art, Southampton Road, Salisbury, Wiltshire, SP1 2PP England, United Kingdom. Telephone: (44X0722) 23711 ext. 275. Contact: Mr. Pitt. Description: Artistic holography; holography education.

ROYAL SUSSEX HOSPITAL. Brighton, England, United Kingdom. Description: Scientific and Medical Holography research; Interferometry.

RUTHERFORD AND APPLETON LABORATORIES. Chilton, Didcot, Oxon OXll OQX, England, United Kingdom. Telephone: (44) (0235) 821900. Contact: Robert Sekulin. Company description: Particle measurements; Holographic non-destructive testing.

S

SAAB-SCANIA. S-581 88 LINKOPING, Sweden. Telephone: (46) (013) 129020. Contact: Sven Malmqvist. Company description: Holographic non-destructive testing; scient ific holography research.

SAGINAW VALLEY STATE UNIVERSITY. 2250 Pierce Road, University Center, Michigan 48710-0001 USA. Telephone: (517) 790-4000. FAX: (517) 790-2717. Contact: Hsuan Chen (Course instruction) or Greg Wright (marketing holographic technology). Description: Research includes HOEs, multiplex and rainbow holography.

SANDIA NATIONAL LABORATORIES. Combustion Research Facility. Livermore, CA 94550 USA. Telephone: (415) 294 3138. Contact: Donald Sweeney. Company description: Scientific holography research; Computer-generated holography; Holographic Optical Elements.

THE SCHOOL OF THE ART INSTITUTE OF CHICAGO - HOLOGRAPHY DEPARTMENT. Columbus Drive and Jackson Boulevard, Chicago, IL 60603 USA. Telephone: (312) 443-3883 (Holography lab). Contact: Ed Wesley, Visiting Artist. Description: The SAlC olTers an MFA degree with concentration in holography, and is equipped with three tables, with one containing a stereogram printer for computer-generated imagery or views from life. One instructor, three graduate assistants.

SCHOOL OF HOLOGRAPHY. 263 Montego Key, Novato, CA 94949 USA. Telephone: (415) 824 3769. Contact: Sharon McCormack. Description: Holography education; workshops

SCHOOL OF HOLOGRAPHY/CHICAGO. 1134 West Washington Boulevard, Chicago, IL 60607 USA. Telephone: (312) 226 1007. Contact: L. Billings. Company description: Museum/Curator;

Holography Teacher Training; Hands-On Workshops; Curriculum development; Educational materials; Independent educational facility.

SCHULTZ & CO. (see E.C. Schultz & Co.) SCIENCE KIT & BOREAL LABORATORIES. 777 East Park Drive, Tonawanda, NY 14150-6784 USA. Telephone: (716) 874 6020. Company description: Suppliers and mail-order cataloguers of holography educational materials including holography kits, books and more.

SCIENCE & MECHANICS INSTRUMENTS. 605 East 59th Street, Dept. H-2, Brooklyn, NY 11234 USA. Telephone: (718) 531 3381. Company description: Manufacture and sell light meters and shutters.

SCIENTIFIC COUNCIL ON EXHIBITIONS OF THE USSR, Academy of Sciences, 30, Varllov Street, Moscow USSR. Telephone: (7) (135) 64 86. Contact: Larisa Nekrasova. Description: Traveling exhibits, artistic holography.

SCOPE OPTICS LTD. Noon Lane, Barnet, Hertfordshire EN5 5ST, England, United Kingdom. Telephone: (44X81) 441 2283. Contact: Mr. T.R. Hollinsworth. Description: Manufacturer of medical and industrial endoscopes; micro-precision optics.

SEE 3 (HOLOGRAMS) LTD. (Also trading as "HOLOGRAMS/3D") Established 1978. 4 Macaulay Road, London, SW4 OQX, United Kingdom. Telephone: 071-622-7729. FAX: 071-622-5308. Contact: Jonathan Ross. Description: Consultant for exhibitions, fine art and commercial applications of holography.

SEVCHENKO RESEARCH INSTITUTE ... (see A.N. Sevchenko Research Institute)

SHIPLEY CHEMICAL CO. 1457 McArthur, Whitehall, PA 18052 USA. Telephone: (215) 820 9777. Contact: Stu Price. Company description: Manufacture holographic photochemistry, film, plates and more.

SIEMENS. 186 Wood Avenue South, Iseling, NJ 08830 USA. Telephone: (800) 827 3334, (201) 321 3400.FAX: (812) 422 2339. Contact: Marketing department. Company description: Manufacturer of lasers and components. .

SIEMENS LTD. Siemens House, Windmill Road, Sunbury-on-Thames, Middlesex, TW16 7HS, England, United Kingdom. Company description: Manufacturer oflasers and components.

SILLCOCKS PLASTICS INTERNATIONAL. (MAIN HEADQUARTERS). Established in 1930. 310 Snyder Avenue, Berkeley Heights, NJ 07922 USA. Telephone: (800) 922 0958. FAX: (201) 665 1856. Contact: Tom Palmentieri, Holographic Consultant. President, John Herslow; Vice-President, Vic Berkowitz; Sales Manager, Jack Fenimore; Holographic Customer Service, Shari Spiro. Subsidiary company: SPI Identification, Inc.; Dayluic, Inc. Company description: Applies embossed foil to sheeted pvc vinyl creating holographic and diffraction foil for credit, promotional, and business cards, buttons, rulers, keytags, calenders, dazers, danglers, tentsigns, with custom printing, laminating, diecutting.

S.I. VAVILOV, State Optics Institute, 199064 Leningrad, USSR. Contact: A. Prostev. Description: artistic holography; recording materials research.

SMITH & McKAY PRINTING CO. INC., Established in 1919. 25 Employees at this address. 96 North Almaden Boulevard, San Jose, CA 95110-2490 USA. Telephone: (408) 292 8901 FAX: (408) 292 0417. Contact: Dave McKay, President. President, Dave McKay; Sales Manager, John Rogers; Customer Service, Stephanie Louie. Parent Company: Holographic Impressions. Company description: Hot stamp foil holograms onto paper products. Dimensional printing and fine lithography. Assist coordination of printing projects featuring embossed holography. Publish educational newsletter. Hold holography seminars for graphic designers.

SOCIETY FOR PHOTO-OPTICAL INSTRUMENTATION ENGINEERS (S.P.I.E.). P.O. Box 10, Bellingham, WA 98227 USA. Telephone: (206) 676 3290. Contact: Sales & Marketing Department. Company description: Seminars, conferences on Holography; publishes news in OE Reports.

SO1 - SOCIETA OLOGRAFICA ITALIA. Via degli Eugeoii 23, 00178 Roma, Italy. Telephone: (39X6) 7180976 FAX: (39) (6) 7185172. Company description: Contact: Luigi Attardi, Public Relations Representative. Company description: produces artistic/commercial holograms of any type.

SORO ELECTRO-OPTICS, S.A. 26 rue Berthollet, 94110 Arcueil, France. Company description: Design & manufacture optics.

SOUTHERN INDIANA HOLOGRAPIDCS. Established 1989. 6841 Newburgh Rd, Evansville, IN, 47751. Telephone: (812) 474-0604. Contact: Larry Johann.

SOVISKUSSTVO vlo MEZHDUNARODNAYA KNIGA. Established in 1968. 20 Employees at this address. Art holography department, 141120 Tryazino, Moscow region, USSR. Telephone: (7) 2384600. FAX: (7) 230 2117. Contact: German B. Avksentjev, Director. Dep't head, Valery A. Vanin. Parent company: Platan. Branch offices: contact for info. Company description: Our firm offers: DCG pendents; lamps with holograms; art reflection holograms (size 102x127 mm, and 180x240 mm.).

SPACE AGE DESIGNS INC., Established in 1984. P.O. Box 72, Carversville, PA 18913 USA. Telephone: (215) 297 8490. Contact: Valli Rothaus, President. Description: Designer, Artistic holography; Marketing consultants.

SPECAC Limited. 6A River House, Lagoon Road, St. Mary Cray, Orpington, Kent, BR5 3Qx, England, United Kingdom. Telephone: (44) 689 873134. Description: Artistic holography.

SPECTRA-PHYSICS INC., (MAIN HEADQUARTERS). 3333 North First Street, San Jose, CA USA. Parent company: CIBAGeigy. Branch offices: following listing.

SPECTRA-PHYSICS INC., LASER PRODUCTS DIVISION. Established in 1961. Over 500 Employees at this address. 1250 West Middlefield Road, Box 7013, Mountain View, CA 94039-7013 USA. Telephone: (415) 961 2550. FAX: (415) 969 4084. Contact: Steve Guggenheimer. General Manager, Henry Massenburg. Parent company: CIBA-Geigy. Company description: Spectra-Physics is the world's largest supplier of CW and pulsed gas and solid state laser systems, including a comprehensive optical accessories line and a worldwide customer service network.Branch offices: below

SPECTRA-PHYSICS AG. (Branch Office). Schweizergasse 39, 4054 basel, Switzerland. Telephone: (41)(061) 541 154. FAX: (41)(061) 541129.

SPECTRA-PHYSICS B.V. (Branch Office). Prof. Dr. Dorgelolaan 20, P.O. box 2264, 5600 Cg Eindhoven, The Netherlands. Telephone: (31)(040) 451855. FAX: (31)(040) 439 922.

SPECTRA-PHYSICS B.V.B.A. (Branch Office). North Trade Build ·

:ng, Noorderlaan 133, 2030 Antwerp, Belgium. Telephone: (32)(03) 5426203.

SPECTRA-PHYSICS GmbH, (Branch Office) Siemensstrasse 20, D-0100 Darmstadt, Federal Republic of Germany. Telephone: (49)(061) 517 080. FAX: (49X061) 51 708 232.

SPECTRA-PHYSICS INC., LASER PRODUCTS DIVISION. ("Branch Office). 366 South Randolphville Road, Piscataway, NJ 08854-4175. Telephone: (800) 6315693. FAX: (201) 981 0029 ..

SPECTRA-PHYSICS K.K. (Branch Office). 15-8 nanpeidai-cho, 3hibuya-lru, Tokyo 150, Japan. Telephone: (81)(03) 770 5411. FAX: 81X03) 770 4197.
SPECTRA PHYSICS LIMITED. (Branch Office). Established in :'965. 50 Employees at this address. Boundary Way, Hemel Hemp-nead,
Herts. HP2 7SH, England, United Kingdom. Telephone: 0442 232 322. FAX: 044268538. Contact: Mrs. Molly Helm, Sales Secre: ary. President, Mr. Jon Tomkins; Sales Manager, Anthony Brown.
SPECTRA PHYSICS PTY. LTD. (Branch Office). 826 Mountain Highway, Bayswater,Victoria, Australia. Telephone: (61X03) 729 5155. FAX: (61X03) 720 4256 ..
SPECTRA-PHYSICS S.A. (Branch Office). Avenue de Scandinavie, Z.A. de Courtaboeuf, BP 28,91941 LES ULIS Cedex, France. Tele? hone: (33X1) 6907 9956. FAX: (33X1) 6907 6093.
SPECTRA-PHYSICS S.R.L. (Branch Office). via Derna 1, 1-20132 ~ano, Italy. Telephone: (39)(02) 2853665. FAX: (39X02) 282 7445.
SPECTRATEK CORPORATION. Established 1980. 10 employees at this address. 1510 Cotner Avenue, Los Angeles, CA 90025 USA. 'elephone (213) 473-4966. FAX: (213) 477-6710. Contact: Mike Wanlass, v.P. Marketing, or Terry Conway, V.P. Finance. Company description: Specialists in high-resolution, low-cost holograms. We have delivered hundreds of millions of holograms in 1989 and again in 1990.
SPECTROGON. Established in 1976. 550 County Avenue, Secau;: us, NJ 07904 USA. Telephone: (201) 867 4888. FAX: (201) 867 2191. Contact: Sam Ponzo. President, Sam Ponzo; Sales Manager, Sam Ponzo. Parent Company: Spectrogon AB, Box 2076, S-18302 Taby, Sweden. Company description: Manufactures & designs interference fllters for the IR, visible & UV spectral regions; narrow and bandpass, long and short-wavepass, isolation-line, & ND filters; atmospheric windows; diffraction gratings; AR & metallic coatings; laser optics.
SPECTROGON AB, (MAIN HEADQUARTERS) Box 2076, S-18302 Taby, Sweden. Subsidiary company: Spectrogon, Secaucus NJ, USA
SPECTROLAB INTERNATIONAL LTD. Established in 1962. 200 Employees at this address. P.O. Box 25, Newbury, Berkshire, RH6 8BQ, England United Kingdom. Telephone: (44)0635 248080 FAX: (44) 0635248745. Contact: Bill Vince. Company description: Manu-facture
holographic systems including laser tables, optics, optical and laser positioning systems.
SPECTRON DEVELOPMENT LABORATORIES. 3535 Hyland, Suite 102, Costa Mesa, CA 92626 USA. Telephone: (714) 549 8477. Contact: Dr. Medhat Azzizy. Company description: Scientific holog-raphy
research; Interferometry; Thermoplastic recording material research; Holographic non-destructive testing.
S.P.I.E. (See Society for Photo-Optical Instrumentation Engineers).
SPINDELER & HOYER GmbH. Postfach 3353, Koenigsallee 23, D-3400 Goettingen, Federal Republic of Germany. Contact: Rainer Lessing. Company description: Manufactures H-1s.
SPOT. Agentur fur Holographie, An St. Katharinen 2, D-50oo Koln Chapter Eleven- Page 117
1, Federal Republic of Germany. Description: Holography marketing consultant,

STANFORD UNIVERSITY--Department of Mechanical Engineer-ing, Building 570 Room 571C, Stanford, CA 94301 USA. Telephone: (415) 723 3243. Contact: Joseph Goodman. Description: Optical di-agnostics.

STARCKE, KY. (Main Headquarters) Established in 1983. 4 Em-ployees at this address. P.O. Box 22, SF-32811 Peipohja, Finland, Europe. Telephone: +(358) (39) 360700. FAX: +(358) (39) 367230. Contact: Mr. Ari-Veli Starcke. Vice-President, Mrs. MillaRitta Starcke; Customer Service, Ms. Mikaela Renvall. Subsidiary: Oy Foil It Ltd. Branch offices: Starcke KY Branch, Helsinki, Finland; Ho-lografia Galleria, Oulu, Finland. Company description: Starcke KY is the leading company selling holograms in Scandinavia

STARCKE, KY, (Branch office). Unionikatu 45111 Floor, SF-00170 Helsinki, Finland. Telephone: (358) 0135 5044. FAX: (358) 0135 5306. Contact: Mr. Ari-Veli Starcke.

STARLIGHT HOLOGRAPHIC INC. (MAIN HEADQUARTERS), Established in 1987. 5 Employees at this address. 73 Stable Way, Kanata, Ontario K2M 1A8, Canada. Telephone: (613) 235 0440. FAX: (613) 592 7647. Contact: Stephen Leafloor, President. Vice Presi-dent, Susan Skrypnyk. Parent company: Starmagic Holographic Gal-lery, Ontario Canada. Company description: Starlight Holographics Inc. is one of Canada's leading representatives of the international holographic community, in both stock and custom holography. Its subsidiaries, "Starmagic Holographic Gallery" are currently opening across Canada.

STARMAGIC HOLOGRAPHIC GALLERY. 47 Clarence Street, Ot-tawa, Ontario K1N 9K1, Canada. Telephone: (613) 235 0440. Branch office of Starlight Holographic Inc., Ontario, Canada. Starmagic will be opening galleries in Tbronto and Montreal in 1991.

STATE EDUCATION DEPARTMENT OF NEW YORK. Bureau of Arts and Music Education. Room 681, Albany, NY 12234. Telephone: (518) 474-8773. FAX: (518) 473-3072. Contact: Bob Reals, Associ-ate. Description: "Holography in the Classroom" program. Includes workshops for teachers, exhibitions of student-created holograms, a six-week summer school for gifted holography students, curriculum materials.

STATE RESEARCH AND PROJECT INSTITUTE OF CHEMICO-PHOTOGRAPHIC INDUSTRY, 125167 Moskow, USSR. Descrip-tion: Materials research.

STEINBICHLER OPTOTECHNIK GMBH, Established in 1987. 9 Employees at this address. Am Bauhof 4, D-8201 Neubeuern, FRG Federal Republic of Germany. Telephone: 8035 1017 FAX: 8035 1010. Contact: Dr. H. Steinbichler. President, Dr. H. Steinbichler; Vice-President, J. Engelsberger; Sales Manager, T. Anzenberger; Customer Service, W. Sixt. Company description: Holographic vibra-tion analyzers, image analysis systems, instant holographic camera processing thermoplastic film, optic contour measuring systems, speckle measuring systems, holographic non-destructive test equip-ment.

STEREO WORLD MAGAZINE, Established 1974. 1 Employee at this address. 5610 SE 71st Street, Portland, OR 97206 USA. Tele-phone: (503) 771-4440. Contact: John Dennis, Editor. President: Gor-don D. Hoffman. Vice President: John Waldsmith. Sales Manager: John Weiler. Company description: Magazine published bimonthly by its parent company, National Stereoscopic Association.

Each issue has 40 or more pages and includes topics ranging from holography to all aspects of stereoscopic imaging_

STEUER KG GmbH & Co. Postfach 100327, Ernst May Strasse 7, D-7022 Leinfelden-Echterdingen, Federal Republic of Germany. Telephone: (49X0711) 753143. Company description: Manufacturer of holographic embossing machines.

STEVE PROVENCE HOLOGRAPHY. 15220 Fern Avenue, Boulder Creek, CA 95006 USA. Telephone: (408) 338 2051. Contact: Steve Provence. Subsidiary company: Active Image, Boulder Creek, CA USA.

STOLTZ AG. Tafernstrasse 15, CH-5405 Baden Dattwil, Switzerland. Telephone: (41X056) 840 151. Contact: Beat Ineichen. Company description: Holographic non-destructive testing.

STUDIO FiiR HOLOGRAPHIE. Established in 1985. 1 Employee at this address. MoosstraBe 27, D-8031 Eichenau, Federal Republic of Germany (FRG). Telephone: 08141 70831. Contact: Dr. Carlo Schmelzer. Company description: Products: mastering and copy services (rainbow/reflection), production of mass-run embossed holograms, open stock images, art-pieces.

STUDIO WEIL-ALVARON. Established in 1933. Ostra Tullgatan 8, 8-211 28 Malm6, Sweden. Telephone: 08141 70831. Contact: Lektor H. Herman Weil. Description: Hans Weil's inventions were made in the period 1933-1937, while Gabor invented holography in 1948, the laser was invented 1962 and the first laser-illuminated hologram was exposed as late as 1964.

SUPERBIN CO. LTD, 6F-3, 792, Tun Hua South Road, P.O. Box 69656, Taipei, Taiwan 10662., Republic of China. Telephone: (886) (02) 733 3920. FAX: (886) (02) 732 6442. Contact: Simon Lin, Sales Engineer. Description: Exclusive representative of Coherent (Active Stabilized Single Frequency Argon, Krypton Laser, Dye Laser); Continuum (ruby Laser, Double Pulse Single Mode nd: YAG laser); Newport (optical tables, components, holographic equipment).

SUPERIOR TECHNOLOGY IMPLEMENTATION. Established in 1988. 4 Employees at this address. Hjortekaersvej 99B, 2800 Lyngby, Denmark. Telephone: (45)(45) 933 358. FAX: (45)(45) 930353. Contact: Knud Banck-Hammerurp, GM. President, Knud BanckHammerum, GM; Vice-President, Per Ibsen; Sales Manager, Knud Banck-Hammerum, GM; Customer Service, Jan Stensborg. Company description: Embossed holography; Artistic holography; Education; HOE manufacturer; Equipment and supplies.

SWEDE HOLOPRINT. Duvhoksgatan 6A, Malmo 21460, Sweden. Telephone: (46) (040) 898 21. Contact: Bjorn Wahlberg. Company description: Artistic holography; Artistic marketing consultant.

SWISS FEDERAL INSTITUTE OF TECHNOLOGY. Laboratory of Photoelasticity, Ramistrasse 101, CH-8092 Zurich, Switzerland. Telephone: (41)(0380 246 000. Contact: Walter Schumann. Description: Holographic non-destructive testing; Scientific and industrial research.

SYDNEY COLLEGE OF THE ARTS. Dept. of Holography, 58 Allen Street, Glebe, Sydney, Australia. Description: Holography courses.

SYNCHRONICITY HOLOGRAMS. Est~blished in 1984. 1 Employee at this address. Box 4235, Lincolnville, ME 04849 USA. Telephone: (207) 763 3182. Contact: Arlene Jurewicz. Company description: Synchronicity Holograms provides outreach educational presenta-tions on the artistic, scientific and perceptual aspects of holography. Presentation and workshops can be geared to all age levels and abilities of audience.

SYSTEMS GROUP OF TRW INC. One Space Park, Redondo Beach, CA 90274 USA. Telephone: (213) 812 4321. Contact: Customer Service Dept. Company description: Scientific holography research; Interferometry; Particle measurement; Industrial research.

T

T.A.1. INCORPORATED. 12021 South Memorial Parkway, Huntsville, AL 35803 USA. Telephone: (205) 881 4999. FAX: (205) 880 8041. Contact: Loy Shreve, Director. Company description: Manufacture and testing of optic and laser inspection equipment.

TECHNICAL MARKETING ASSOCIATES, 6695 Millcreek Drive Unit 1#1, Mississauga, Ontario, Canada L5N 5M5. Telephone: (416) 5670390. FAX: (416) 826 8225. Branch office of Newport Corporation, Fountain Valley, CA USA.

TECHNICAL UNIVERSITY OF BUDAPEST. Institute of Precision Mechanics and Optics, Applied Biophysics Laboratory, H-1621 Budapest, Hungary. Contact: Pal Greguss. Description: Medical holography research.

TECHNICAL UNIVERSITY OF EINDHOVEN. Faculty of Architecture, Calibre Institute, P.O. Box 513, Eindhoven, NL-5600MB, The Netherlands. Contact: Geert T. A. Smelzer. Description: Academic research, computer generated holograms

TECHNICAL UNIVERSITY OF WROCLAW. Institute of Physics, Wybrvzeze Wyspianskiego 27, PL-50-370 Wroclaw, Poland. Contact: Henryk Kasprzak. Description: Academic, Scientific research.

TECHNOEXAN/SEMICON, INC. 2300 Drew Road, Mississauga, Ontario L5S 1H4 Canada. Telephone: (416) 672 7271. FAX: (416) 677 7299. Contact: Director. Company description: The Soviet company Technoexan, a joint venture of the Physical Technical Institute JotTe (FTI) in Leningrad and Semicon GmbH-Austria, makes Denisyuk holograms available for western customers.

TECHNOEXAN/SEMICON. Morellenfeldgasse 41, A-8010 Graz, Austria. Telephone: (43)(0316) 38 25 41. FAX: (43) (0316) 38 24 03. Contact: Christian Haydvogel, Director. Company description: The Soviet company Technoexan, a joint venture of the Physical Technical Institute JotTe (FTI) in Leningrad and Semicon GmbH-Austria makes Denisyuk holograms available for western customers. '

TEXTILE GRAPHICS, INC. (Branch office) 201 North Fruitport Road, Spring Lake, MI 49456 USA. Telephone: (616) 842 5626. Contact: Jan Bussard.

TEXTILE GRAPHICS, INC. (MAIN HEADQUARTERS). Established in 1987. 10884 South Street, P.O. Box 68, Nunica, MI 49448 USA. Telephone: (616) 837 8048. Contact: Jan Bussard, President. Branch office: Spring Lake, MI USA. Company description: integration of holographic art into silk screen printing on textiles for creation of a total graphic, with patented bonding process (licensed geographically) using heat and pressure to withstand 100 washings/dryings.

THIRD DIMENSION ARTS INC. Established in 1983. 8 Employees at this address. 1241 Andersen Drive, Suites C & D, San Rafael, CA 94901 USA. Telephone: (415) 4851730. FAX: (415) 4850435. Contact: Tim LaDuca. President, Dara Haskell; Sales Manager, Tim

LaDuca; Customer Service, Shirley Patterson. Parent company: Holocrafts of Canada. Company description: Third Dimension Arts Inc. manufacturers of dichromate jewelry, gifts, and 3-D arts trademark hologram watches. Suppliers to: the gift, jewelry, museum, and entertainment industry; (licencing) markets. Custom designs welcome!

THREE DIMENSIONAL IMAGERY, LTD. 3031-K Nihi Street, Honolulu, III 96819 USA. Telephone: (808) 373 1810. Company description: Artistic holography.

THREE-D LIGHT GALLERY. 107 The Commons, Ithaca, NY 14850 USA. Telephone: (607) 273 1187. Contact: Sheryl Cinkow. Company description: Artistic holography; holography gallery.

TJING LING INDUSTRIAL RESEARCH. 130.Keelung Road, Section III, Taipei, Taiwan. Telephone: (86X20 704 1856. Description: Fine art originals.

TNO INSTITUTE OF APPLIED PHYSICS. Department of Optics, P.O. Box 155, NL-2600 AD Delft, The Netherlands. Contact: Ruud L. van Renesse. Description: Academic and industrial holography research.

TOKA! UNIVERSITY. Department of Electro Photo Optics, 1117 Kitakaname Hiratsuka City, Kanagawa 259-12, Japan. Contact: Hidetoshi Katsuma. Description: Artistic holography; scientific research.

TOKYO UNIVERSITY, Medical Division, 2nd Surgical Research, Toyko, Japan. Description: Medical holography

TOPAC GmBH HOLOGRAPHY. Established 1985. 6 Employees at this address: Aufm Eickholt 47, 4830 Gutersloh 1, West Germany. Telephone: 0 52 41 580192. FAX: 0 52 41 58549. Contact: Dr. Werner Schumacher. President: Udo Thalemann. Vice President: Gerold Lipka. Company description: Embossed holography full service. Product and marketing concepts, creation of artwork and modelling. Production of holograms and application to specific products. Packaging and distribution service.

TOPPAN PRINTING CO. LTD., Central Research Institute. 5, 1-chome Taito, Taito-ku, Tokyo 110, Japan. Contact: K. Ohnuma.

TOPPAN PRINTING CO. LTD. (Branch office). 1 Embarcadero Center, Suite 2106-MP, San Francisco, CA 94111 USA. Telephone: (415) 982 7733. FAX: (415) 956 2551. Contact: Mr Ozawa, Sales & Marketing. Company description: Full holographic services, design, execution, print mass-manufacturing, embossing, application onto products.

TOUCHWOOD HOLOGRAPIIlCS, Established in 1985. 1 Employee at this address. 50 Sugworth Lane, Radley, Abingdon, Oxon, OX14 2HY, England, United Kingdom. Telephone: 0865 735874. Contact: Mr. George W. Clare. Chairman, Mr. G.W. Clare. Parent company: Touchwood Sports Ltd. Main headquarters: 426 Abingdon Road, Oxford, OX1 4XN England. Branch Offices: 4 High Street, Abingdon; 107 St. Aldates, Oxford. Company description: Small experimental laboratory investigating holographic applications for advertising and promotions in the normal sports trade. One-off experimental commissions can be undertaken. Sorry, no long runs or repeat work.

TOWNE LABORATORIES, INC. Established in 1956. 65 Employees at this address. P.O.Box 460-HM, One U.S. Highway 206,Somerville, NJ 08876-0460 USA. Telephone: (908) 722 9500 FAX: (908) 722-8394. Contact: Charles Ondrejik, Sales Manager. President,

J.J. Obzansky; MA. Obzansky; Sales Manager, C. Ondrejik. Company description: Towne Laboratories is a producer of fine quality precision holographic photo plates with or without a sub-layer of IRON-OXIDE and precision spun striation free photoresist in sizes to 18"x 18".

TRANSFER PRINT FOILS INC. P.O.Box 518, 9 Cotters Lane, East Brunswick, NJ 08816 USA. Telephone: (908) 722 9500 FAX: (908) 722-8394.Contact: Charlie Yetka. Manufacture holographic foils; embossing of foils.

TREND. Miramarska 85, 41000 Zagreb, Yugoslavia. Telephone: (38) (041) 511 426. Contact: Dalibor Vukicevic. Company description: Gallery.

TRIDIMENSIONALE HOLOGRAMAS. Alberto, Alcocer, 38-2D, 28016 Madrid, Spain. Telephone: (34)(481) 290 745. Contact: Daniel Weiss. Company description: Artistic holography; Pulsed portraiture.

TRI-ESS SCIENCES:STUDENT SCIENCE SERVICE. 1020 West Chestnut Street, Burbank, CA 91506-1623 USA. Telephone: (818) 247 6910. Company description: Educational materials and laser equipment supplier.

U

U.K. GOLD PURCHASERS DBA HOLOGRAMS UNLIMITED .. 7907 NW 53 Street, Miami, FL 33166 USA. Subsidiary company: U.K. Gold Purchasers, Inc. dba Holograms Unlimited, Corpus Christi TX, USA.

U.K. GOLD PURCHASERS, INC. DBA HOLOGRAMS UNLIMITED. Established in 1979. 13 Employees at this address. 20131 Highway 59, Unit 2216, Humble, Texas 77338 USA. Telephone: (800) 722-7590 or (512) 993 5211. FAX: (512) 993-0467. Cont act: Marvin Uram, President. President, Marvin Uram; Vice-President, Amy Uram; Sales Manager, Marvin Uram; Customer Service, Amy Uram. Parent company: U.K. Gold Purchasers Inc., Miami FL USA. Company description: If its holographic, we have it, or want it! One stop wholesale distributor for varied products the public can afford and will buy. All items at factory-to-you prices.

UK OPTICAL SUPPLIES. Established in 1980. 20 Employees at this address. 84 Wimborne Road West, Wimborne, Dorset BH21 2DP, England, United Kingdom. Telephone: (44)(0202) 886 83l. FAX: (44)(0202) 742 236. Contact: Ralph Cullen, Director. President, R. Cullen. Company description: Supplying probably the world's largest selection of Holographic/Optical components which are: Best Quality; Best Value; Designed by Experienced Hologrsphers. Plus component selection and laboratory/studio set-up advice freely available.

ULTRAFINE. 16 Foster Road, Chiswick, London W4 4NY, England, United Kingdom. Telephone: (44)(01) 995 2303. Company description: Holographic non-destructive testing; scientific and industrial research.

UNIPHASE VETREIDS-GMBH. Established in 1984. 5 Employees at this address. LeiI3str. 8, 8152 Feldkirchen-Westerham, Federal Republic of Germany. Telephone: (49)08063/9036. FAX: (49) 08063/ 7663. Contact: Werner Bleckwendt, Managing Director. Company description: Manufacturer of lasers

UNITED TECHNOLOGIES RESEARCH CENTER. 1100 Employees at this address. Silver Lane, East Hartford, CT 06108 USA. Telephone: (203) 727 7060. FAX: (203) 727 7852. Contact: Dr. Karl A. Stetson. Parent company: United Technologies Corporation. Company description: UTRC electronic holography systems display real-time fringe patterns on TV comparable to photographic holography. They also provide data output for quantitative analysis. Complete systems or retrofits are available.

UNIVERSIDADE DO PORTO. Laboratorio de Fisica, Praca Gomes Teixeira, P-4000 Porto, Portugal. Contact: Oliverio Soares. Description: Holographic non-destructive testing; Academic holography research.

UNIVERSITA Dr ROMA-LA SAPIENZA. Dipartimento di Fisica, Piazzale Aldo Moro, 2, 1-00185 Roma, Italy. Contact: Paolo De Santis. Description: Scientific research.

UNIVERSITAT ERLANGEN - NURNBERG. Physikalisches Institute, Erwin-Rommel Strasse 1, D-8520 Erlangen, Federal Republic of Germany. Telephone: (49)(09131) 857 408. Contact: Adolf Lohmann. Description: Scientific holography research; HOE; computer generated holography.

UNIVERSITe DE FRANCHE-COMTE. Laboratoire d'Optique, L.A. 214, UFR Sciences et des Techniques, F-25030 Besancon Cedex, France. Description: Scientific holography research.

UNIVERSITE DE NEUCHATEL. Institut de Microtechnique, 2, rue A L Breguet, CH-2000 Neuchatel, Switzerland. Telephone: (41) (038) 246 000. Contact: Rene Dandliker. Description: Industrial research.

UNIVERSITe DE PARIS-SUD. Institute d'Optique, F-91405 Orsay, France. Telephone: (33X9) 416 750. Contact: Serge Lowenthal. Description: Scientific holography research.

UNIVERSITE LAVAL. Dept. Physique-COPL, Pavilion Vachon, Ste-Foy, Quebec, G1K 7P4 Canada. Contact: Roger A. Lessard. Description: Holography education; workshops.

UNIVERSITe LOUIS PASTEUR. Photonics Systems Laboratory. 7 rue de L'Universite, 67000 Strasbourg, France. Telephone: (88) 355750. Contact; P. Meyrueis. Description: Scientific holography research.

UNIVERSITY ESSEN. Fachbereich 7lPhysik, Universitatsstrasse 2, D-4300, Essen 1, Federal Republic of Germany. Telephone: (49) (0201) 183 3019. Contact: Detlef Leseberg. Description: Scientific holography research; HOE; computer generated holography.

UNIVERSITY GENT, Workshop Holography. Established in 1968. 2 Employees at this address. 41 St. Pieters Nieuwstraat, B 9000, Gent, Belgium. Telephone: (32)(0) 91 233 821, ext. 2465. FAX: (32)(0) 91 237 326. Contact: Pierre M. Boone, PhD. Parent company: Center for Applied Research in Art and Technology, Gent, Belgium. Description: We are a small optical group in a 70-person Strength of Materials Laboratory, mainly doing holographic interferometry, but also commissioned work for display, museums, etc.

UNIVERSITY OF ABERDEEN, Dept. of Engineering, Kings College, Aberdeen AB9 2UE, Scotland, United Kingdom. Description: Holographic non-destructive testing, industrial research.

THE UNIVERSITY OF ALABAMA IN HUNTSVILLE, Center for Applied Optics, Huntsville, AL 35899 USA. Telephone: (205) 895 6030. FAX: (205) 895-6618. Contact: John Caulfield. Description: Scientific holography research, NDT.

UNIVERSITY OF ALICANTE. Department of Applied Physics, Centro de Holografia, Facultad de Ciencias, Alicante Apdo 99, Spain. Telephone: (34X566) 1200 ext 1147. Contact: A. Fimia. Description: Artistic holography; HOEs; workshops.

THE UNIVERSITY OF ARIZONA, Optical Science Center, Tucson, AZ 85721 USA. Telephone: (602) 621 6997. Contact: Robert Shannon. Description: Industrial and scientific Holography research; Holographic interferometry; Holographic non-destructive testing.

UNIVERSITY OF BOLOGNA. via Fiumazzo 347, 1-48010 Belricetto (RA), Italy. Contact: Pier Luigi Capucci. Description: Artistic holography research & education.

UNIVERSITY OF CALIFORNIA, BERKELEY. See LAWRENCE BERKELEY LABORATORIES.

UNIVERSITY OF CALIFORNIA. SAN DIEGO. Dept. Electrical and Computer Engineering, La Jolla, CA 92093 USA. Contact: Sing Lee. Description: Holography research

UNIVERSITY OF CALIFORNIA, SANTA BARBARA. Advanced Imaging Center, College of Engineering, Santa Barbara, CA 93106 USA. Telephone: (805) 892-4341. FAX: (805) 893-3262. Holography Project Contact: John Landry or Fred Unterseher. Description: Holography research.

UNIVERSITY OF DAYTON, Research Institute. 300 College Park, Dayton, OH 45469 USA. Telephone: (513) 229 3221. Contact: Lloyd Huff. Description: Scientific research, Industrial research; courses.

THE UNIVERSITY OF MICHIGAN. College of Engineering, Chrysler Center, Ann Arbor, MI 48109-2092 USA. Telephone: (313) 763 5464. Contact: Peter M. Banks. Description: Holographic interferometry; Particle measurement; Holographic scientific & industrial research; Holographic non-destructive testing.

UNIVERSITY OF MICHIGAN, Department of Electrical Engineering and Computer Science; Room 1108, EECS Building, Ann Arbor, MI 48109-2122 USA. Telephone: (313) 764 9545. Contact: Emmet Leith. Description: Scientific holography research; Design H.O.E.s; Courses on holography.

UNIVERSITY OF MUNICH. Institute of Medical Optics, Thresienstrasse 37, D-8000 Munich 2, Federal Republic of Germany. Contact: R. Rohler. Description: Medical holography; scientific holography research.

UNIVERSITY OF MUNSTER. Ear, Nose and Throat Clinic, Kardinal von Galen Ring 10, D-4400 Munster, Federal Republic of Germany. Telephone: (49)(0251) 836 861. FAX: (49)(0251) 836 960. Contact: Gert von Bally. Description: Medical holography; Interferometry.

UNIVERSITY OF NORTH CAROLINA, Department of Cell Biology, CB 7090, 108 Taylor Hall, UNC-Chapel Hill, North Carolina 27599. Telephone: (919) 966-2941. Contact: Dr. Ammasi Periasamy. Description: Medical holography.

UNIVERSITY OF OXFORD--HOLOGRAPHY GROUP, Department of Engineering Science, Parks Road, Oxford OX1 3PJ, England, United Kingdom. Contact: Paul Hubel. Description: Scientific holography research; workshops.

UNIVERSITY OF ROCHESTER. Institute of Optics, Rochester, NY 14627 USA. Telephone: (716) 275 5248 .. Contact: Dr. Duncan Moore. Description: Scientific and industrial holography research;

interferometry; particle testing & measurement.

UNIVERSITY OF SOUTHERN CALIFORNIA. Department of Physics, University Park, Los Angeles, CA 90089-0484 USA. Telephone: (213) 740 1134. Contact: Jack Feinberg. Description: Scientific holography research; Interferometry.

UNIVERSITY OF STRATHCLYDE. Mechanical Engineering Group, Glasgow, Scotland, United Kingdom. Contact: P. Waddell. Description: Scientific research; industrial research.

UNIVERSITY OF STImGART. Institute of Applied Optics, PCaffenwaldring 9, D-7000 Stuttgart 80, Federal Republic of Germany. Telephone: (49)(0711) 685 6075. Contact: Hans Tiziani. Description: Scientific holography research; interferometry.

UNIVERSITY OF TOKYO. Faculty of Engineering, Hongo 7-3-1, Bunkyo-ku, Toyko, Japan. Contact: T. Uyemura. Description: Scientific and Medical holography research; Interferometry.

UNIVERSITY OF WISCONSIN, College of Engineering. 432 North Lake Street, Madison, WI 53706 USA. Telephone: (608) 263 7427. FAX: (608) 263 3160. Contact: Francis P. Drake, Program Director. Description: Courses on Laser System Design. Covers: Laser operation, techniques for using & modifying laser output; types of lasers combined with material on scanning, modulation, & detection of laser radiation; designing practical laser systems.

UNIVERSITY OF ZAGREB. Institute of Physics, Bijenicka 46, 41000 Zagreb, Yugoslavia. Telephone: (38X041) 271 211. Description: Industrial, Holographic non-destructive testing; medical holography.

V

VINCENNES UNIVERSITY. 1002 North First Street, Vincennes, IN 47591 USA. Telephone: (812) 885-5294. Contact: Richard Duesterberg. Description: Offering holography workshops for high school teachers, & college level courses in holography. We have 4 research-grade optical tables, as well as argon and krypton lasers. Call for more details

VINTEN ELECTRO OPTICS LTD. Unit 28 Ashfield Way, Whetstone, Leicester, LE8 3NU, England, United Kingdom. Telephone: (44)(533) 867 110. Description: ManufacturelDesign mirrors and optics.

VISIONS OF THE FUTURE. 1040 N. Kings Highway, Suite 717, Cherry Hill, N.J. 08034 USA. Telephone: (609) 667-5622. FAX: (609) 667-6289. Contact: Alan Schorr. Company description: Distributors for all of the holographic merchandise that you will ever need. Prompt, knowledgeable service is our specialty. Direct from factory prices. There is no need to call anyone else.

VNlKTK KULTURA. ul. Intusiastov 34, 1051118 Moskow, USSR. Contact: O.B. Serov. Description: Marketing consultant.

VOLVO-FLYGMOTOR. S-461 81 Trollhattan, Sweden. Telephone: (46) (0520) 94471. Contact: Robert Frankmark. Company description: Holographic non-destructive testing.

VOLKSWAGEN AG. Forschung und Entwicklung, Messtechnik D-3180, Wolfsburg 1, Federal Republic of Germany. Telephone: (49) (05) 361 221. Contact: Armin Felske. Company description:

Industrial research, Interferometry; Holographic non-destructive testing.

W

WASEDA UNIVERSITY. Dept. of Applied Physics, School of Science and Engineering, 3-4-1, Ohkubo, Shinjuku-Ku, Tokyo 160, Japan. Telephone: (81X03) 209 3211. Description: Medical holography research.

WAVEFRONTS. Established in 1974.2428 Judah Street, San Francisco, CA 94122 USA. Telephone: (415) 664 0694. Contact: Louis Brill, Director. Company description: Involved in developing & expanding market & sales efforts for holographic retail/wholesale product lines. Assist in preparation of promotions and collateral sales materials, identify potential sales markets & implementation of sales.

WAVE MECHANICS. 1535 North Ashland Avenue, Second Floor, Chicago, IL 60622 USA. Telephone; (312) 384 4860. Contact: Deni Drinkwater-Welch. Description: Artistic holographer; silver halide transmission and reflection; consultant.

WENTWORTH LABORATORIES LTD. Sunderland Road, Sandy, Bedfordshire, SG19 1RB, England, United Kingdom. Telephone: (44) (0767) 81221. FAX: (44)(0767) 291 951. Contact: M.F. Horgan. Company description: Manufacturer of holography Lab equipment, including labware and isolation tables.

WHILEY FOILS LIMITED (formerly GEORGE M. WHILEY LIMITED) (MAIN HEADQUARTERS). Established in 1783. 182 Employees at this address. Firth Road, Houston Industrial Estate, Livingston, West Lothian EH54 5DJ, Scotland, United Kingdom. Telephone: (0506) 38611 FAX: (0506) 38262. Contact: B J Sitch, Technical Director. President, J.B, McCleery; Vice-President, B.J. Sitch. Parent company: API Group pIc, Silk House, Park Green, Macclesfield, Cheshire SK11 7NU, England, United Kingdom. Branch offices: Auckland, New Zealand Company description: Whiley Foils Limited is a long-established manufacturer of stamping foils. We have developed special base materials for Holographic embossing and market these and Holographic foils worldwide.

WHITE LIGHT WORKS, INC. P.O. Box 851,Woodland Hills, CA 91365 USA. Telephone: (818) 703 1111. FAX: (818) 703 1182. Contact: Jerry Fox. Company description: White Light Works is a full service holographic production company specializing in low cost embossed holograms and Multiplex "People Stopper" holographic deisplays for trade shows and POP applications.

WHITE TIGER HOLOGRAMS. Johannes Verhulststraat 45, 1071 MS Amsterdam, The Netherlands. Telephone: (31)(20) 797 270. FAX: (31)(20) 790 896. Contact: Neil Walker. Creative Director: Ian Cinn. Company description: Partners' experience 30 years in advertising, and 12 in holography. Established 1986. White Tiger Holograms has completed major projects in embossed, dichromate and silver halide, as well as security and production consultancy.

WHOLE HOGRAPHY. Established in 1987. 1 Employee at this address. 4142 Bellefontaine Street, Houston, TX 77025-1105 USA. Telephone: (713) 667 3325. Contact: Michael E. Crawford. Company description: Broker for silver halide holograms

THE WHOLE PICTURE, A GALLERY OF HOLOGRAPHY. Established in 1987. 4 Employees at this address. 634 Parkway, Gatlinburg, TN 37738 USA. Telephone: (615) 436 3650. Contact: Jim Kelly, Owner. Company description: Gallery

WISE INSTRUMENTS. Unit 9, Hollins Business Centre, Marsh Street, Stafford, ST16 3BG, England, United Kingdom. Telephone: (44)(0785) 223 535. Contact: Peter Wise. Company description: Manufactures optics

WOBER DESIGN HOLOLAB AUSTRIA .. Established in 1985. KahlenbergstraBe 6, A-3042 Wurmla, Austria. Telephone: (43) 02275 8210. FAX: (43) 22 75 82105. Contact: Irmfried Wober. Parent company: Amblehurst Ltd., England. Company description: Our Holography Laboratory, founded in 1985, is the first in Austria. We are the biggest hologram producers in town. We organize exhibitions in Austria and Germany. and sell embossed holograms.

WOLVERHAMPTON POLYTECHNIC INSTITUTE, HOLOGRAPHY UNIT. Art and Design Building, Molineux Street, Wolverhampton, West Midlands, WVl ISB, England, United Kingdom. Contact: Graham Saxby, Tutor in Holography. Telephone: (44) 902 313004. FAX: (44) 902 25015. Description: The Holography Unit offers consultancy services in applied and display holography, production of silver halide holograms, projects for students on degree and post-degree courses in arts and science/technology.

WONDEROUS UGHT GALLERIES (formerly Light Fantastic) Established in 1987. 3 Employees at this address. Tabor Center, Bridge Mart, 1201 16th Street, Denver, CO 80202 USA. Telephone: (303)399 2603. Contact: Richard B. Osada, Owner. Company description: Retail shop and gallery; Buying & selling holograms; holography education.

WONDERS OF HOLOGRAPHY GALLERY. Established in 1986. 25 Employees at this address. P.O. Box 1244, Jeddah 21431, Saudi Arabia. Telephone: (966)(2) 652 0052. FAX: (966)(2) 6511325. President, M.A. Baghdadi; Vice-President, A.M. Baghdadi; Sales Manager, M.A. Salam; Customer Service, Walled Saddeq. Company description: We are the first gallery in the Arab World, and the first live laser show company .. Please contact us - we are distributors for several laser and holographic companies all over the world.

WORCESTER POLYTECHNIC INSTITUTE. Mechanical Engineering Department, 100 Institute Road, Worcester, MA 01609-2280 USA. Telephone: (508) 831 5536. FAX: (508) 831 5483. Contact: Ryszard Pryputniewicz. Description: Scientific, Medical & Industrial holography research; Interferometry; Holographic non-destructive testing.

WORLD ART PROJECT. 10247 40th Street West, Webster, MN 55088 USA. Telephone: (507) 744 2913. Company description: Artistic holography-artistic environments- mixed media; architectural holography; international collaborations.

WOTAN LAMPS LTD. 1 Gresham Way, Durnsford Road, London, SWI9, England, United Kingdom. Telephone: (44)(l) 947 1261. Company description: Equipment for holography; H-l producers.

WRIGHT PATTERSON AIR FORCE BASE. WRDCIFIBG, Dayton, OH 45433 USA. Thlephone: (513) 255-5159. Contact: Gene Maddux. Description: Scientific and holography research; holographic nondestructive testing; interferometry.

WYKO Corporation. 2650 East Elvira Road, Tucson, AZ 85706-7123 USA. Telephone: (602) 741-1044. FAX: (602) 294-1799. Contact: James Wyant. Company description: Scientific holography research; Interferometry and analysis.

X, Y, Z

X-IAL. Established in 1987.6 Employees at this address. "Les Al-

gorithmes", Parc d'Innovation, F-67400 Illkirch, France.Telephone: (33) 88 67 44 90. FAX: (33) 88 67 40 66. Contcat: Dr. Christian D. Liegeois, Director. Parent company: Pirelli. Company description: Stereograms for embossing; HOEs designed and manufactured

YORK UNIVERSITY. Department of Physics, 4700 Keele St., North York, Ontario, Canada M3J 1P3. Contact: Dr. S.B. Joshi. Description: Holography courses.

ZARDON HOLOGRAPHIC. 3654 North Dancer Road, Dexter, MI 48130 USA. Telephone: (313) 426-0452. Company description: Holographic non-destructive testing, production display, consulting, artistic limited editions. Contact: Charles Lysogorski.

3

3D GALLERY. Established in 1987.4 Employees at this address. 207 Queen's Quay West, Toronto, Ontario, M6J 1A7 Canada. Telephone: (416) 359 0417. Contact: Brian Postnikoff, President. President, Brian Postnikoff; Vice-President, Avi McCaffrey. Company description: 3D Gallery is Canada's premier exponent of the 3D cult. From holographic pop-art, to the laser fish hook, 3D Gallery is the 3D solution to the 2D blues.

3DI. Established 1985.49 Upper Woodford, Salisbury, Wilshire SP4 6NW England. Telephone 0722-73471. FAX: 0722-73373. Contact: Caroline Palmer. Description: Commercial range of holograms.

3D MEDIA. Kyrkvagen 24, S-91O 36 SAVAR, Sweden. Telephone: (46)(090) 98141. Contact: Hans Hallstrom. Company description: Artistic holography consulting.

3M--OPTICS TECHNOLOGY CENTER. 1331 Commerce Street, Petaluma, CA USA. 94954 USA. Telephone: (707) 765 3240. Contact: Sharon. Company description: Manufacturer of Holographic Optical Elements and other optics.

12

Businesses By Category

This section lists the businesses and individuals from the main directory, separated into the sectors of the industry in which they operate. We have made every effort to ensure that the categories to which each company claims to belong are accurate: we telephoned more than 90% of the US, UK, German and French businesses.

~le cannot prevent, in all cases, companies from mis:- representing themselves. As with any business deal:::

1gs you undertake, it is best to remember the phrase

-caveat emptor" Qet the buyer beware).

We've made every effort to update these charts accurately. If any of the information about your business is incorrect, please let us know and we'll amend it in the fourth edition.

There are 11 headings in this section:
1. Hologram Mastering and Pre-Production
2. Mass-Manufacturing
3. Marketing
4. Embossing
5. Education
6. Medical

7. HOEs and optical components
8. Lasers
9. Laboratory equipment and supplies
10. Non-destructive testing
11. Brokers and consultants.

Each heading has several sub-categories (i.e. Hologram Mastering and Pre-Production has sub-categories like "dichromate mastering ", "model maker", etc.). Each of these sub-categories has a corresponding letter of the alphabet and when the businesses are listed, they are followed by letters that correspond to their categories.

Hologram Mastering And Pre-production

a. Stereogram Master
h. Pulsed Portrait Master
c. Dichromate Master
d. Color Master
e. Silver Halide Reflection Master
f. Silver Halide Transmission Master
g. Fine-art Ltd Edition
h Model Maker
i. Pulsed Laser In House
j. C.W. Laser In House
k. Animator
l. Photopolymer Master

ADEL ROOTSTEIN: h.
ADVANCED HOLO CORP: c, h.
ADVANCED HOLO LTD: e, f, g.
ADVANCED ENV. RES.: d, e,f, gj
AITES LIGHTWORKS: e, f, g.
AKS HOLO: b, e, f, g, h, k.
AMAZON: a, g.
AMERICAN BANK NO~: b, d, f, g, h.
AMHERST MEDIA: g.
ANGSTROM INDUSTRIES: a, e, g, h, k.
APPLIED HOLOGRAPHICS CORP: d, e, f, h.
ARCHEOZOIC INC: b, c, e, f, g, h,j.
ART FREUND HOLO: e, f.
ARTKITEK: f, h.
ASOC ESP DE HOLO
ASTI: d, e, f, g, h,
ASCOT LASER PICT: e, f, g.
ATELIER HOLO: d, e, f, g, h, k.
ATLANTA GALLERY OF HOLO: g.
AUSTRALIAN HOLO: e, f, g.
BOB MADER PHOTO: b.
BRIDGESTONE: c, d, j.
BRODEL HOLO: a, b, g.
BURNS HOLO: a, d, e, g, k.
BURTON HOLMES !NT: g.
CASDIN-SILVER HOLO: a, c, d, e, f, g, k.
CENTER FOR APPLIED RES: b, e, f, h.
CHERRY OPTICAL: e, f, g.
CHROMAGEM INC: d, e, f, k.
CISE SPA: a, c, e, f, g, j.
CHERRY OPT: d, e, f, g, j, k.
DEEP SPACE HOLO: c, e, f, g, h, j, k.
DEUTSCHER HOLO
DIALECTICA
DUSTON HOLO: a, d, e, f, g,j, k.
DUTCH HOLO: a, e, f, g,j, k.
ElF PROD: g.
ERBA
FOCAL IMAGE: d.
FRED UNTERSEHER & .. : b, c, d, e, f, g, h, k .
FRINGE: b, e, f.
FTI
GERALD MARKS : g.
GRAY SCALE: h
HICKMOTT & AUST
HOECHST
HOLAGE: e, f, g.
HOLAR:g.

HOLAXIS: I.
HOLICON: b, e, f.
HOLOARP:g.
HOLOCOM:g.
HOLOCOM HOL: g.
HOLOCRAFTS EURO: c.
HOLODESIGN: g.
HOLO-DIMEN: g.
HOLOFAR
HOLOFAX: e.
HOLOFLEX: e, f.
HOLO GMBH: a, d, e, f, g,j.
HOLOGOS: b, e, f, g, j.
HOLOGRAFIE: g.
HOLOGRAMA LAB: b.
HOLOGRAMAS DE MEX: c, d, j.
HOLOGRAPHIC ART: h.
HOLOGRAPHIC CONCEPTS, IN (MN): e, f.
HOLOGRAPHIC CREATIONS: f.
HOLOGRAPHIC DESIGN, INC.(PA): c, h,j.
HOLOGRAPHIC DESIGN SYS: a, e, f, g, h,j.
HOLOGRAPHIC DIMENSIONS: a, e, f, g, j, k.
HOLOGRAPHIC IMAGES: f, g, h.
HOLOGRAPHIC PRODUCTS: c, d, g, h.
HOLOGRAPHICS NORTH: b, e, f, g, h.
THE HOLOGRAPHIC STUDIO: e, g,j.
HOLOGRAPHIC STUDIOS.: a, e, f, g, h, j.
HOLOGRAPHICS (UK) LTD: b, c, g.
HOLOGRAPHIE KONZEPT: g.
HOLOGRAPHY DEVELOP GP: g, h, j, k.
HOLOGRAPHY INST: a.
HOLO-IMAGES
HOLO-LASER: d, e, f, g.
HOLOMAT:f.
HOLOMEDIA INC: c, e, f, g, h.
HOLOMORPH: e, f, g, h.
HOLO-SERVICE: g.
HOLO/SOURCE : a.
HOLOTEC CC.: g.
HOLOTEC PLC: g.
HOLOTEK, S.A.: c, e, f, g,h, j.
HOLTRONIC: b.
IAN GINN HOLO: e, f, j .
IBERO GESTA.O: f.
IBOU: f, g, h.
ICON: g.
IDHOL: e, f, g.
IMAGES COMPANY SO: a, j.
IMAGING & DESIGN: a, d, e, f, g, h ..
INSTITUTE OF ART & DES: b, f, g.
INTERFERENS: g, h, j.
KYOTO TECH: b, e, f.
LABOR FUR HOLO: g.
LAMINEX
L.A.S.E.R. CO: g.
LASERFILM: g.
LASERGRAFICS: f.
LASER GRAPHICS: e, f, g.
LASERGRUPPEN: g.
LASERLABBET.
LASER LIGHT EXPRESSIONS: d, e, f, g, j, k.
LASER LIGHT LTD: g.
LASER SMITH
LASERWORKS
LASmIS: e, f.
LAZA HOLOGRAMS: a, b, d, e, f, g, h.
LAZART HOLO: e, f.
LES PRODUCTIONS: g.

UGHT CONSTRUCTION: g.
LIGHT FANTASTIC PLC: a.
UGHT IMPRESSIONS, INC: d, h.
LINDA LAW HOLOGRAPlllCS: g.
MACSHANE HOLO: e, C, g, h, j.
MAN ENVIRONMENT: g.
MARKEM SYSTEMS: e, C, g.
MARWELL AB: g.
MATT HANNIFIN CO: g.
MEDIA INTERFACE: g.
MESSERSCHMITT: d.
METAMORFOSI: c.
MGD PRODUCTIONS: g
MIRAGE HOLOGRAMS: e, C.
MOELLER WEDEL: g.
MULTI ·PURPOSE HOLO: b, c, e, C, g, h, i,j, k.
MUNDAY SPATIAL IMAGING: a, b, e, C, j, k.
NATIONAL HOLOGRAPHIC STUDIOS: e, f, g.
NEOVISION: g.
NEWBOLD WELLS: e, C.
NEW YORK HOLO LAB: g.
NORTH AMERICAN HOLO: b.
ODHNER HOLO: e, C.
OP ·GRAPHICS: a, b, e, C, g, h.
OPTICAL LAB: g.
OPTICAL SURFACES: g.
ORIEL CORP: g.
PACIFIC HOLO: e, C, g.
PERCEPTION HOLO: g.
PENN STATE U: d.
PHOENIX HOLO: b, e, C, g, j.
PHOTON LEAGUE: g,j.
PHYSICAL OPTICS CORP: c, d, h, j.
PHYSICS INSTITUTE. LATV: e,j.
POINT OF VIEW: a.
POLYMER HOLO: a, e, C, g,j, 1.
PORTS ON: b, d, e, f, g, 1.
RANDY JAMESIHOLO: e, C, g, 1.
RAVEN HOLO: c, d.
REAL IMAGE: g.
REEL IMAGE: g.
RICHARD BRUCK HOLO: d, e, C, g, j.
RICHMOND HOLO: b, d, e, C, g.
RITCHER HOLOGRAMS: g.
SCHOOL OF THE ART INSTI: a, e, f, g, h, j.
SOl: all categories.
SPINDELER & HOYER: g.
STUDIO FUR HOL: e, C, g.
SUPERBIN: a, b, c, d, e, j.
SUPERIOR TECHNOLOGY: a, e, C, k.
SWEDE HOLO: g.
TECHNOEXAN: b, g.
TJING LING IND: g.
TOUCHWOOD HOLO: c, e, C.
TRIDIMENSIONAL: b, g.
UNIVERSITY GENT: b, e, C, h.
WAVE MECHANICS: e, C.
WeBER DESIGN HOLOLAB: e, f, g.
WORLD ART PROJECT
X ·IAL: a.
3D GALLERY: a, d, e, C, g.
3DI: c, e, C, g.

Individual Artists

ABRAMS, C.: g.

ALEXANDER: g.
ALLON, G: b, e, f, g.
BENYON, M.: b, e, C, g.
BERKHOUT, R. C, g.
BOISSANT, P: g.
BOYD, P: a, b, e, C, g.
BUNTS,F:g
BYRNE,K:g.
CARLSSON, T: g.
CARLTON, D: g.
CONNER, A: g.
COSSETTE, M.: e, C, g.
COWLES, S: g.
DEEM, R: e, C, g.
DEFREITAS, F: e, g.
DIDIER, L: d, e, C, g, h.
DINSMORE, S: c, d, e, f, g.
DUBOV,P:g.
DYENS, G:g.
FEFERMAN, B: g.
FEROE,J:g.
FISCHER, J: g.
FOLTZ, S:g.
FORNARI, A: e, C.
FRIGON, J: g.
GILLES, J: g.
HANSEN, M: g.
HARRIS, N: a, e, C.
HEATON, L: g.
HOWLETT, G: g.
ISHII, S: c, e, g.
JACKSON SMITH, R: g.
JUNG,D:g.
KAC, E: e, f, g.
KAUFMAN, A: g.
LAFRENIERE, A: g.
LESAR, C:g.
LEVINE, C: g.
LIGHT,G:g.
LIJN, L:g.
McCLEAN, e, g, h.
McINTYRE, J: e, C, g, h.
MENNING, M: g.
MERRICK, M: e, f, g.
MERRILL, M: g.
MULHEM, D: a, b, c, d, e, C, g, h.
MUSKOVITZ, A: g.
MYRE,R:g.
NAIMARK, M: g.
NAKAMURA, I: e, C, g, h.
NEMTZOW, S: g.
NEWMAN, P: b, e, C, g.
PADNOS, W: e, f, g.
PALMER, C: g.
PETERSEN, J: g.
PETICOV, A: g.
PIZZANELLI, D: a, b, e, g.
PYE, T:g.
REILLY, J: g.
REUTERSWARD, C: g.
ROSEWELL, M: g.
ROY,J:g.
SENECHAL, J: g.
STALLARD, P: g.
TURNADINE, G: g.
UNTERSEHER, F: a, b, c, d, e, C, g.
VILA, D: e, f.

VOLKER, M: g.
WELBY, D: g.
WELLS, S: g.
ZEMAN,L:g.

Mass Manufacturing

a. Photoresist Master
b. Silver Halide
c.Embossed
d. Photopolymer
e. Dichromate
f. Packaging (Frames, Jewelry Etc.)
g. Portrait

ADVANCED ENV. RES.: a,b, c, d, e.
A.H. PRISMATIC, Inc.: f.
AKS HOLO: a, b.
ALPHA FOILS, Inc. c.
AMBLEHURST LTD. c.
APPLIED HOLOGRAPHICS CORP: a., b, c.
ARCHEOZOIC INC: a, b, d, e.
ASTI: a, c, d, e, f.
BEDDIS KENLEY: c.
BRIDGESTONE: c, d, f.
BURNS HOLO: a, b, f.
CENTER FOR APPLIED RES: b.
CHERRY OPT:
CISE SPA: a, b, c, d, e.
COSSE'ITE, M: b.
COBURN:c.
DIFFRACTION CO: c.
DUTCH HOLO: a, b, c.
DZ CO: f.
EDMUND SCI: f
EMPAQUES: c, f.
FLEXcon: e, f.
FOREIGN DIM: e, f.
FRED UNTERSEHER &: a, b, d, e.
HIGH TECH: c.
HOLAXIS: c.
HOLICON:g.
HOLO 3: a, b, g.
HOLOCRAFTS: e, f.
HOLOGRAMAS DE MEX: a, b, c, e.
HOLOGRAPHIC ART: f.
HOLOGRAPHIC DESIGN, INC.(PA): a, d, e, f.
HOLOGRAPHIC DESIGN SYS: a, b, c, d, e, f.
HOLOGRAPHIC DIMENSIONS: a, b, c, f.
HOLOGRAPHIC PRODUCTS: e, f.
HOLOGRAPHICS NORTH: b.
HOLOGRAPHIC STUDIOS: b, g.
HOLO·LASER: a, b, c, d.
HOLO·MORPH: b.
HOLOPRODUCTION: b.
HOLO/SOURCE : c, f.
HOLOTEK, S.A.: b.
IAN GINN HOLO: b.
IMAGING & DESIGN: a, b, c, f.
KOLBE·DRUCK: c.
LASER FARE: d.
LASERGRAFICS: f.
LASER LIGHT EXPRESSIONS: a, b, c, d.
LASIRIS:c.

LAZA HOLOGRAMS: b.
LIGHT IMPRESSIONS, INC: c.
MACSHANE HOLO: b, f.
MARKEM SYSTEMS: c.
MULTI·PURPOSE HOLO: b, d.
MUNDAY SPATIAL IMAGING: b.
ODHNER HOLO: b.
OP.GRAPHICS : b, f, g.
OPTICAL IMAGES: d, e, f.
PACIFIC HOLO: a, c.
PHANTASTICA: c, f.
PHYSICAL OPTICS CORP: a, b, d, e, f.
PILKIN"GTON: e.
POLAROID: d.
POLYMER HOLO: d, f.
PORTS ON: a,b, c, d, e, f.
RAINBOW SYMPHONY: f.
RAVEN HOLO: e, f.
REGAL PRESS: c.
RICHARD BRUCK HOLO: b.
RICHMOND HOLO: b, f.
SOVISKUSSTVO: b, f.
STUDIO FUR HOL: a.
SUPERBIN: a, c, d.
SUPERIOR TECH: a, c.
THIRD DIMENSION ARTS: e, f.
UNIVERSITY GENT: b, f.
WHITE LIGHT WORKS: c.
3D GALLERY: b, f.

Individual Artists

BERKHOUT, R: b.
COSSE'ITE, M: b.
DINSMORE, S: b.
MULHEM, D: a, b.
UNTERSEHER: a, b, d, e.

Marketing

a. Retail Shop
b. WholesalerlDistributor
c. Catalog/Mail Order
d. Gallery/Museum Shop
e. Private Collection
f. Hologram Display Units
g. Lighting/Framing
h. Jewelry/Novelties
i. Silver Halide Holograms
j. Embossed Holograms
k. Dichromate Holograms
ACTIVE IMAGE: b,j.
ADVANCED HOLO CORP: b, f, k.
AD 2000: b, d.
A.H. Prismatic, Inc: a, b, c, d, f, h.
A.H. Prismatic, Ltd.: b, f, h, i, j, k.
AKS HOLO: i, j.
Alpha Foils, Inc: j.
AMAZING WORLD OF HOLO: all except (c)
AMAZON: d.
APPLIED HOLOGRAPHICS CORP: b, f, g, i, j.
ARCHEOZOIC INC: c, e, f, g, h, i, j, k.

ARTIGLIOGRAPHY: d, h.
ASTI: c, d, f, g, k.
ATELIER HOLO: d, g, i.
ATLANTA GALL: a, d, h.
AUSTRALIAN HOLO: g.
BENYON, M. i.
BERKHOUT, R. e.
BURNS HOLO: h, e, i, j.
CASDIN-SILVER HOLO: h, e, i, j, k.
CHERRY opr: h, c, e.
COSSETTE, M: d, e, f, i.
DEEP SPACE HOLO: a, h, e, f, g, h, i,j, k.
DIE DRITTE DIM: a, d, h, i, j, k.
DREAM IMAGES: d.
DUSTON HOLO: e, i.
DZ CO: h, d, h.
E.C. SCHULTZ: e.
EDMUND SCI: a, h.
ELUSIVE IMAGE: d.
FOREIGN DIM: h, f, g, h, k.
FOUNDATION IDEE: d.
F UNTERSEHER &: f, i, j, k.
FRINGE: d.
GENERAL HOLO: h, f, h, i.
HALO POWER: g.
HOL3: a, d.
HOLAXIS: f, k.
HOLO ARP. h, d.
HOLOCOM:j.
HOLOCRAFT !NT
HOLOCRAFTS EURO: h, c, f, g, h, i,j, k.
HOLOGOS: h, i.
HOLOGRAFIA: a, d.
HOLOGRAFIA OU: a, d.
HOLOGRAFIE: h.
HOLOGRAMAS DE MEX: j, k.
HOLOGRAM EURO: a.
HOLOGRAM LAND: a, c, d, g, h, i,j, k.
HOLOGRAMM WERK: h, d, e, f, i.
HOLOGRAM ROADSHOW: a.
HOLOGRAMS AND OTHER: a, c, h.
THE HOLOGRAM SCHOPPE: a.
HOLOGRAM WORLD: a, h, c, f, g, h, i, j, k.
HOLOGRAPHIC APPL: h, f, g, h, i,j, k.
HOLOGRAPHIC ART: a, h, c, e, f, g, h, i, j, k.
HOLOGRAPHIC DESIGN, INC.(PA): h, h, k.
HOLOGRAPHIC DIMENSIONS: h, d, i,j.
HOLOGRAPHIC INDUSTRIES: a, h, d, e, g, h, i, j, k.
HOLOGRAPHIC MARKETING: h, c, h,j, k.
HOLOGRAPHIC PRODUCTS: a, h, c, d, e, f, h, k.
HOLOGRAPHIC STUDIOS: a, h, c, d, f, i, j, k.
HOLOGRAPHIE LABOR: a, i, k.
HOLOGRAPHY DEVELOP GP: h, f, g, i, k.
HOLOGRAPHY INST: h, c, e, f, g, j, k.
HOLOGRAPHY LTD
HOLO-LASER: e, f.
HOLO-LASER TECH: a, h, i, k.
HOLOMEDIAAB: h, d, i.
HOLOMEDIA INC: h, c, i, k.
HOLO-MORPH: c.
HOLOPRINT: h.
HOLOS ART GALERIE: a, d.
HOLOS GALLERY: a, d.
HOLO-SPECT: h, i, k.
HOLOTEK, S.A.: a, h, c, e, f, g, h, i, j.
HOLOVISION: a, i.
IAN GINN HOLO: h, c, i.

IBERO GESTAo: h, i.
IBOU: all except g.
IMAGEN HOLO: h.
IMAGES COMP: h, c, g.
IMAGES COMPANY SO: h, c, g.
IMAGING & DESIGN: h, g, h, i, j, k.
INFOTECH: h.
INTERFERENS: d, e, h, i, j, k.
KOLBE-DRUCK: h, c, k.
LASART: h, g, h.
LASER ARTS: h, e.
LASER HOLOGRAPHICS: h.
LASER INTERNATIONAL: d.
LASER LIGHT DESIGNS: a, h, d.
LASER LIGHT EXPRESSIONS: h, d, e, f, g, h, i, j, k.
LASER LIGHT IMAGE: all except c.
LASmIS:j.
LAZER WIZARDRY: h, c, h, i,j.
LET THERE BE NEON: a, d.
LICHT-BLICKE: d.
LIGHT ENGINEERING: a, d.
LIGHT FANTASTIC PLC: d, f, i, j, k.
LIGHT IMPRESSIONS, INC: d, e, h, j.
LIGHT WAVE GALLERY: a, d.
MACSHANE HOLO: h, c, h, i.
MAGIC LASER: a, h, d.
MAGIC LIGHT: d.
McCAIN MARKETING: c, d, e.
METAMORFOSI: h, e, h, j, k.
MULTI-PuRPOSE HOLO: h, d, e, f, g, h, i, k.
MUNDAY SPATIAL IMAGING: a, d, e, i, j, k.
MU8eE DE L'HOLOGRAPHIE: d, e.
MUSEUM FUR HOLOGRAPHIE: all categories except (g)
MUSEUM OF HOLOGRAPHY: c, d.
MUSEUM OF HOLOGRAPHY/CHICAGO: a, h, c, d.
MUSEUM OF THE FINE ARTS RES: d.
NEW CLEAR IMPORTS: a, d.
NEW DIMENSION HOLO: a.
NEWPORT HOLOGRAMS: a, d.
NEW YORK HALL OF SCI: d.
ONTARIO SCIENCE CENTRE: d.
OXFORD HOLO: a, h, g, h, i, j, k.
PALCO MARKETING: a, h, h.
PHANTASTICA: a, h, j.
PHOENIX HOLO: d, i, j, k.
PHOTON LEAGUE: d, e.
PHYSICAL OPrICS CORP: h, i, k.
PHYSICS INSTITUTE. LATV: h.
PORTSON: h, i,j, k.
QU~ SERA: h, k.
RAINBOW SYMPHONY: h.
RALPH CULLEN HOLO: a, h, e, f, g, i, j, k.
RAVEN HOLO: h, c, h, k.
RICHARD BRUCK HOLO: i.
RICHMOND HOLO: h, e, f, i.
SOl: h, e, f, h, i, j, k.
STARCKE, KY: a, h, d, i, j.
STARLIGHT HOLO: a, h, d, g, h, i, j, k.
STUDIO FUR HOL: i, j.
SUPERBIN: f, g. j, k.
TECHNOEXAN: c, d.
THREE-D LIGHT: d.
TREND:d.
U.K. GOLD: h, c, d, f, g, h, i,j, k.
VISIONS OF THE FUTURE: a, h, c, d.
WAVE MECHANICS: d, i.
WHITE TIGER HO LO: j, k.

WHOLE PICTURE: d.
WOBER DESIGN HOLOLAB: b, d, g, j.
WONDEROUS LIGHT: a, c, d, h, i, j, k.
WONDERS OF HOLO: a, b, d, e, h, i, j .
3D GALLERY: all categories except (c)

Individuals

ALLEN, J: b, c, e, f, h, i, j.
ANDREWS, M: b, e, i.
BENYON,M:i.
COSSETTE, M: i.
DIDIER, L: i
DINSMORE, S: i
MC CLEAN, J: i.
NAKAMURA, I: d, e, f, g.
UNTERSEHER, F: f, i, j, k.

Embossed Holography

a. Model Maker
b. Color Master
c. Photoresist Master
d. Shim Master
e. Holographic Embossing
f. Application To Paper
g. Application To Toys/Clothing
h. Application Of Stickers
i. Hot Stamping
j. Pulsed Laser
k. CWLaser

ADVANCED ENV RES.: a, b, c, d, e, k.
AMBLEHURST LTD.: a, b, c, d, i.
AMERICAN BANK NOTE: a, b, c, d, e, f, h, i.
APPLIED HOLOGRAPHICS CORP: a, b.
ASTI: a , f, h, i
BEDDIS KENLEY: c.
BURNS HOLO: c, d.
CHROMAGEM: c, d, e.
CISE SPA: b, c, d, k.
CROWN ROLL: d, e
COBURN: d, e.
CREATIVE LABEL: f, g, h, i.
DENNISON: e, f, i.
DIFFRACTION CO: e.
DUTCH HOLO: b, k.
EMPAQUES: e, f.
FLEXcon: e, f, i.
FRED UNTERSEHER &: a, b.
GRAY SCALE: a.
HOLAXIS a, b, e, i.
HOLOCOR I.B.F: a, d, e, f.
HOLOCRAFTS EURO: g, h.
HOLO GMBH: b, c, k.
HOLOGRAMAS DE MEX: a, b, c, d, e, f, g, h, i, k.
HOLOGRAM. IND: c, d, e, h, k .
HOLOGRAPHIC DESIGN, INC. (FA): a, c, k.
HOLOGRAPHIC DESIGN SYS: a, b, c, f, h, k.
HOLOGRAPHIC DIMENSIONS: b, c, d, k.
HOLOGRAPHIC LABEL: f, g, h.
HOLOGRAPHIC STUDIOS: a.
HOLOGRAPHY INST: a, b.
HOLO ·LASER: a, b, c.

HOLO/SOURCE: b, e.
HOLOTEK, S.A.: g.
IMAGES COMP: a .
IMAGING & DESIGN: a, b.
INGENIEUR BuRO: a, b, d, e.
JAEGER GRAPHIC: f, i.
JAYCO HOLO: e.
KOLBE ·DRUCK: d, e, f, g, h, i.
LASERGRAFICS: i.
LASER LIGHT EXPRESSIONS: a, b, c, d, f, g, i, k.
LASIRIS: c, d, e, f, h, i.
LETTERHEAD PRESS: g, i.
LIGHT FANTASTIC PLC: a, b, c, d, e, f, h, i, k.
LIGHT IMPRESSIONS, INC: a, b, c, d, e.
LOUGHBOROUGH UNIV: d, e.
MACSHANE HOLO: a.
MARKEM SYSTEMS: a, c, d, e.
MARTINSSON ELEKTRONIK: c.
MULTI.PURPOSE HOLO: g.
MYERS PRINTING: f, i.
PHYSICAL OPTICS CORP: b, c, k .
PHYSICS INSTITUTE. LATV: i, j, k.
PORTS ON: a, c, e.
REGAL PRESS: f, i.
REYNOLDS METALS: f, i.
SILLCOCKS PLASTICS: f, i.
SMITH & McKAY: f, i.
SOl: a, b.
SPECTRATEK: e.
STARLIGHT HOLO: f, h.
STEVE PROVENCE HOLO
STUDIO FUR HaL: c.
SUPERBIN: all categories.
SUPERIOR TECH: c, d.
TOPAC GmBH: a, b, c, d, e, f, h, i, j, k .
TaPPAN PRINT: b, c, d, e, f, i.
TRANSFER PRINT FOIL: i.

Embossing Equipment

a. Foil Manufacturer
b. Photoresist Mftr.
c. Shim Mftr.
d. Embossing Machine Mft.
e. Hot Stamp Machine Mft.
f. Label-applying Machine Maker
g. Electroplating Mft.

ALPHA FOILS, INC: a, b, c, d. f
ANGSTROM INDUSTRIES: b.
APPLIED HOLOGRAPHICS CORP: a, b, c.
BEDDIS KENLEY: b
BOBST GROUP: f.
BRANDTJEN & KLUGE: e.
BURNS HOLO: b, c.
CHROMA GEM: b, c.
CROWN ROLL: a, c.
COBURN: c, d, g.
E.C. SCHULTZ: b.
GLOBAL IMAGES: d.
HOLAXIS: b, c.
HOLOGRAMAS DE MEX: a, c, d, g.
HOLOGRAPHIC DIMENSIONS: b, c.

HOLO·LASER: b, c.
HOLO-LASER TECH: b, c.
HOLO/SOURCE: b, c.
INGENIEUR BuRO: c.
JAEGER GRAPHIC: a.
JAMES RIVER: d.
JAYCO HOLO: c, d.
LASmIS: b, c.
LENOX LASER: a, c.
LIGHT IMPRESSIONS, INC: c.
LOUGHBOROUGH UNN: c.
MARKEM SYSTEMS: b, c, e.
METAPLAST: c.
PORTS ON: a, b.
SPECTRATEK
STEUER KG: d.
TRANSFER PRINT FOIL: a.
WHILEY FOILS: a.
WHITE LIGHT WORKS: b.

EducationIResearch

a. College And University Classes
h. Independent Schools With Regular Schedule
c. Independent Schools: By Appointment
d. Touring Exhibitions
e. Academic/industrial Research

ALPHA FOILS INC.: e.
APPLIED HOLOGRAPHICS CORP: e.
ARCHEOZOIC INC: c, e.
ASTI:b, d.
ASCOT LASER PICT: c.
AT&T BELL LAB: e.
BELJING INST: a.
BELJING NORMAL UNI: e.
BIAS: e.
BROOKHAVEN NAT LAB: e.
CASDIN-SILVER HOLO: c, e.
CENTER FOR APPLIED RES: c, e.
CENTRAL MICHIGAN UNN: a.
CHERRY OPT: c.
CHIBA UNN: e.
CINEMA & PHOTO RES: e.
CISE SPA:e.
COSSETTE, M: c, e.
COLLEGE OF MANUFACTURING: e.
DEUTSCHE GESEL: c, e.
DREAM IMAGES: c.
DUSTON HOLO: e.
ECOLE NAT: e.
EDMUND SCI (ed. materials)
ED WESLYHOLO: c, e.
EOCS:c,e.
ENVIRONMENTAL RES: e.
ERBA:c.
EVE RITSCHER: c.
F.A.S.T.
FRED UNTERSEHER & .. : c, e.
FREE UNIVERSITY: a, e.
FUJI PHOTO: e.
HOLOARP:c.
HOLO GMBH: c.
HOLOGOS: c, e.

HOLOGRAMAS DE MEX: e.
HOLOGRAM ROADSHOW: d.
HOLOGRAPHIC CONCEPTS, IN (MN): c.
HOLOGRAPHIC DESIGN, INC.: c.
HOLOGRAPHICS INT.
HOLOGRAPHICS NORTH: c.
HOLOGRAPHY DEVELOP GP: c.
HOLOGRAPHY INST: c, e.
HOLOGRAPHY YEARBOOK
HOLO ·LASER: c, e.
HOLOMAT:e.
HOLOPRODUCTION: c, e.
HOLOPUBLIC: c, e.
IBERO GESTAO: c, e.
ILLINOIS INST: e.
ILLINOIS VALLEY: e,
IMAGES COMP: c.
IMPERIAL COLL: e.
ING. ·AGENTUR: e.
INSTITUTE OF ART & DES: a, e.
INSTITUTE OF NUCLEAR PHYSICS: e.
INSTITUTE OF OPTICAL SCI: e.
INSTITUTE OF PHYSICS: e.
INSTITUTE OF PLASMA PHYS: e.
INTERCHANGE STUDIO: c.
JOHNS HOPKINS: e.
KRAFTWERK: e.
LABOR FUR HOLO: c.
LAKE FOREST COLL: a, e.
LASER AFFILIATES: c.
LASER INSTITUTE: c.
LASER LIGHT EXPRESSIONS: e.
LAWRENCE BERKELEY LAB: e.
LCPC: e.
LENINGRAD SUBS: e.
LIGHT CONSTRUCTION: c.
LINDA LAW HOLOGRAPHICS: c., e
L.LR.: e.
LA SCHOOL OF HOLO: c.
LOUGHBOROUGH UNN: a, e.
LULEA UNN: e.
LUND INSTITUTE: e.
LURE:e.
MACSHANE HOLO: c, e.
MARKEM SYSTEMS: e.
M.LT: a.
McCAIN MARKETING: c.
MESSERSCHMITI: e.
METRO LOGIC INSTRUMENTS: e.
MITSUBISHI: e.
MOSCOW PHYSIC ENG. INST.
MUSeE DE LIIOLOGRAPHIE: c, e.
MUSEUM FUR HOLOGRAPHIE: c.
MUSEUM OF HOLOGRAPHY: c, d, e.
MUSEUM OF HOLOGRAPHY/CHICAGO
MUSEUM OF THE FINE ARTS RES: c.
NASA:e.
NATIONAL PHYSICAL LAB: e.
NEWCASTLE UPON TYNE: a, e.
NEW YORK HALL OF SCI: c.
NEW YORK HOLO LAB: c.
NEW YORK INSTITUTE OF TECH: e.
NORGES TEKNISK: e.
NORTHERN ILLINOIS UNN: e.
NORTHWESTERN: a, e.
ODHNER HOLO: c.
ONTARIO COLLEGE: a.

ONTARIO HYDRO: e.
ONTARIO SCIENCE CENTRE: c.
OPTICAL IMAGES: e.
OPTISCHE FENOMENEN (newsletter)
OXFORD UNIVERSITY: a, e.
PENN STATE UNIV: e.
PHOENIX HOLO: c.
PHOTONICS DIR (magazine)
PHOTON LEAGUE: c.
PHOTONICS SYSTEMS: a, e.
PHYSICS INSTITUTE. LATV: e.
QU~SERA:c.
RALCON:e.
RECONNAISSANCE: c, e.
ROCHESTER INST: a, e.
ROSS BOOKS (publisher)
ROTTENKOLBER: e.
ROWLAND INST: e.
ROYAL COLLEGE OF ART: a, e.
ROYAL INST TECH: c, e.
ROYAL PHOTOGRAPHIC SOC: c.
SAAB:e.
SAGINAW VALLEY: a, e.
SANDIA NATIONAL LAB: e.
SCHOOL OF ART INSTI: a.
SCHOOL OF HOLOGRAPHY: c.
SCHOOL OF HOLOGRAPHY/CHI: c.
SCIENTIFIC COUNCIL: c, d.
S.I. VAVILOV: e.
SMITH & McKAY: (seminars, newsletter)
S.P.I.E
SOl: c, e.
SPECTRON DEV: e.
STATE ED DEPT NY: c.
STATE RESEARCH: e.
STUDIO FuR HOL: c.
SUPERB IN: c.
SUPERIOR TECH: c.
SWISS FEDERAL INST: e.
SYDNEY COLLEGE: a.
SYNCHRONICITY HOLO: c, d, e.
SYSTEMS GROUP OF TRW: e.
TECHNICAL UNIV ·EIND: e.
TECHNICAL UNIV ·WROC: e.
TNOINST:e.
TOKA! UNIV: e.
ULTRAFINE: e.
UNIVERSIDADE DO PORTO: e.
UNIVERSITA DI ROMA: e.
UNIVERSITAT ERLANGEN: e.
UNIVERSIT~ DE FRANCHE: e.
UNIVERSITE DE NEUCHATEL: e.
UNIVERSIT~ DE PARIS: e.
UNIVERSrrE LAVAL: a.
UNIVERSIT~ LOUIS PAS: e.
UNIVERSITY ESSEN: e.
UNIVERSITY GENT: a, e.
UNIVERSITY OF ABERDEEN: e.
UNIVERSITY OF ALABAMA: e.
UNIVERSITY OF ALICANTE: e.
UNIVERSITY OF ARIZONA: e.
UNIVERSITY OF BOLOGNA: a, e.
UNIVERSITY OF CAL (SD): e.
UNIVERSITY OF CAL (SB) a, e
UNIVERSITY OF DAYTON: a, e.
UNIVERSITY OF MICHIGAN: a, e.
UNIVERSITY OF MUNICH: e.

UNIVERSITY OF MUNSTER: e.
UNIVERSITY OF N.C.: e.
UNIVERSITY OF OXFORD: a, e.
UNIVERSITY OF ROCHESTER: e.
UNIVERSITY OF SO CAL: e.
UNIVERSITY OF STRATHCLYDE: e.
UNIVERSITY OF STU'ITGART: e.
UNIVERSITY OF TOKYO: e.
UNIVERSITY OF WISCONSIN: e.
UNIVERSITY OF ZAGREB: e.
VINCENNES UNIV: a.
WASEDA UNIV: e.
WOLVERHAMPI'ON POLY: a.
WORCESTER POLY: e.
WRIGHT PATTERSON: e.
WYKO CORP: e.
YORK UNIV: a.

Individuals

ANDREWS,M.
CAPUCCI,P.
COSSETTE, M.
DEEM,R.
DEFREITAS, F.
DIDIER, L.
DINSMORE, S.
HARRIS,N.
KAC,E.
MEANS,M.
NEWMAN,P.
PEPPER,A.
STOCKLER, L
UNTERSEHER, F.
VILA, D.

Medical Holography

a. Dental
h. Endoscopic
c. Time-sequenced
c. Plasma Research
d. Microscopic

CENTER FOR APPLIED RES: a, b, d.
COLUMBIA UNIV: b.
ELANBIO
EOCS: a.
FOCAL IMAGE: b.
HOLOGRAPHY INST: c.
HOLO ·LASER: b, d.
HOLOPRODUCTION: a, b, c, d, e.
INSTITUTE OF OPTHAL: b.
INSTITUTE OF OPTICAL SCI
LASER LABS: b.
NOVATOR RESEARCH
ROYAL SUSSEX HOSP.
TECHNICAL UNIV ·BUD (research)
UNIV GENT: b, d.
TOKYO UNIV (research)
UNIVMUNICH
UNIV MUNSTER
UNIVTOKYO
WASEDAUNIV

Holographic Optical Elements/ Optical Components

a. Gratings
b. Heads-up Display
c. Lenses/prisms/windows
d. Mirrors
e. Beamsplitters/steerers
f. Collimators
g. Filters/polarizers
h. Coating
i. Cutting/grinding/polishing
j. Software For Opt Sys. Design
k. Optics Testing
l. Spatial Filters

ADVANCED OPrICS, INC.: c., d., e., i.
ALPHA FOILS INC.: 1 (shims)
AMERICAN HOLO INC.: a, k.
AMITY PHOTONICS CO.: c, d, e., , g., i. (D, B)
ApA OPrICS: b, c, k.
APOLLO LASER: d, e,1.
ASHLAND ELECTRIC: c, g.
CENTER FOR APPLIED RES: c, f.
CISE SPA: a, b, c, d, j, k.
COBURN:c.
CONTINENTAL OPrICAL: a, c, d, e, g, h
CORION CORP: c.
COULTER opr: c, e.
CSI: c, d, e, f, g, h, i, k.
CVI LASER: c., d, e.
DELL OPrICS : c, d, g, i.
DIE DRI'ITE DIM: c, d (D) for both
DUSTON HOLO: c.
EALING ELECTRO-OP: b, c, d, e, g, 1.
ELECTECH: c, e, 1.
ELECTRO OPrICAL IND: f, g, k.
ELECTRO OPrICS DEV: d, e.
FISHER SCI: c, d, e.
FRESNEL: c, d, f.
FUJI PHOTO:, c, d, e.
GALVOPrICS: c, d, e, f, g, h, i.
HOLOGRAMAS DE MEX: a, c, e, f, h, i.
HOLO-LASER TECH: e, 1.
HOLO-OR: c, d, j.
HOLOTEK LTD: c, d.
HOWARD SMITH: c, d.
IBM:a,e.
ICI: a, c, d.
IMAGES COMP: d, e, 1. (D)
INFRARED opr: c, d, e, g, h, i, k, 1.
INRAD: d, e, f, h, i.
INSTITUTE OF ELE: c, d, e, f, g.
JODON:e,g.
KREISCHER opr: c, d, e, f, g, h, i.
LAMBDAfI'EN: c, d, f, g.
LASER ELECTRONICS: c, d, e, i, 1.
LASER OPrICS: c, d, e, f, h, i, k.
LASER TECH INDUSTRIES: c, d, e, h, i.
LAZAP INC: c, d, e,j, k,1.
LENOX LASER: d, f, g, 1.
L.I.R.: d.

MARKEM SYSTEMS: c, f.
MELLES GRIOT: c, d, e, f, g, h, i, j, k, 1.
MITUTOYO: k.
MODERN OPrICS
MULTI-PURPOSE HOLO: a, d.
NEWPORT CORP: c, d, e, f, g, h, j, 1.
OPrICAL IMAGES: a, b, c, d.
OPrICAL WORKS LTD: c, d, e, f, g, h, i, k, 1.
OPrICS PLUS: b, c, e, f, h, i, k.
OPTILAS B.V.: a, b, c, d, e, f, g, 1.
PHOTONICS SYSTEMS: a, h, c, d.
PHYSICAL OPrICS CORP: all except 1.
PHYSICS INSTITUTE. LATV: a, c, d, e, f, h, i, k.
PILKINGTON: h, c, d, e, f, g, h, i, j, k.
RALCON: a, c, d, e, i.
RAVEN HOLO: a, h, h, i.
ROCHESTER PHOTONICS: a, c, f, g.
SCOPE OPrICS
SORO ELECTRO
SPECTRA-PHYSICS: c, d, e, f, g, h, i, k, 1.
SPECTROGON: a, c, d, e, f, k, 1.
SPECTROLAB: all except h.
STEINBICHLER: c, d, e, f, 1.
SUPERBIN: h, c, d, e, f, g, j.
SUPERIOR TECH: c, k.
T.A.1. INC: h.
TOWNE LAB: a, c, d, e, f, g.
UK OPrICAL: a, c, d, e, f, i, 1.
VINTEN ELECTRO: d.
WISE INSTRUMENTS: a, h, c, d, e, f, g, k.
X-1AL: a, h, c, d, e, f, g, h, k.
3M--OPrICS: c, d.

Lasers: Manufacturers and Distributors (D)

A Argon
B.Krypton
C. Helium Cadmium
D. Helium Neon
E.Ruby
F. Krypton Chloride
G. Krypton Fluoride
H.Xenon
I. Xenon Chloride
J. Carbon Monoxide
K Carbon Dioxide
L. Nitrogen
M. Nitrogen Flouride
N. Hydrogen
O. Rods
P. 'lUbes
Q. Diodes
R. Collimators
S. Etalons/etalon Filters
T. Phase Conjugators
U. Cleaning Equipment
V. Used Laser Sales/repairs
W. Pulsed
X. Other.

.ADLAS G.m.h.H. :w.
AMERICAN LASER CORP: a, h.
APOLLO LASER: e, k, 0, p, r, s, t, u.
AEROTECH: d,(D) k (D) 0, p.
BURLEIGH INSTR: s, x (color-center).
CAMBRIDGE LASERS: a, h.
COHERENT: a, d, f, ,a, p, s, x (dye).
CVI LASER: s.
DIE DRITIE DIM: a, d, q (D)
EALING ELECTRO-OP: a, d, r.
EALING SCI: a, d.
ELECTECH: 0, p, r, s.
EVERGREEN: v.
EXCITEK: p, v.
FISHER SCI: a, d, e, q, r, w.
FUJI ELECT: k, w.
HOLO-SPECT: v.
HUGHES AIRCRAFT: k, w.
IMAGES COMP: d, p.(D)
IMAGES COMPANY SO: 0, p, v.
INSTITUTE OF ELE: e, s, t, w.
ION LASER: a.
JODON: d.
LASER ELECTRONICS: a, h, c, d, g, i, j, k, la, p, q, r, u.
LASER IONICS: a, h, p, s.
LASERMETRICS
LASER RESALE: v.
LASER TECH INDUSTRIES: o. p, s, v.
LAZAP INC: c, d, k, s, v, w, x (chemical)
LEXEL: a, h, w.
LICONIX: a, h, c, p, w.
LINE LITE LASER: j, k, 0, p, v, w, x (Nd: YAG)
LUMONICS INC: e, k, w, x (Nd: YAG)
MELLES GRIOT: d, k, p, q, r, w.
MEREDITII INSTRUMENTS: a, d, s, v.
METRO LOGIC INSTRUMENTS
MWK INDUSTRIES: a, d (D for both)
NEWPORT CORP: a, c, f, d.
OMNICHROME: a, c.
OPTILAS B.V.: a, h, c, d, e, f, g, h, s, v ..
PHASE-R
SIEMENS: a, d.
SPECTRA-PHYSICS: a, h, d, g, p, s,u.
SPECTROLAB: all categories.
SUPERBIN: a, f, j, k, u, v.
UK OPTICAL: a, h, c, d, e, k, 0, p, q, r, s, t, u.
UNIPHASE: a, d.
WONDERS OF HOLO: a, h, c, d (D for all)

Lab Equipment

A. Chemicals
B. Film/recording Material
C. Lamps/isolation Tables
D. Plates/emulsions
E. Safety Equipment
F. Holography Kits

D= distributor

AGFA-GEVAERT: a, h, d.
ALPHA PHOTO: a, h, d, e (D for all)
APOLLO LASER: h, c, d, f.
BRIGHTON: h.

C.ITOH & CO: h, d .
CITY CHEMICAL: a, d.
DUTCH HOLO: h.
EASTMAN KODAK: a, h, d.
E.I. DUPONT: a, h, d.
G.M. VACUUM: d.
HOLOFAX: a, c.
HOLO-LASER TECH: a, c (D for both)
HOLOMEX: f.
ILFORD: a, h, d.
IMAGES COMP: a, h, c, e, f. (D for all)
IMAGES COMPANY SO: a, h, c, e, f. (D for all)
IMAGING & DESIGN: a, h, c, d, e (D for all)
INSTITUTE OF ELE: a.
INTEGRAF: a, h, d, f (D for all)
KEYSTONE: a, h, d, f. (D for all)
KINETIC SYS: c.
LASER ELECTRONICS: c, e (D for hath), f.
LASER FARE: h, d.
LASER TECH INDUSTRIES: c.
MELLES GRIOT:.c.
NEWPORT CORP: h, c, f.
NORLAND: a.
OPTILAS B.V.: c, e.
PHOENIX HOLO: f.
PHOTOGRAPHERS FORMULARY: a, h, f.
POLAROID: h.
PORTSON: h, d.
RESIST MASTERS: d.
SCIENCE KIT: f.
SCIENCE & MECHANICS: c.
SHIPLEY CHEMICAL: a, h, d.
STEINBICHLER: f.
SUPERB IN: a, h, c (D for all)
TOWNE LAB: a, d.
TRI-ESS: f.
UK OPTICAL: h, c, d, e, f.
WENTWORTH LAB: c.
WOTAN LAMPS: c.

Non-destructi ve Testing

A.CW
B. Pulsed
C. Real Time
D. Time Average
E. Double Exp.
F. Flow Visualization
G. Vibration Visualization
H. Implact Pulse Probe
I. Load Or Thermal Strain
J. Pressure Cycling Or Effects
K. NDT Particle Testing
L.DCG
M. Silver Halide
N. Thermoplastic
O. Photopolymer Monobath
P. Video

ABBOTI LAB: h, c, i.
AEROSPATIALE: a, h, c, d, e, g, i,
AMERICAN BANK NOTE: a, h, c, d, e.
AMERICAN LASER CORP.: a.

BEIJING NORMAL UNI: c, f, k, p.
BIAS: a, b, c, d, e, g, i, j, k, m, n, p.
BRITISH AERO: a, b, c, d, e, f, g, i, I, m, p.
CENTER FOR APPLIED RES: all except k, 1.
CHERNOVTSY STATE UNIV: a, b, c, d, e, f, k, m, p.
CHIBA UNIV: a, d.
CISE SPA: a, c, d, e, f, g, i, j, I, m, o.
COLLEGE MANUFACTURING: b, c.
CITRoEN INDUSTRIE: a, c, h, i, p.
ECOLE NAT: a, d, k, m, n.
ELECTECH: a, c, d, e, f, g, i, j, n.
EOCS: a, b, c, d, e, g, i, j, k, m, n, p.
EUROLAUNCH: a, f.
FORD: a, f, j, k.
HOLO-LASER: a, b, c, d, e, g, m, p.
HOLOMETRIC: a, d,
HOLOPRODUCTION: all except I, p.
HOLOTEK, S.A.: a, h, i, k.
IBERO GESTAo: a, c, d, e, g, i, m, n, p.
ILLINOIS INST: b, c, i.
ING.-AGENTUR: b, f, i.
INSTITUTE OF ART & DES: a, b, c, e, f, m.
INSTITUTE OF ELE: b, d, e, f, g, h, i, j.
JODON:m.
KRAFTWERK: a, k, m.
LABOR DR. STEIN: all except I, o.
LASERMET LIMITED
LITTON SYSTEMS: b, j.
LULEA UNIV: b, c.
MACSHANE HOLO: a, c, e, m.
MAZDA MOTOR CORP
McMAHAN ELECTRO: a, b, c, d, e, g, k, p.
METRO LASER: all categories except I, o.
MICRAUDEL: a, b, c, c, e, f, g, h, i,j,l, n, p.
MITSUBISHI: b, i.
NASA
NATIONAL PHYSICAL LAB: b, c.
NEWPORT CORP: a, c, e, g, i, I, m, n, p.
NEW YORK INSTITUTE OF TECH: b, d.
NORGES TEKNISK: a, c.
NORTHERN ILLINOIS UNIV
NORTHWESTERN: b, c.
ODHNER HOLO: a, c, d, e, g, i, j.
PHOTONICS SYSTEMS: all except I, 0, p.
PHYSICAL OPTICS CORP: a, c, d, e, I, o.
PHYSICS INSTITUTE. LATV: a, b, c, n.
RALCON: a, b, c, i, m.
ROCHESTER INST: a, c, d, e, g, i, m, n.
ROLLS-ROYCE PLC: all except k, m, o.
RUTHERFORD AND APPLETON
SPECTRON DEV: b, d, e, n.
STEINBICHLER: all except I, o.
STOLTZ AG: b, d, n.
SYSTEMS GROUP OF TRW: b, k.
UNIVERSITY GENT: a, b, c, d, e, g, h, i, j, m, n, p.
UNITED TECH: b, c, d, e, f, g, j, k, m, p.
VOLVO
VOLKSWAGEN
WONDERS OF HOLO: a, c.
WRIGHT PATTERSON
ZARDONHOLO

Brokers/consultants
(area of specialty in parentheses)

ACME HOLO (artistic)
AG PRISMATIC (artistic)
ALPHA FOILS(embossed holograms)
AMITY PHOTONICS(HOEs)
ARTIGLIOGRAPHY (embossed)
ARTPLAY (design)
ASTI (embossed holograms)
BRIDGESTONE (marketing/packaging)
CAMBRIDGE STEREOGRAPHICS (optics, imaging)
CASDIN-SILVER HOLO (art, marketing)
CENTER FOR APPLIED RES (Ed, NDTlMedical)
CHERRY OPT (framing, education)
CHIMERIC IMAGES (lighting/framing)
Coherent (medical)
DEEP SPACE HOLO (Marketing, lighting)
DIFFRACTION CO (Marketing, lighting)
DREAM IMAGES: (Marketing, lighting, masters)
DUSTON HOLO (HOE, Defense, Ed.)
E.C. SCHULTZ (Embossing) .
ELECTECH: (NDT)
EOCS (marketing, lasers)
EVERGREEN (lasers)
FOREIGN DIM (embossing, artistic)
FRED UNTERSEHER & (artistic, embossing)
GALVANOART (artistic)
GARDENER PROMO (artistic)
GRAY SCALE (embossing)
HOECHST (marketing)
HOLO GMBH (marketing)
HOLOGOS (lasers/optics)
HOLOGRAMAS DE MEX (optics, embossing)
HOLOGRAPHIC APPL (artistic, education)
HOLOGRAPHIC CONCEPTS (MA) (artistic)
HOLOGRAPHIC CONCEPTS, INC. (MN) (artistic, industrial)
HOLOGRAPHIC MARKETING (artistic)
HOLOGRAPHIC SERVICE. (marketing)
HOLOGRAPHICS NORTH (artistic)
HOLOGRAPHY DEVELOP GP(embossing)
HOLOGRAPHY INST (artistic)
HOLOGRAPHY NEWS (marketing, medical)
HOLOPUBLIC (education)
IAN GINN HOLO (marketing)
IAN M. LANCASTER HOLO (marketing, art)
IBERO GESTAO (marketing)
IBOU (marketing)
IMAC INT (marketing)
IMAGEN HOLO (apparel)
JODON(NDT)
JR HOLOGRAPHICS (marketing)
KREISCHER OPT (optics)
LASART: (artistic)
LASER AFFILIATES (lighting, framing)
LASER ARTS: (education, marketing)
L.A.S.E .R. CO (lighting)
LASER HOLOGRAPHICS (embossing)
LENOX LASER (lasers, optics)
LEONARD KURZ GMBH (embossing)
MACSHANE HOLO (marketing, framing)
McCAIN MARKETING (marketing, education)
McMAHAN ELECTRO (medical)
MEDIA INTERFACE (artistic)
METAMORFOSI (marketing)

MIND'S EYE (marketing)
MULTI ·PURPOSE HOLO (lab equip)
MU's LASER WORKS (light shows)
NATIONAL HOLOGRAPHIC STUDIOS (marketing)
OPTICAL IMAGES (NDT)
PACIFIC HOLO (Lighting, framing)
PHOTONICS SYSTEMS (NDT, optics)
POLYMER HOLO (HOEs)
RALPH CULLEN HOLO (HOE)
RECONNAISSANCE (marketing, education)
RICHMOND HOLO (Technology transfer)
RICHTER ENTERPRISES (optics, lab equip)
ROBERT SHERWOOD HOLO (artistic, marketing)
ROBINSON HOLO (lighting)
ROLLS ·ROYCE PLC (NDT)
ROYAL PHOTOGRAPHIC SOC (artistic)
SEE 3 (exhibits, embossing)
SPACE AGE DESIGNS (marketing)
SPOT (marketing)
SWEDE HOLO (marketing)
TEXTILE GRAPHICS (embossing/clothing)
U.K. GOLD (marketing)
VNIKTK KULTURA (marketing)
WAVEFRONTS (marketing)
WAVE MECHANICS (art)
WHITE TIGER HOLO (commercial finishing)
WHOLE HOGRAPHY (silver halide)
3D MEDIA (art)

13

Individuals

A

ABENDROTH, Detlev. President, AKS Holographie Galerie GmbH. Potsdamer Strasse 10, 4300 Essen I, Federal Republic of Germany.

ABOUCAR, Nathalie. Vice President, The Foreign Dimension Ltd., Suite B, 8th Floor, Central Mansion, 270-276 Queen's Road Central, Hong Kong. Telephone: (852) 542- 0282. FAX: (852) 541-6011.

ACKROYD, A. Customer Service, Amblehurst Ltd., 52 Invincible Road, Farnborough, Hants GU14 7QU, England, United Kingdom: Telephone: (44) 0252 520052. FAX: (44) 0252 373871.

ADAM, Fani. Contact for Multi-Purpose Holograms. 6 rue de l'orme, 75019 Paris France. Telephone: (331) 42009866. FAX: (331) 47434744.

ADAMI, Catlin. Customer Service, AH Prismatic, 1792 Haight Street, San Francisco, CA94117.

AGEHALL, Christer. Contact for High Tech Network, Skeppsbron 2, S-211 20 Malmo, Sweden. Telephone: (460(040) 350750. FAX: (46)(040) 237667.

AINSLEY-BROWN, Stuart. President, Raven Holographics Ltd, Old Saw Mills, Nyewood, Near Petersfield, Hampshire, GU3 15HX, England, United Kingdom. Telephone: 0730-821612. FAX: 0730-821260.

AITES, Edward. Contact for Aites Lightworks. 2148 North 86th Street, Seattle, WA 98103 USA. Telephone: (206) 526 5752.

AKBARI, Homaira. Contact for The Johns Hopkins University. Dept of Physics and Astronomy, Baltimore, MD 21218 USA. Telephone: (301) 3387385.

ALBRECHT, Gerd. Contact for Phantastica, Hohenweg 14, 5760 Amsberg I, Germany. Telephone: (49) 2932-81917. FAX: (49) 2932-29441.

ALEXANDER. 1323 14th Street, Apt. L, Santa Monica, CA 90404 USA. Telephone: (213) 393 9846. FAX: (213) 451

5291. Deacription: Holographic artist.

ALLEN, Jeffrey. Established in 1970. Contact for Envision Enterprises, P.O. Box 2003, Sausalito, CA 94965 USA. Telephone: (415) 459 8232. Deacription: All types, full service consulting. Twenty years experience, including direct involvement with the first dichromate and embossed holograms. Designed,produced,manufactured & marketed: fine art multiplexes, environments, jewelry, gifts, etc.

ALLEN, Keith. Contact for Imaging & Design. 1101 Ransom Road, Grand Island, NY 14072-1459 USA. Telephone: (716) 773 7272.

ALLON, Gerard. Holographic srtist. 4446 Saint-Laurent Blvd., Montreal, Canada H2W lZ8. Telephone: (514) 289-9841. Also at 5 Louis Pasteur St., Jaffa Tel-Aviv, Israel. (973) 3 830-538. Deacription: Studio: pulsed and CWo Origination on silver halide (mastering), pulsed. Large editions, limited editions. Advertising applications. Holoposter, portrait service, consultancy.

ALTON, Mary. Coordinator, Photon League. 110 Sudbury Street-Unit B, Toronto, Ontario M6J 1A7 Canada. Telephone: (416) 531 7087. FAX: (416)536-4291

ALVA, Edwina. Contact for Multi-Purpose Holograms. 6 rue de l'onne, 76019 Paris France. Telephone: (331) 42009866. FAX: (331) 47434744.

ANAIT. 1685 Fernald Point Lane, Santa Barbara, CA 93108 USA. Telephone: (805) 969 5666. Contact: Anait Stephens. Description: Artist in hands-on holography; reflection holography; portraiture; commissions.

ANDERSEN, Chad. Contact for Meredith

Instrumenta. 6403 North 59th Avenue, Glendale, AZ 85301 USA.

ANDERSON, Michael & Susan. President, Vice-President, Holomex. 4 Borrowdale Avenue, Harrow HA3 7PZ, England, United Kingdom. Telephone: (44)(01) 4279685.

ANDERSON, Steve. Proprietor, ArtKitek.122 Myrtle Avenue, Cotati, CA 94931 USA.

ANDREWS, John. Contact for Advanced Holographies, LTD. 243 Lower Mortlake Rd.,Unit 11, Richmond, Surrey TW9 2LL, England, United Kingdom.

ANDREWS, Matthew, M.A. (RCA). 7A Brunswick Park, Camberwell, London, SE5 7RH, England, United Kingdom, Telephone: 703 1254. Description: Freelance Holographer

ANGELSKY, Oleg V. Contact for Chernovtsy State University, 2 Kotsyubinsky Str., 274012 Chernovtsy, USSR.

ANTHONY, S. Lee. Vice-President, Cambridge Stereographies Group. P.O. Box 159, Kendall Square Station, Cambridge, MA 02142-0002 USA.

ANZENBERGER, T. Sales Manager, Steinbichler Optotechnik GmbH. Am Bauhof 4, D-8201 Neubeuern, FRG Federal Republic of Germany. Telephone: (0049) 8035 1018. FAX: (0049) 80351010.

APRILE, Eva, Sylvia. President, Vice-President, Metamorphosi Olografia Italia SRL. Via Lecco 6, 20124 Milano, Italy. Telephone: (39)(2) 204 9943. FAX: (39)(2) 204 1625.

ARANDA, Frank. Customer Service, Newport Corporation, 18235 Mt. Baldy Circle, P.O. Box 8020, Fountain Valley, CA 92708-8020 USA. Telephone: (714) 963 9811. FAX: (714) 963 2015.

ARKIN, R. Contact for Holo-Spectra. 7742-B Gloria Avenue,Van Nuys, CA 91406 USA. Telephone: (818) 994 9577. FAX: (818) 994 4709.

ARMSTRONG, Frank. Sales Manager, Aerotech Inc., Electro Optical Division, 101 Zeta Drive, Pittsburgh, PA 15238 USA.

ATTARDI, Luigi. Contact, Societa Olografica Italia. Via degli Eugenii 23, 00178 Roma, Italy. Telephone: (39)(6) 7180976 FAX: (39) (6) 7185172.

AUSTIN, M. Contact for Hickmott & Austin Holograms, 11 Castelnau, London SW13 9RP, England,United Kingdom. Telephone: (44)(01) 486 5811.

AVKSENTJEV, German B. Director, Soviskusstva vlo mexhdunarodnaya Kniga. Art holography department, 141120 Tryazino, Moscow region, USSR. Telephone: (7) 238 4600. FAX: (7) 230 2117.

AZZAZY, Dr. Medhat. Contact for Spectron Development Laboratories, 3303 Harbor Blvd., Suite 0.3, Costa Mesa, CA 92626. Telephone: (714) 549 8477.

B

BACK, Johnathan. Contact for Holosystems Inc.P.O.Box 6810, Ithaca, NY 14850 USA. Telephone: (607) 2731187.

BAG DAN, Szilvia. Customer Service, Artplay Holographic Studio. H-1191 Budapest, Ady Endre ut. 8, Hungary.

BAGHDADI, A.M. Vice-President, Wonders of Holography Gallery, P.O. Box 1244, Jeddah 21431, Saudi Arabia. Telephone: (966)(2) 652 0052. FAX: (966)(2) 651 1325.

BAGLEY, Sheila. Customer Service, A.H. Prismatic, LTD. New England House, New England Street, Brighton, BN1 4GH England, United Kingdom. Telephone: (44) 0273 686 966. FAX: (44) 0273676692.

BAILEY, Jean. Contact for Architectural Glass & Holography, 1 South Street, Great Waltham, Chelmsford, Essex CM3 1 OF, England, United Kingdom.

BAINS, Sunny. Contact for Holographies International, BCM Holographies, London WC1N 3XX, England, United Kingdom. Telephone: (44)(01) 584 4508.

BALLARD, Marie Vice-President, American Laser Corporation, 1832 South 3850 West, Salt Lake City, Utah 84104 USA. Telephone: (801) 972-1311. FAX: (801) 972-5251.

BALOGH, Tibor. President, Artplay Holographic Studio. H-1191 Budapest, Ady Endre ut. 8, Hungary.

BANCK-HAMMERUM, Knud. GM, Superior Technology Implementation. Hjortekaersvej 99B, 2800 Lyngby, Denmark. Telephone: (45) (45) 933 358. FAX: (45)(45) 930 353.

BANFANTI, F. Sales Manager, CISE SpA Technologie I nnova tive. via Reggio Emilia, 39, 20090 Segrate, Milano, Italy. Mailing address: P.O. Box 12081,1-20134 Milano, Italy.

BANKS, Peter M. Contact for The University of Michigan, College of Engineering holographic interferometry. Telephone: (313) 763 5464.

BAR, Edgar. Contact for Holo.Service, Neuensteinerstrasse 19, CH-4153 Basel, Switzerland. Telephone: (41)(061) 502287.

BARATON, Alan. Contact for Holocom Holographie. 13, rue Charles V., Facul te des Sciences et des Techniques, F- Paris, France. Telephone: (33)(1) 948) 04 0058.

BARKER, Graham. Contact for Architectural Glass & Holography, 1 South Street, Great Waltham, Chelmsford, Essex CM3 1 OF, England, United Kingdom.

BARKIN, Wendy. Vice-President, Point of View Dimensions, Ltd. 45-2903 River Drive South, Jersey City, NJ 07310 USA. Telephone: (201) 6268844.

BARNES, Don. Sales Manager, Lenox Laser.1 Green Glade Court, Phoenix, MD 21131 USA. Telephone: (301) 592-3106. FAX: (301) 592-3362.

BARNEY, Lynn. President, Ion Laser Technology Inc. 263 Jimmy DolittJe Road, Salt Lake City, UT 84116 USA. Telephone: (801) 537 1587. FAX: (801)5371590.

BARNHART, Donald. Contact for Holoflex Company, RR 3, Box 381, Urbana, IL 61801 USA. Telephone: (217) 684 2321.

BARRE, Pascal. Contact for Holos Art Galerie. 4 Place Grenus, 1201 Geneva, Switzerland.

BASLER, Dieter. Contact for Holtronic. Melchior-Huber Strasse 25, 0-8011 Ottersberg, Post P1iening, Federal Republic of Germany. Telephone: (49X08121) 81005.

BASSIN, Jim and Barry. Contacts for Infrared Optical, 120 Secatogue Ave, P.O. 292, Farming-dale, NY 11735 USA. Telephone: (516) 694-6035. FAX: (516) 694-6049.

BAUCCIO, Candice. Sales Manager, Melles Griot,1770 Kettering Street, Irvine, CA 92714 USA Telephone: (714) 261-5600. FAX: (714) 261-7589.

BAUGN, Drurey. Contact for James River Products. 5420 Distributor Drive, Richmond, VA 23225 USA. Telephone: (804) 233 9145. F~ (804) 231 7891

BAZARGAN' Kaveh. Contact for Focal Image Ltd. P.O.Box 1916, 1 Kelvin Court, London W11 3QR, England, United Kingdom. Telephone: (44) (071) 229 0107. FAX: (44)(071) 727 3438.

BEAUREGARD, Alain. President, Lasiris Inc. 3549 Ashby, Ville St. Laurent, Qu~, Canada H4R 2K3. Telephone: (514) 335 1005. FAX: (514) 335 4576

BEECK, Manfred-Andreas. Contact for Volkswagen AG. Forschung Messtechnik, Postfach 0-3180, Wolfsburg 1, Federal Republic of Germany.

BEKKER, A.M. Contact for Institute of Nuclear Physies, Leningradska obI., 188350 Gatchina, USSR.

BELK, Joseph. Holography consultant. Telephone: (415) 776-0581. 2416 Polk Street, San Francisco, CA 94109 USA.

BENTON, Stephen. Contact for Massachusetts Institute of Technology. M.I.T. Media Laboratory, Spatial Imaging Group, 20 Ames Street, El5-416, Cambridge, MA 02139 USA.

BENYON, Margaret. Established in 1968. Holography Studio, 40 Springdale Avenue, Broadstone, Dorset, BH18 9EU, England United Kingdom. Telephone: (44)(0202) 698 067. FAX: (44) (0202) 694161. Description: Independent holographic artist.

BERG, Marcia. Contact for Hughes Aircraft Co. Laser Products. 6155 EI Camino Real, Carls-bad, CA 92008 USA. Telephone: (619) 9313252.

BERKHOUT, Rudie. 223 West 21st Street, Apt. B, New York, NY 10011. Telephone: (212) 255 7569. Description: Artist.

BERKOWITZ, Vic. Vice-President, Sillcocks Plastics International. 310 Snyder Avenue, Berkeley Heights, NJ 07922 USA. Telephone: (800) 9220958. FAX: (201) 6651856.

BERNARD, B. Sales Manager, Laser Applications, Inc. 12722 Research Parkway, Orlando, FL 32826 USA. Thlephone: (407) 380 3200. FAX: (407) 3819020.

BERNIER, Jean-Robert. President, Holocor I.B.F. Printing Inc. 95 des Sulpiciens, L'Epiphanie, Quebec JOK 1JO, Canada.

BERSON, Martin. President, Holaxis Corporation. 499 Farmington Ave, Hartford, CT 06105 USA. Thlephone: (203) 232 2030. FAX: (203) 236-3767.

BILLINGS, R. or L. Contact for Museum of Holography/Chicago; School of Holography/ Chicago. 1134 West Washington Boulevard, Chicago, IL 60607 USA. Telephone: (312) 829 2292.

BIRCH, C. Customer Service, Kendall Hyde Ltd. Kingsland Industrial Park, Stroudley Road, Basingstoke, Hants., RG24 OUG, England, United Kingdom. Telephone: (44) 0256 840 830. FAX: (44)0256 840 443.

BJELKHAGEN, Dr. Hans. Contsct for Holicon Corporation. 906 University Place, Evanston, IL 60201 USA; Northwestern University, Biomedical Engineering, Evanston, IL 60208 USA. Thlephone: (708) 491 4310. FAX: (708) 491 7955.

BLECKWENDT, Werner. Contact for Uniphase Vetreibs GmbH. LeiBstr. 8, 8152 Feldkirchen-Westerham, Federal Republic of Germany. Telephone: (49)0806319036. FAX: (49) 08063n663.

BLYTH, Jeff. Contact for Brighton Imagecraft. 7 Bath Street, Brighton, East Sussex, BN1 3TB, England, United Kingdom. Telephone: (44)(0273) 202 069.

BOBECK, Paula. Contact for E.I. DuPont De Nemours & Company, Optical Element Venture, Experimental Station Laboratory, P.O. Box 80352, Wilmington, DE 19880-0352 USA. Thlephone: (302) 695 4893. FAX: (302)695 9631.

BOHAN, Brian. Contact for Cambridge Lasers, Inc., 43216 Christy Street, Fremont, CA 94538 USA. Thlephone: (415) 651-0110. FAX: (415) 651-1690.

BOHLMANN, Claus-Peter. President of KolbeDruck, Coloco GmbH & Co. KG. 1m. Industrigell1nde 50, Postfach 1103, D-4804 Versmold, Federal Republic of Germany. Thlephone: (05423 2431 (-5). FAX: 05423 41230.

BOISSONNET, Philippe. 5294 De L'Esplanade, Montreal H2T 2Z5 Quebec, Canada. (514) 270-1840. Description: holographic artist, limited editions =and multi-media installlations.

BONFANTI, F. Customer Service, CISE SPA Thehnologie Innovative, Reggio Emilia 39, 20090 Segrate, Milano, Italy. Thlephone: (39) (2) 2167 2634. FAX: (39)(2) 2167 2620.

BONNELL, Jean. MarketinglLaser Sales, 1832 South 3850 West, Salt Lake City, Utah 84104. Thlephone: (801) 972-1311. FAX: (801) 972-5251.

BOONE, Pierre. Contact for University Gent, Workshop Holography. 41 St. Pieters Nieuwstraat, B 9000, Gent, Belgium. Telephone: (32) (91) 626384. FAX: (32) (91) 237326.

BOTOS, Stephen A. President, Aerotech Inc., Electro Optical Division, 101 Zeta Drive, Pittsburgh, PA 15238 USA. Thlephone: (412) 963-7470, FAX: (412) 963-7459,

BOULTON, J.B.Contact for Hologram Europe sprl. Avenue Voltaire 137, 1030 Brussels, Belgium Telephone: (32)(2) 242 7284.

BOUSIGUE, Jacques. Contact for IDHOL. Boite Postale 7, F. 89340 Saint-Agnan, France Thlephone: (33)(16) 8696 1929.

BOWMAN, Jim. Owner, L.A.S.E.R. Co. 1900 Grove Drive, Haymarket, VA 22069 USA. Telephone: (703) 754 2526.

BOXALL, Ossie. President, Applied Holographics PLC, Braxted Park, Great Braxted, Malden, Essex CM8 3XB, England, United Kingdom.

BOYD, Patrick. Established in 1985. 1 Employee at this address. 18 Whiteley Road, London SE19 1JT, England United Kingdom. Telephone: (44)(081) 670 4160. FAX: (44)(71) 284 0437. Parent company: Space Time Holographic Applications. Description: Creative holographer specialising in large fonnat pulsed holography of fashion imagery (commissions for Zandra Rhodes in London & Gallerias Preciados in Madrid) United States Representative: Elusive Image, Dallas, TX.

BRADSHAW, Roy. Owner, Reel Image, P.O. Box 566, Pacifica, CA 94044 USA. Thlephone: (415) 355 8897.

BRAGINTON, Mary. Contact, Coulter Optical Company. P.O. Box K Dept. MP, 54121 Pinecrest Road, Idyllwild, CA 92349 USA. Thlephone: (714) 6592991.

BRANDIGER, F.J. Vice-President, CVI Laser Corporation, CVI West. 470 Lindbergh Avenue, Livermore, CA 94550 USA. Thlephone: (415) 449-1064.

BRANDTJEN , Henry A., Hank A. President, Vice-President, Brandtjen & Kluge, Inc, 539 Blanding Woods Road, St. Croix Falls, WI 54024 USA. Thlephone: (715) 483 3265. FAX: (715) 483-1640.

BRECHEISEN, Robert. Sales Manager, Polaroid Corporation, 730 Main Street-lA, Cambridge, MA 02139 USA.

BRILL, Louis. Director, Wavefronts. 2428 Judah Street, San Francisco, CA 94122 USA. Telephone: (415) 664 0694.

BROADBENT, Donald. Owner, Optical Images. 1309 Simpson Way, Suite J, Escondido, CA 92025 USA. Telephone: (619) 746 0976. FAX: (619) 746 6141.

BRODEL, D. Contact for Ascot Laser Picture Studio, 27 Upper Village Road, Sunninghill, Ascot, Berkshire SL5 7AJ England, United Kingdom. Telephone: (202) 667 6322. FAX: (202) 861 0621.

BROEDERS, Jan M. Owner, Optische Fenomenon. Nederlandse Stichting Voor, Waameming & Holografie, Warenarburg 44, NL 2907 CL

Capelle aid Ijssel, The Netherlands.

BROWN, Anthony. Sales Manager, Spectra-Physics, Boundary Way, Hemel Hempstead, Herta. HP2 7SH, England, United Kingdom. Thlephone: 0442232 322. FAX: 0442 68538.

BROWN, George. Contact for Barr and Stroud Ltd., Caxton Street, Anniesland, Glascow, Scotland G13 1HZ, UK. Thlephone: (44)(41) 954 9601

BROWN, Gordon. Contact for Ford Scientific Labs. 2000 Rotunda Drive, Dearborn, MI 48121 USA. Thlephone: (313) 323 1539.

BROWN, K.C. Contact for K.C. Brown Holographics. 22 St. Augustine's Road, Camden Town, London NW1 9RN, England, United Kingdom. Telephone: (44)(1) 482 2833.

BROWN, Kerry J. Contact for Artigliography Co. 7130 Mohawk West Drive, Indianapolis, IN 46236 USA; holographic artist.

BROWN, Kevin. Contact for Holographic Dimensions, Inc., 9235 SW 179 Thrrace, Miami, FL 33157 USA. Thlephone: (305) 255 4247.

BRUCK, Richard. Owner, Richard Bruck Holography. 33i2 West Belle Plaine 112, Chicago, IL 60618 USA. Thlephone: (312) 267 9288.

BRYNGDAL, Olof. Contact for University of Essen. Pherdemarkt 4, D-4300 Essen, Federal Republic of Germany.

BUELL, D. Marketing Department, Applied Holographies Corp., 1721 Fiske Place, Oxnard, CA 93033 USA. Thlephone: (805) 385-5670. FAX: (805) 385-5671.

BUNTS, FrankIFlatiron Studio. 15 West 24th Street 7th Floor, New York, NY 10010 USA. Thlephone: (212) 645 5173. Company description: Fine Art using interference patterns to produce a sense of depth and movement. - Work in the collections of The Museum of Holography, N.Y.; The Philadelphia Museum of Art; The Library of Congress.

BURCH, J.M. Contact for College ofManufacturing, Cranfield Institute of Technology, Cranfield, Bedford MK43 OAL, England,United Kingdom.

BURGER, Philippe. Directeur Technique, Micraudel. 93, rue AdelshofTen, F-67300 Schiltigheim, France. Thlephone: (33)88813293.

BURGMER, Brigitte. Contact, Deutsche Gesellschaft fUr Holografie e.V. LerchenstraBe 142 a, D-4500 Osnabr!lck, Federal Republic of Germany.

BURNS, Joseph. Contact for Burns Holographics, P.O. Box 377, Locust Valley, NY 11560 USA .. Description: Holographic artist. Thlephone: (516) 6743130.

BURRIDGE, Mike. Contact for Perception Holography. Thornton Marketing Ltd. Aketon Close, Haggs Lane, Follifoot, Harrogate, North Yorks., England HG3 1A2 United Kingdom. Thlephone: (44)(93) 782 323.

BUSSARD, Jan. Contact for Textile Graphics, Inc., 201 North Fruitport Road, Spring Lake, MI 49456 USA. Thlephone: (616) 837 8048.

BUSSAUT, Laurent. Research Director, Art, Science and Technology Institute--Holography Society. 2018 R Street, N.W.,Washington D.C. 20009

USA_ Telephone: (202) 667 6322_ FAX: (202) 861 0621.

BUTTERISS, Tony_ Contact for New Dimension Holographies_ 65-72 Pier One, Hickson Road, Sydney, New South Wales 2000, Australia. Telephone: (61)(2) 247 6063. FAX: (61) 2 419 8670.

BYRNE, Kenneth G. 77 8th Avenue, Brooklyn, NY 11215 USA. Telephone: (718) 789-5854. Description: Holographic artist.

C

CANTOS, Brad D. Contact for Holage. 1881 Eighth Avenue, San Francisco, CA 94122 USA. Telephone: (415) 664 1840.

CAPUCCI, Pier Luigi. Established in 1984. 1 Employee at this address. University of Bologna, via Fiumazzo 347,1-48010 Belricetto (RA), Italy. Telephone: (39) 0545 77 296. Contact: Dr. Pier Luigi Capucci, Visual Communications. Description: My services are : Art and communication exhibitions curator, consultant and critic. Expert in display/ embossed commercial images for subject, structure, disposition in the final setting. Media languages expert for the best image making/ employing.

CARLSSON, Torgny E. Vickervagen 4, 14569 Norsborg, Sweden. Telephone: (46) (753) 85659. Description: Holographic artist.

CARLTON, David. 7153 Cavalry Drive, Warrenton VA 22186 USA. Telephone: (703)361 9443. Description: Artist.

CAROLA, Maggie. Customer Service, Crown Roll Leaf, Inc. 91 Illinois Ave., Paterson, NJ 07503 USA.

CASDIN-SILVER, Harriet. Contact, Casdin-Silver Holography. 99 Pond Avenue, Suite D403, Brookline, MA 02146 USA. Telephone: (617) 4234717.

CASE, Steven. Contact for University of Minnesota, 4-174 Electrical Engineering/Computer Science Building, 200 Union Street, S.E., Minneapolis, MN 55455 USA.

CASSEN, J.Jay. President, Diversified Graphics, Ltd. 5433 Eagle Industrial Court, Hazelwood, MO 63042 USA.

CAULFIELD, John. Contact for University of Alabama in Huntsville, Center for Applied Optics, Huntsville, AL 35899 USA. Telephone: (205) 895 6030.

CAUSSIGNAC, Jean-Marie. Contact for LCPC- Lab Central des Ponts et Chaussees. 58 Boulevard Lefebvre, F-75015 Paris, France. Telephone: (33)(l) 4532 3179.

CHALK, Kenneth. Contact for Holomorph Visuals, Inc. 273 de la Gauchetiere W., Montreal, Quebec H27 lC7 Canada. Telephone: (819) 872-3622.

CHANG, B.J. Contact for Kaiser Optical Systema, Inc. P.O. Box 983, 371 Parkland Plaza, Ann Albor, MI 48106 USA. Telephone: (313) 665 8083

CHANNER, David. Vice-President, Images Company. P.O. Box 313, Jamaica, NY 11419

USA. Telephone: (718) 706-5003.

CHAPPELLE, Mike. Contact for National Holographic Studios, Technology Development Center, Kresson-Gibbsboro Road, Voorhees, NJ 08043.

CHEN, David. Vice-President, Acme Holography, 12 Sunset Road, West Somerville, MA, Boston Area. USA.

CHERRY, Greg. Contact, Cherry Optical Company, 2047 Blucher Valley Road, Sebastopol, CA 95472 USA. Telephone: (517) 774 3025.

CHIA, Ching-Wat. Divisional Manager for Electech Distribution Systema, PTE. Ltd. 605- A, MacPherson Road, #03-03, Citimac Industrial Complex, Singapore 1336. Telephone: (65)2869933. FAX: (65) 284 3256.

CHI'ITIM, Kevin. Vice-President, Melles Griot,1770 Kettering Street, Irvine, CA 92714 USA Telephone: (714) 261-5600. FAX: (714) 261-7589.

CHOUDRY, Amar. Contact for University of Alabama in Huntsville, Center for Applied Opties, Huntsville, AL 35899 USA.

CHRISTAKIS, Anne-Marie. Manager, Magic Laser. Quartier de L'horloge, 4 rue Branttlme, 75003, Paris France; Manager, Musee de L'holographie. Telephone: (33)(l) 4274 3357.

CHRISTE-WILTON, Lars. Contsct for High Tech Network, Skeppsbron 2, 8-211 20 Malmo, Sweden.

CHRISTIE, Robin. Sales Manager, The DZ Company. P.O. Box 5047-R, 181 Mayhew Way, Suite E, Walnut Creek, CA 94596 USA.

CIFELLI, Dan. President, The DZ Company. P.O. Box 5047-R, 181 Mayhew Way, Suite E, Walnut Creek, CA 94596 USA. Telephone: (415) 9354856. FAX: (415) 9354660.

CINKOW, Sheryl. Contact for Three-D Light Gallery, 107 The Commons, Ithaca, NY 14850. Telephone: (607) 273 1187.

CLARE, G.W. Chairman, Touchwood Holographies, 50 Sugworth Lane, Radley, Abingdon, Oxon, OX14 2HY, England, United Kingdom. Telephone: 0865 735874.

CLARKE, Walter. President, Global Images, Inc. 509 Madison Avenue, Suite 1400, New York, NY 10022 USA. Telephone: (212) 759 8606. FAX: (604) 734 2842.

CLASSEN, Walter. Contact for Licht-Blicke-Buro. Bornemannstrasse 10, D-6000 Frankfurt 70,Federal Republic of Germany.

CLAUS, Patrick. Customer Service, Micraudel. 93, rue Adelshoffen, F-673oo Schiltigheim, France. Telephone: (33) 8881 3293.

CLAY, Burton. Contact for Holographix Inc. 87 Second Avenue, Burlington, MA 01803 USA.

CLAYTOR, Linda H. Contact for Fresnel Technologies Inc. 101 West Morningside Drive, Fort Worth, TX 76110 USA. Telephone: (817) 926 7474. FAX: (817) 9267146.

COBLIJN, Alexander. Owner, Que Sera Sera. P.O. Box 29, 9700 AA, Groningen, The Netherlands.

COBURN, Joe. President, Coburn Corporation. 1650 Corporate Road West, Lakewood, NJ

08701 USA. Telephone: (908) 367-5511. FAX: (908) 367-2908.

COHN, Dr. Gerald. Contact for Abbott Laboratories. Department 93F, (Building AP-9), Routes 43 and 137, Abbott Park, IL 60064 USA.

COLLIER, Richard G. Contact for Ion Laser Technology Inc. 263 Jimmy Dolittle Road, Salt Lake City, UT 84116 USA. Telephone: (801) 537 1587. FAX: (801) 5371590.

COLLINS, Jonathan. Vice-President, Atelier Holographique de Paris, 13, Pa88age Courtois, F-75011 Paris, France. Telephone: (33)(l) 43 79 6918.

COMINELLI, Francesca. Customer Service Metamorphosi Olografia Italia SRL. Via Lecco 6, 20124 Milano, Italy. Telephone: (39)(2) 204 9943. FAX: (39)(2) 204 1625.

CONLAN, J. Contact for Ashland Electric Products-Visual Electronies Corp., 8O-39th Street, Brooklyn, NY11232 USA. Telephone: (718) 855 3319.

CONNER, Arlie. 1514 South East Salmon, Portland, OR 97214 USA. Telephone: (503) 239 0545. Description: Research and development, large display format, artistic and commi88ions. Expert in liquid crystal devices as well.

CONNORS, Betsy. Contact for Acme Holography. 12 Sunset Road, Somerville, MA 02144 USA.

CONTE, Linda. Contact for Massachusetts Institute of Technology, MIT Media Laboratory, Spatial Imaging Group, 20 Ames Street, El5-421, Cambridge, MA 02139.

CONWAY, Terry. Contact for Spectratek Corporation. 1510 Cotner Avenue, Los Angeles, CA 90025 USA. Telephone (213) 473-4966. FAX: (213) 477-6710.

COOK, Richard. Contact for Lazap Inc., 2614 N. Sweetbriar Drive, Claremont, CA 91711, USA. Telephone: (714) 624-9923.

COOKE, D.J. Contact for Oxford University. Department of Engineering Science, Parks Road, Oxford OX1 3PJ England, United Kingdom. Telephone: (44) (0865) 273 805.

COOMBES, Angela. Contact for Dovecote Studio, Witham Friary, Frome, Somerset , England, United Kingdom. Telephone: (44)(74) 985 691

COOPER, Nick. Contact for Oxford Holographics. 71 High Street, Oxford OX1 4BA, England, United Kingdom. Telephone: (44)0865 250 505. FAX: (44) 0865 790 353.

CORNELIUS, Mark. Customer Service, Advanced Holographies Corp. 2469 East Fort Union Blvd. Suite 108, Salt Lake City, UT 84121 USA.

COSSETTE, Marie Andree. Artistic holographer. 1145 Avenue des Laurentides, Apt 2. Quebec, Quebec, GIS 3C2, CANADA. Telephone: (418) 687-2985. FAX: (418) 656-7305. Description: One-offsllimited editions: continuous wave. Holographic exhibitions: private gallery, permanent exhibition. Holographic education:

private tutoring, specialist college course. Research: artistic, academic, educational. Studio rental; consultancy.

COWLES, Susan Ann. Educational Director, Museum of Holography, 11 Mercer Street, NY, NY 10013. Telephone: (212) 925 0581 or 925-0526. FAX: (212) 334-8039.

CRANFORD, Robert. Contact for Infotech International, Holography Division, 3607 West Magnolia Blvd., Suite 2, Burbank, CA 91505 USA. Telephone: (818) 845-7997. FAX: (818) 845-0312.

CRAFT, Steven. Contact for Laser Labs, Inc. 8000 W.ll0th Street, Suite U15, Overland Park, KS 66210 USA. Telephone: (913) 383 7639.

CRAWFORD, Michael E. Contact for Whole Hography, 4142 Bellefontaine Street, Houston, TX 77025-1105 USA. Telephone: (713) 667 3325.

CRENSHAW, Melissa. Contact for The Holographic Studio, Ltd. 2525 York Avenue, Vancouver, British Columbia V6K lE4 Canada. Telephone: (604) 7341614. FAX: (604) 734 2842.

CRIST, William. Contact fOT Holocrall International, P.O. Box 152, Lake Forest, IL, 60045-0152 USA. Telephone: (708) 234-7625. FAX: (708) 615-0835.

CROSBY, Paul. Sales & Marketing Dir. ,Coherent. 3210 Porter Drive, Palo Alto, CA 94306 USA. Telephone: (415) 493-2111. FAX: (415)858-7631.

CROSS, Lloyd. P.O. Box 145, Point Arena, CA 95468 USA. Description: Holography Consultant

CRUMLEY, Elizabeth. Contact for F.A.S.T. Electronic Bulletin Board, P.O. Box 421704, San Francisco, CA 94142-1704 USA. Telephone: (415) 845 8306. FAX: (415) 8416311.

CUCHE, Denis. Contact for Ciba-Geigy AG. Research Center P&A, CH-1701 Fribourg, Switzerland.

CUELI, Manuel. Vice-President, Crown Roll Leaf, Inc. 91 Illinois Ave., Paterson, NJ 07503 USA. Telephone: (201) 742-4000. FAX: (201) 742-0219.

CULLEN, Ralph. Director, UK Optical Supplies; Ralph Cullen HolographieslUKOS. 84 Wimborne Road West, Wimborne, Dorset BH21 2DP, England, United Kingdom. Telephone: (44)(202) 886 831. FAX: (44)(202) 742236.

CULLEN, Gary & Karoline. President, Managing Director, Holographic Developments LTD. Box 1035, Delta, B.C., V4M 3T2 Canada. Telephone: (604) 948-1926. FAX: (604) 948-1648.

CURCI, J. Vice-President, Jaeger Graphic Technology, J.G.T- Holofoil S.A. 20 Avenue des Desirs,B-114O Bru.ssels, Belgium Telephone: 00-322-7359551. FAX: 733 1035.

CUSSEN, Arthur J . President, Electro Optical Industries, Inc. 859 Ward Drive, Santa Barbara, CA 93111 USA. Telephone: (805) 964 6701, sales ext. 280. FAX: (805) 967 8590.

CVETKOVICH, Thomas J . Owner, Chromagem. 573 South Schenley, Youngstown, OH 44509 USA. Telephone: (216) 799 0323. FAX: (216) 747 9371.

D

DA-HSIUNG, Hsu. Contact: Beijing institute of posts and Telecommunications. Department of Applied Physies, Holography Laboratory, Beijing 10080, People's Republic of China.

DAINTY, J. Contact for Imperial College of Science. Opties Section, Blackett Laboratory, London SW7 2BZ, England, United Kingdom. Telephone: (44)(71) 589 5111.

D'AMATO, Salvatore, President. American Bank Note Holographies, Inc. 500 Executive Boulevard, Elmsford, New York 10523 USA. Telephone: (914) 592-2355. FAX: (914) 5923248.

DANDLIKER, Rene. Contact for Universite de Neuchatel, Institut de Microtechnique, 2, rue A L Breguet, CH-2ooo Neuchatel, Switzerland. Telephone: (41)(038) 246 000.

DARCY, L. Customer Service, Laser Electronies Pty., Ltd. P.O. Box 359, Southport, Queensland, 4215, Australia. Telephone: 61 75 53 2066. FAX: 61 7553 3090.

DAVIS, Frank. President, Angstrom Industries, Inc. 3202 Argonne, Houston, TX 77098 USA. Telephone: (713) 526 0006.

DAYUS, Ian. Sales Manager, A.H. Prismatic, LTD. New England House, New England Street, Brighton, BNl 4GH England, United Kingdom. Telephone: (44) 0273 686 966. FAX: (44) 0273676 692.

D'CRUZ, A. Wasy. Director, E.I. Dupont De Nemours & Co. Inc Optical Element Venture, Experimental Station, P.O.Box 80352, Wilmington, DE 19880-0352 USA. Telephone: (302) 695 4893. FAX: (302)695 9631.

DeBITETTO, David. Contact for Holo-Images, 167 Washburn Road, Briarcliff Manor, NY 10510. Telephone: (914) 9418811.

DEEM, Rebecca. 709 112 West Glenoaks Blvd., Glendale, CA 93105. USA. Telephone and FAX: (818) 549-0534 Description: Fine art works incorporating holography-single sculptural and installation pieces combining reflection holograms with mixed media-limited edition wall mounted pieces-curatorial, catalog and lecture services.

DEFREITAS, Frank. Publisher, The Hologram. P.O. Box 9035, Allentown, PA 18105 USA. Telephone: (215) 434 8236. Description: Limited edition holograms; workshops; publishes "l'he HoloGram' newsletter (free); global networking of enthusiasts, collectors and practitioners; bibliographies and title searches (current, rare, out-ofprint books/P/Patentsl manuscripts).

DEGAIRE, Bridgette. Sales Manager, Hologram Industries, 42-44 Rue de Trucy, 92140 Fontenay Sous Bois, France. Telephone: 1 4394 1919. FAX: 143940032.

DEKENS, Joop. Contact for Qu~ Sera Sera, P.O. Box 29, 9700 AA, Groningen, The Netherlands. Telephone: (31)(050) 140417. FAX: (31) (050) 144142.

DELEPIERE, Marc. President, Micraudel. 93, rue Adelshoffen, F-67300 Schiltigheim, France. Telephone: (33)88813293.

DEL-PRIETE, Sandro. Contact, Gallerie III-

usoria, Schwarztorstrasse 70, CH-3007 Bern, Switzerland

DEMICHER, Charles. Contact for Laser Holographies, Inc. 1179 King St. West, Unit 111, 'lbronto, Ontario, M6K 3C5 Canada. Telephone: (416) 531 4656. FAX: (416) 229 6724.

DEMMER, Richard. President, Laser Ionies Inc. 701 South Kirkman Road, Orlando, FL 32811 USA. Telephone: (407) 298 1561. FAX: (407) 297 4167.

DENISYUK Yo. Contact for FTI Joffe, Politechnicheskaya 26, Academy of Sciences of the USSR, 194021 Leningrad, USSR.

DENIZOT, Stephane. Customer Service. The Foreign Dimension, Suite B, 8.1F1., Central Mansion, 270-276 Queen's Road Central, Hong Kong.

DENNIS, John. Editor, Stereo World Magazine. 5610 SE 71st Street, Portland, OR 97206 USA. Telephone: (503) 771-4440.

D'ENTREMONT, Joseph. President, Lenox Laser. 1 Green Glade Court, Phoenix, MD 21131 USA. Telephone: (301) 592-3106. FAX: (301) 592-3362.

DE ROOS, Marc. President, Deep Space Holographies, 1328 Dunsterville Avenue, Victoria, British Columbia, Canada V8Z 2Xl. Telephone: (1)(604) 384-3927.

DE SANTIS, Paulo. Contact for Universita di Roma--La Sapienza,. Dipartimento di Fisica, Piazzale Aldo Moro, 2, 1-00185 Roma, Italy.

DESCHAUMES, H. Sales Manager, Agfa Gevaert N.V. Holography Film Dept, Septestraat 27, B2510, Mortsel, Belgium. Telephone:(03) 444 2111. FAX: (03) 444 7094.

DEUTSCHMANN, Sven. Product ManageI' Holography, Kolbe-Druck, Coloco GmbH & Co. KG. 1m. Industrigellinde 50, Postfach 1103, D- 4804 Versmold, Federal Republic of Germany. Telephone: (05423 2431 (-5). FAX: 05423 41230.

DEWAR, David.39 Rue D'Oradour, Luxembourg 2266, Luxembourg. Description: Artist

DE WINNE, R. Product Manager, Agfa Gevaert N.V. Holography Film Dept, Septestraat 27, B2510, Mortsel, Belgium Telephone: (03) 444 2111. FAX: (03) 444 7094.

DE WOLF, E. President, Agfa Gevaert N.V.Holography Film Dept, Septestraat 27, B2510, Mortsel, Belgium Telephone:(03) 444 2111. FAX: (03) 444 7094.

DIKMAK, Lorraine. Sales Program Manager, Polaroid Corporation. Holography Business Group,730 Main Street-lA, Cambridge, MA 02139 USA. Telephone: (800) 237 5519 or (617) 577-4307. FAX: (617) 577 5434.

DICK, Jim. Contact for Myers Printing,Inc. 914 East Gier Street, Lansing, MI, 48906, USA. Telephone: (1) (800) 333-3645. FAX: (517) 482-0550.

DIDIER, Leduc. 152 rue de la Raquette, 75011 Paris, France. Telephone: 43-79-52-48. Description: Artist Independent

DIETRICH, Edward. 2036 West Haddon, Chicago, IL 60622 USA. Telephone: (312) 292 0770. Description: Consultant.

DIETZ, Andrew, AI. Vice-President, Sales Manager, Excitek, Inc. 277 Coit Street, Irvington, NJ 07111 USA. Telephone: (201) 372 1669. FAX: (201)372 8551.

DINSMORE, Sydney. Artist I Educator. The Holographic Studio. 2525 York Avenue #1, Vancouver, B.C. V6K 1E4, Canada. Telephone: (604) 734 1614. FAX: (604) 734 2842. Description: Specialize in multicolour reflection transfers from pulse masters. Extensive experience as curator and project co-ordinator for fine art holography.

DIRTOFT, Ingegard. Contact for Holometric AB, Bjornasvagen 21, 8-113 47 Stockholm, Sweden. Telephone: (46)(08) 790 9780.

DISHMAN, Rick. Contact for American Holographic. P.O. Box 1310, 521 Great Road, Littleton, MA 0 1460 USA. Telephone: (914) 592-2355. FAX: (914) 592 3248.

DOBRANYI, Zsuzsa. Vice-President, Artplay Holographic Studio. H-1191 Budapest, Ady Endre Ilt 8., Hungary.

DOMINGUEZ, Carmenza. Schweizerstrasse 77,6000 Frankfurt 70, Federal Republic ofGermany. Description: Artistic holographer.

DORAISWAMY, Krishna C. Marketing Manager, E.I. Dupont De Nemours & Co. Inc Optical Element Venture, Experimental Station, P.O.Box 80352, Wilmington, DE 19880-0352 USA. Telephone: (302) 695 4893. FAX: (302) 695 9631.

DOREe, R. Sales Manager, Jaeger Graphic Technology, J.G.T-Holofoil S.A. 20 Avenue des Desirs,B-114O Brussels, Belgium Telephone: 00-322-7359551. FAX: 733 1035.

DOTY, Jim. Sales Manager, Newport Corporation, 18235 Mt. Baldy Circle, P.O. Box 8020, Fountain Valley, CA 92708-8020 USA. Telephone: (714) 963 9811. FAX: (714)963 2015.

DOVGALENKO, Georgi. Contact for Novator Research Center, PI. Lesi Ukrainki I, Kiev-196, 252196 USSR. Telephone: (7X296)8048.

DOWLEY, M.W. President, LiCONjx, 3281 Scott Boulevard, Santa Clara, CA, 95054 USA.

DRAKE, Francis P. Program Director, University of Wisconsin, College of Engineering. 432 North Lake Street, Madison, WI 53706 USA. Telephone: (608) 2637427. FAX: (608) 263 3160.

DRINKWATER-WELCH, Deni. Contact for Wave Mechanics. 1535 North Ashland Avenue, Second Floor, Chicago, IL 60622 USA. Telephone; (312) 384 4860.

DRISCOLL, John. Contact for New York Hall Of Science. 47-01 11lth Street, Corona, NY 11368 USA Telephone: (718) 699 0005.

DROK, P. Vice-President, Optilas B.V. P.O. Box 222, 2400 AE Alphen, AID Rijn, The Netherlands

DRUCKER, Dennis & Harriet. President, VicePresident, Holograms and Other Strange Things_ 3200 West Oakland Park Boulevard, Lauderdale Lakes, FL 33311 USA. Telephone: (305) 739 9634.

DUBOV, Philip. 5401 Bee Cave Road .. Austin,

TX 78746 USA. Telephone: (512) 327 5961. Description: Artist; fine art holography.

DUDDERAR, Thomas. Contact for AT&T Bell Labs-Materials Group. 600 Mountain Avenue, Murray Hill, NJ 07974-2070 USA.

DUESTERBERG, Richard. Contact for Vincennes University. 1002 North First Street, Vincennes, IN 47591 USA. Telephone: (812) 885- 5294.

DUFFEY, William. Contact for Regal Press Inc., Regal Holographics Division, 129 Guild Street, Norwood, MA 02062 USA. Telephone: (617) 769 3900.

DUNNIGAN, Paul. Contact for Newcastle Upon Tyne Polytechnic, Department of Physics, Ellison Building, Newcastle upon Tyne, NE1 SST. England, United Kingdom. Telephone: (44)(091) 2358453. FAX: (44)(091) 235 8017.

DUSTON, Deborah A. President, Duston Holographic Services. 90 Sherbrooke Avenue, Ottawa, Ontario, K1Y lR9 Canada. Telephone: (613) 722-9004.

DUTTON, Keith. Contact for Laser International, 19 Normanton Rise, Holbeck Hill, Scarborough, N Yorks Y011 2XE, England, United Kingdom. Telephone: (44X0723) 366096.

DYENS, Georges M. 1293 Rue de la Visitation, Montreal, Quebec H2L 3B6 Canada. Telephone: (514) 598 8860. Description: Artistic holographer.

DYER, Jack. Managing Editor, Journal of Laser Applications, 4143 Merriweather Road, '!bledo, OH 43623 USA. Telephone: (419) 885 4803. FAX: (419) 885-5895.

E

EDWARDS, A. Vice-President, Kendall Hyde Ltd. Kingsland Industrial Park, Stroudley Road, Basingstoke, Hants., RG24 OUG, England, United Kingdom. Telephone: (44) 0256840 830. FAX: (44) 0256 840 443.

EIMERS, Tilman. Sales Manager, Phoenix Holograms. Trendelbuscher Weg #1, 2875 Ganderkesse 2, Federal Republic of Germany. Telephone: (49)044 89 15198

EIZYKMAN, Claudine. Contact, ElF Productions. Eizykman I Fihman. 19 Rue Jean Jacques Rousseau, F- 75001 Paris, France. Telephone: (33)(1) 4236 0631.

EMMONS, Larrimore B. Contact for Armstrong World Industries. P.O. Box 3511. Lancaster, PA 17604 USA.

ENGELSBERGER, J . Vice-President, Steinbichler Optotechnik GmbH. Am Bauhof 4, D- 8201 Neubeuern, FRG Federal Republic of-Germany. Telephone: (0049) 8035 1018. FAX: (0049) 80351010.

ENIVOI, Ruth. Sales Manager, Images Company. P.O. Box 313, Jamaica, NY 11419 USA. Telephone: (718) 706-5003.

ENNOS, Tony. Contact for U1trafin. 16 Foster Road, Chiswick, London W4 4NY, England, United Kingdom.

EPSTEIN, Max. President, Holicon Corporation, Inc. 3 Warwick Lane, Lincolnshire, IL 60069 USA; Northwestern University, Dept. of Electrical and Biomedical Engineering, Technological Institute, Evanston, IL 60208 USA. Telephone: (708) 491 4310. FAX: (708) 4917955.

EPSTEIN, Michael. Customer Service, Holographic Industries, Inc. 3 Warwick Lane, Lincolnshire, IL 60069 USA. Telephone: (708) 680 1884. FAX: (708) 680-0505.

ERF, Robert. Contact for United Technologies Research Center, Optics and Acoustics 86, Silver Lane, East Hartford, CT 06108 USA.

ERIKSON, Ron. Contact for Media Interface, Ltd. 167 Garfield Place, Brooklyn, NY 11215 USA

EVANS, John. Vice-President, British Aerospace PLC. Sowerby Research Centre, FPC: 267, P.O. Box 5, FUton, Bristol BS12 7QW, England, United Kingdom. Telephone: (44) (0272) 366 842. FAX: (44)(0272) 363733.

F

FAGAN, William. Contact for Lasermet Limited, Five Oaks, Sway Road, Brokenhurst, Hants S04 27RX, England, United Kingdom. Telephone: Lymington (0590) 23075.

FAKLIS, Dean. Vice-President, Rochester Photonics Corporation. 67 Nettlecreek Street, Fairport, NY 14450 USA. Telephone: (716) 377-7990. FAX: (716) 377-7913.

FARGION, Daniele. Contact for Universita di Roma--La Sapienza,. Dipartimento di Fisica, Piazzale Aldo Moro, 2, 1-00185 Roma, Italy.

FARRELL, David. President, Burleigh Instruments, Burleigh Park, Fishers, NY 14453, USA. Telephone: (716) 924-9355. FAX: (716) 924- 9072.

FEENEY, Michael. Sales Manager, Physical Optics Corporation, 2545 W 237th Street, Suite B, Torrance, CA 90505 USA. Telephone: (213) 530-1416. FAX: (213) 530-4577.

FEFERMAN, Bennett J . 13 Woodstream Drive, Norristown, PA 19403-3754 USA. Description: Artistic holographer.

FEINBERG, Jack. Contact for University of Southern California. Department of Physics, University Park, Los Angeles, CA 90089-0484 USA Telephone: (213) 740 1134.

FELSKE, Armin. Contact for Volkswagenwerk AG, Forschung und Entwicklung, Messlechnik D-3180, Wolfsburg I, Federal Republic of Germany. Telephone: (49)(05) 361 221

FENIMORE, Jack. Sales Manager, Sillcocks Plastics International. 310 Snyder Avenue, Berkeley Heights, NJ 07922 USA. Telephone: (800) 9220958. FAX: (201) 665 1856.

FEROE, James, 1420 45th Street, Studio 35-A, Emeryville, CA 94608 USA. Telephone: (415) 658 9787. Description: Artistic holographer.

FIMIA, A. Contact for University of Alicante. Department of Applied Physics, Centro de Holografia, Facultad de Ciencias, AIicante Apdo 99, Spain. Telephone: (34X566) 1200 ext 1147.

FINe, Hartmut. Contact for Holographic Art, WerderstraBe 1173, 2BOO Bremen 1, BR DeutschlandlFederal Republic of Germany. Telephone: (49)(421) 555 690. FAX: (49)(421) 556 202.

FISCHER, Julian. K1ugstrasse 49, 8000 Munchen 19, Federal Republic of Germany. Telephone: (49)(83) 1572 682. Description: Artist.

FISHER, Gary. Contact for Man Environment, Inc. P.O. Box 25959, 2041 Sawtelle Boulevard, Los Angeles, CA 90025 USA

FITZPATRICK, Colleen. Contact for Spectron Development LaboratorieslElectro-Optic Consulting, 3303 Harbor Boulevard, Suite G-3, Costa Mesa, CA 92626 USA. Telephone: (714) 964-0324. FAX: (714) 536-7729.

FOLTZ, Susannah D. 1824 Silver S.E. Albuquerque, NM 87106 USA. Telephone: (505) 277 2616. Description: Artist.

FORD, J. Contact for Spectrolab Ltd. P.O. Box 25, Newbury, Berkshire, RH6 8BQ, England United Kingdom. Telephone: (44)0635 248080. FAX: (44) 0635 248745.

FORNARI, Arthur David. 813 Eighth Avenue, Brooklyn, NY 11215 USA. Telephone: (718) 965 3956. Description: Artistic holographer; silver halide transmission & reflection holograms.

FORSBERG, Mona. Contact for HoloMedia ABI Hologram Museum. P.O. Box 45012, Drottninggatan 100, 10430 Stockholm, Sweden. Telephone: (46)(08) 105 465. FAX: (46)(08) 107638.

FOSTINl, Phil. President, Evergreen Laser Corporation, Unit B, Commerce Circle, Durham, CT 06422 USA. Telephone: (203) 349-1797 (413) 731-8813.

FOULKES-WILLIAMS, Paula. Marketing & Sales Coordinator, Light Fantastic PLC. 4EIF Gelders Hall Road, Shepshed, Leicestershire LE12 9NH, England, United Kingdom.

FOURNIER, Jean-Marc. Contact for Rowland Institute for Science. 100 Cambridge Parkway, Cambridge, MA 02142 USA. Telephone: (617) 497 4657.

FOX, Jerry. Contact for White Light Works, Inc.; L.A. School of Holography, P.O. Box 851,Woodland Hills, CA 91365 USA Telephone: (818) 703 1111. FAX: (818) 7031182.

FRANKMARK, Robert. Contact for Volvo-Flygmotor, S-461 81 Trollhattan, Sweden. Telephone: (46) (0520) 94471.

FRANZ, T. Sales Manager for Labor Dr. Steinbichler, Am Bauhof 4, D-8201, Neubeuern, Federal Republic of Germany. Telephone: (0049) 80351018. FAX: (0049) 8035 1010.

FREUND, Art. Owner, Art Freund Holography. 124 Brookwood Drive, Santa Cruz, CA 95065 USA.

FRIES, Urs. Contact for Holo-Service.Fries, Eulerstrasse 55, CH-4051 Basel, Switzerland.

FRIGON, Jacques. 2375 Fullum, Montreal, Quebec, Canada H2K 3P3. Telephone: (514) 521 4270. Description: Artist.

FRISK, E.O. Managing Director, Optical Works Ltd., Ealing Science Centre, Treloggan Lane, Newquay, Cornwall TR7 1HX, England. Telephone (44) 0637-877222. FAX: (44) 0637-877211.

FRUIT, Vikki. Vice-President, Angstrom Industries, Inc. 3202 Argonne, Houston, TX 77098 USA. Telephone: (713) 526 0006.

FUCHTENBUSCH, A. Contact for Labor fUr Holografie, Am Forst 38, Wesel D-4230, Federal Republic of Germany.

G

GAFUROVA, N.S. Contact for Minchimprom, 101851 Moscov, USSR.

GAGNON, Brigitte.Sales Manager, !BOU Inc. CP 214, Cap-de-Ia-Madeleine, Quebec, GaT 7W2 Canada. Telephone: (819) 295 5229. FAX: (819) 2955229.

GALANTOWICZ, Tom. President, Newport Corporation, 18235 Mt. Baldy Circle, P.O. Box 8020, Fountain Valley, CA 92708-8020 USA. Telephone: (714) 963 9811. FAX: (714) 963 2015.

GALLOWAY, Lory. Venture Manager, E.I. Dupont De Nemours & Co. Inc Optical Element Venture, Experimental Station, P.O.Box B0352, Wilmington, DE 19880-0352 USA. Telephone: (302) 695 4893. FAX: (302) 695 9631.

GARCIA, Julio Ruiz. President, Holotek, S.A. Carretera de Santander, Granda 47, 33199 Granda-Siero, Asturias, Spain. Telephone: (34) (985) 79 35 26. FAX: (34)(85) 27 1853.

GARCON, Thierry. Contact for Holodesign. I, Boulevard de la Republique, F-95600 Eaubonne, France. Telephone: (33)91) 39 593954.

GARDENER, John. President, Gardener Promotion Marketing. 4165 Apalogen Road, Philadelphia, PA 19144 USA. Telephone: (215) 849 4049.

GARRETT, Scott. Contact for Advanced Holographics Corporation, 2469 East Fort Union Blvd., Suite lOB, Salt Lake City, UT 84121 USA.

GASKILL Jack, Contact for The University of Arizona, Optical Sciences Center, Tucson, Arizona 85721 USA.

GAUCHET, Pascal. President, Atelier Holographique de Paris, 13, Passage Courtois, F- 75011 Paris, France. Telephone: (33)(1) 43 79 69 18.

GAUTHIER, Henry. President, Coherent. 3210 Porter Drive, Palo Alto, CA 94306 USA. Telephone: (415) 493-2111. FAX: (415)858-7631.

GEIGER, Thomas. Contact for Ingenieur Btl.ro Geiger. Dieding 7, D-BO 17 Ebersberg, Federal Republic of Germany. Telephone: (08092) 6583. FAX: (08092) 31656.

GERSONDE, Michael. Contact for Elan Bio-Medical. Holography Laboratory, 411 Lewis Road,II417, San Jose, CA 95111 USA.

GERVOLINO, Frank. President, Dell Optics Company, 25 Bergen Blvd., Fairview, NJ 07022, USA. Telephone: (201) 941-1010. FAX: (201) 941- 9524.

GIBSON, Dr. J.A. Contact for AG Electro-Optics Ltd., 29 Forest Road, Tarporley, Cheshire, CW6 OHX England, UK.

GILBERG, Gunilla. Contact for Hoechst Cela-

nese Corporation. 86 Morris Avenue, Summit, NJ 07901 USA. Telephone: (201) 5227816.

GILLES, Jean. 14 Rue des Quatre Vents, 25000 Besançon, France. Telephone: (33) (81) 503 139. Description: Artistic holographer.

GILLESPIE, John, Mike. President, Vice-President, Jodon Inc. 62 Enterprise Drive, Ann Amar, MI 48103 USA. Telephone: (313) 761 4044. FAX: (313) 761 3322.

GINN, Ian. Contact far Ian Ginn Holography, Postbus 116, 5070 AC Udenhout, The Netherlands. Telephone: (31) 42414358. FAX: (31) 4241 4368.

GIPP, R. Sales Manager, Applied Holographics Corp., 1721 Fiske Place, Oxnard, CA 93033 USA.

GITTOES, George. 54 Brighton Street, Bundeena, NSW 2230 Australia. Telephone: (61X2) 523197. Description: H-1 Maker.

GLEASON, 'Ibm. Sales Manager, Laser Technology, Inc. 1055 West Germantown Pike, Norristown, PA 19403 USA Telephone: (215) 6315043.

GLEN, Gerald. Contact for Laser Institute of America, Education Division, 5151 Monroe Street, Toledo, OH 43623 USA.

GOCHA, Dexter. Sales & Marketing Manager, Apollo Lasers, Inc. P.O. Box 2730, Chatsworth, CA 91311 USA.

GOETTEN, John. Contact for Optics Plus Inc. 1369 East Edinger Avenue, Santa Ana, CA 92705 USA. Telephone: (714) 972 1948. FAX: (714) 835-6510.

GOLDBERG, Bruce. 32 Farragut Avenue, San Francisco, CA 94112 USA. Telephone: (415) 584 1197. Description: Consulting and education services.

GOLDSTEIN, Robert. President, Laser Applications, Inc. 12722 Research Parkway, Orlando, FL 32826 USA. Telephone: (407) 3BO 3200. FAX: (407) 3819020.

GOLDSTEIN, Robert. Contact for Lasermetrics, Inc., 196 Coolidge Avenue, Englewood, NJ 07631 USA. Telephone: (201) 894-0550.

GONçALVES, Figueroa. Vice-President, Ibero GestAo Integrada e Tecnologica LDA. Apartado 1267,4104 Porto Codex, Portugal.

GOODMAN, Joseph. Contact, Stanford University-- Department of Mechanical Engineering. Building 570 Room 571C, Stanford, CA 94301 USA. Telephone: (415) 723 3243.

GORGLIONE, Nancy. Contact for Cherry Optical Co.; Laser Affiliates; 2047 Blucher Valley Road, Sebastopol, CA 95472 USA Telephone: (517) 774 3025. Nancy Gorglione produces unique fine art reflection and transmission hologram composites. Created from multiple masters copied onto glass plates, these large multicolored hologram composites create scenes oflasting sensory appeal.

GREEN, Keith. Contact for LaminexlHigh Tech UK Ltd., Bromfield Industrial Estates, Mold, C1wyd CH7 1JR, England, United Kingdom. Telephone: (44)(0352) 58444.

GREENAN, Jay. President, Advanced Holographics Corp_ 2469 East Fort Union Blvd. Suite lOS, Salt Lake City, UT 84121 USA.

GREGUSS, Plil. Institute of Precision Mechanics and Optics, Applied Biophysics Laboratory,Technical University Budapest, H-1621 Budapest, Hungary_

GREISAS, Nisaim. Sales Manager, Holo-Or Ltd. P.O. Box 1051, Rehovot 76110, Israel.

GROSMANN, Michel. Contact for L.I.R. E.R.A.12, rue Libergier, F-51110 Reims, France ..

GROSSINGER, I. President, Holo-Or Ltd. P.O. Box 1051, Rehovot 76110, Israel..

GUEGUEN, Thierry. Sales Manager, Magic Laser. Quartier de L'horJoge, 4 rue Brantbme, 75003, Paris France ..

GUGGENHEIMER, Steve. Marketing for Spectra-Physics Inc. Laaer Products Division. 1250 West Middlefield Road, Box 7013, Mountain View, CA 94039-7013. Telephone: (415) 961 2550. FAX: (415)969 4084.

GUIGNARD, M. Contact for Aerospatiale-Division Helicopteres, 2 avenue Marchel-Cachin, F- 93126 La Courneuve Cedex, France ..

GUISE, Rodney. Contact for Hologram World, 64 The Promenade, Blackpool, Lancashire, England, United Kingdom.

GULBRANSEN, Keith. Customer Service, Advanced Holographics Corp. 2469 East Fort Union Blvd. Suite 108, Salt Lake City, UT 84121 USA_.

GULLAMON, Antoni Pinol. Anselmo Clave 74, Oleaa de Montaerrat, 08640 Barcelona, Spain. Telephone: (34)(3) 778 2299. Description: Holographic artist.

GUNTHER, Roy. Vice-President, Gardener Promotion Marketing. 4165 Apalogen Road, Philadelphia, PA 19144 USA. Telephone: (215) 849 4049.

GUSTAFSSON, Jonny. Contact for Lasergrupperi Holovision AB. Osthammarsgatan 69, S - 11528 Stockholm, Sweden. Telephone: (46)(08) 663 9908. .

GUTEKUNST, Horst. Director, Hologramm Werkstatt &; Galerie, Gallerie Fur Hologramme. Via Principale 30, CH - 7649 Castesegna, Switzerland Telephone: 082 417 18. FAX: 0S2 412 68.

H

HABER, Gary. Vice-President, Holaxis Corporation. 499 Farmington Ave, Hartford, CT 06105 USA. Telephone: (203) 232 2030. FAX: (203) 236-3767.

HAHN, Dr. Yu. H. President, CVI Laser Corporation, CVI West. 470 Lindbergh Avenue, Livermore, CA 94550 USA. Telephone: (415) 449-1064.

HAIDLE, Rudy. Contact for Northwestern University, Fiberoptic Laboratory, Dept of Electrical Engineering, Technological Institute, Evanston, IL 6020S USA. Telephone: (708) 491 4427.

HAJIME, Yamashita. Contact for Mazda Motor Corp. 3-1, Shinchi, Fuchu-cho, Aki-gun, Hiro-

shima, Japan.

HALIVA, Mauric. Contact for New York Institute of Technology. Center for Optics, 100 Glen Cove Avenue, Glen Cove, NY 11542 USA. Telephone: (516) 686 7863 .

HALL, G.G. President, George M. Whiley Limited. Firth Road, Houston Industrial Estate, Livingston, West Lothian EH54 5DJ, Scottland, United Kingdom. Telephone: (0506) 386ll FAX: (0506) 38262.

HALLSTROM, Hans. Contact for 3D Media, Kyrkvagen 24, S-910 36 SAVAR, Sweden. Telephone: (46)(090) 98141.

HAMEIRI, Shimon. Contact, Holography Ltd. 21 Hakomemiut Str., Herzlia Pituah, Israel.

HANNIFIN, Matt. Owner, Matt Hannifin Co. P.O. Box 4574, Austin, TX 78765 USA.

HANSEN, Matthew E. 741 East Gorham Street, Madison, WI 53703 USA. Telephone: (608) 255 3580. Description: Holographic artist.

HARDY, Nick. Contact for Op-Graphics (holography) Ltd. Unit 4: Technorth, 7 Harrogate Road, Leeds LS7 3NB, England, United Kingdom. Telephone: (44)(0532) 628 687. FAX: (44) (532) 374 182,.

HARKEY, Joyce C. Customer Service, E.I. Dupont De Nemours &; Co. Inc Optical Element Venture, Experimental Station, P.O.Box 80352, Wilmington, DE 19880-0352 USA.

HARIDAS, P. Contact for Massachusetts Institute of Technology, Laboratory for Nuclear Science, Cambridge, MA 02139 USA.

HARIHARAN, Parameswaran. Contact for CSIRO, Division of Chemical Physics, P.O. Box 218, Lindfield 2070, Australia.

HARRIS, Kenneth. Contact for Light Impressions Europe Pic., 5 Mole Business Park 3, Off Station Road, Leatherhead, Surrey KT22 7BA, England, United Kingdom.

HARRIS, Nick. 7ll East 13th Street, Houston, TX 77008 USA. Telephone: (713) 861 2865. Description: Artistic holographer; portraits and integrals; consulting.

HARRISON, Susan. Contact for Hologram World, 1212 112 Dixon Boulevard, Cocoa, FL 32922 USA. Telephone: (407) 631 3615 and 1 (800) 882-4656.

HASHIMOTO, Daijo. Contsct for'Ibyko University, Medical Division, 2nd Surgical Research, 'Ibyko, Japan.

HASKELL, Dara. President, Third Dimension Arts, Inc. 1241 Andersen Drive, Suites C &; D, San Rafael, CA 94901 USA. Telephone: (415) 485 1730. FAX: (415) 485 0435.

HAYDVOGEL, Christian. Director, Technoexanl Semicon. Morellenfeldgasse 41, A-80lO Graz, Austria. Telephone: (43X0316) 38 25 41. FAX: (43) (0316) 382403.

HAYS, Joel Todd. President, Hologram World, 1212 112 Dixon Boulevard, Cocoa, FL 32922 USA. Telephone: (407) 631 3615 and 1 (800) 882-4656.

HEATON, Lorna. 3998 Rue de Bullion, Mon-

treal, Quebec, Canada H2W 2E4. Telephone: (514) 845 5403. Description: Artist

HEIDT, Marina. Customer Service, Holographic Products, Inc. 755 South 200 West, Richmond, UT 84333 USA.

HEIN, Mrs. Elke. Sales Manager, die dritte dimension. Frankfurter StraBe 132 -134, D-6078 Neu Iaenburg, Federal Republic of Germany. Telephone: (49)(06102) 33367. FAX: (49) (06102) 36709.

HEINRICH, Philip. President, Laser Optics, Inc. III Wooster St., Bethel, CT 06801. (mailing address) P.O. Box 127, Danbury, CT 06813 USA. Telephone: (203) 744-4160. FAX: (203) 798-7941.

HELFAND, Cindy. Contact for Light Wave, North Pier, 435 East Illinois Street, Chicago, IL 60611 USA.

HELM, Molly. ContactlSales Secretary, Spectra-Physics, Boundary Way, Hemel Hempstead, Herta. HP2 7SH, England, United Kingdom. Telephone: 0442 232 322. FAX: 0442 68538. Contact:

HENNESSEY, Sigi. Customer Service, Melles Griot,l770 Kettering Street, Irvine, CA 92714 USA Telephone: (714) 261-5600. FAX: (714) 261-7589.

HERFORTH, Pamela. Sales Manager, Crown Roll Leaf, Inc. 91 Illinois Ave., Paterson, NJ 07503 USA. Telephone: (201) 742-4000. FAX: (201) 742-0219.

HERSLOW, John. President, Sillcocks Plastics International. 310 Snyder Avenue, Berkeley Heights, NJ 07922 USA. Telephone: (800) 922 0958. FAX: (201) 6651856.

HESLOP, Lynne. Contact for Mirage Holograms Ltd. Unit 2 Brook Lane, Business Centre, Brook Lane North, Brentford, Middlesex TW8 OPP England, United Kingdom. Telephone: (44)(1) 568 2454.

HESS, Cecil. Partner, Metrolaser. 18006 Skypark Circle, Suite lOS, Irvine, CA 92714-6428 USA. Telephone: (714) 553-0688. FAX: (714) 553-0495.

HEWITT, Brian. Contact for Spycatcher Limited. 2 Croft Cottages, Wattisfield, Diss Norfolk 1P22 INS, England, United Kingdom.

HICKMAN, Hope. Contact for Bob Mader Photography. P.O. Box 796728, Richardson, TX 75080 USA. Telephone: (2 14) 690 5511

HICKMOTT, A. Contact for Hickmott &; Austin Holograms, 11 Castelnau, London SW13 9RP, England,United Kingdom. Telephone: (44)(01) 4865811.

HIGGINS, Dr. Jim. Contact for Lumonics Inc. 105 Schneider Road, Kanata, Ottawa, Ontario K2K 1Y3, Canada.

HIGGINS, Sam. Contact for Spectral Images, 15 Wanderdown Way, Ovingdean, Brighton BN2 7BX, England, United Kingdom.

HILDEBRAND Percy. Contact for Spectrnn Development Laboratories, 3303 Harbor Boulevard, Suite G-3, Costa Mesa, CA 92626 USA.

HILLIARD, Bill. Contact for Holographic Visions. 300 South Grand Avenue, Los Angeles, CA 90071 USA; contact for Neovision Productions, P.O. Box 74277, Los Angeles, CA 90004 USA Telephone: (213) 387 0461.

HINTZ, Jürgen. Contact for Phoenix Holograms, Traubenstr. 41, D-2913 - Apen, Federal Republic of Germany. Telephone: (49)0448915198

HOBART, James. CEO, Coherent. 3210 Porter Drive, Palo Alto, CA 94306 USA. Telephone: (415) 493-2111. FAX: (415) 858-7631.

HODGES, Dean. Vice-President, Newport Corporation, 18235 Mt. Baldy Circle, P.O. Box 8020, Fountain Valley, CA 92708-8020 USA. Telephone: (714) 9639811. FAX: (714) 9632015.

HOFFMAN, Gordon. President, Stereo World Magazine. 5610 SE 71st Street, Portland, OR 97206 USA. Telephone: (503) 771-4440.

HOFFMANN, John. Contact for Museum of the Fine Arts Research & Holographic Center. 1134 West Washington Boulevard, Chicago, IL 60607 USA. Telephone: (312) 2261007.

HOFMANN, Martin. Contact for Holografie - Hofmann Labor. Carl-Hermann-Gaiserstrasse 20, 7320 Groppingen, Federal Republic of Germany.

HOGMOEN, Kare. Contact for Det Norske Veritas, Research Division, P.O. Box 300, N-1322 Hovik, Oslo Norway.

HOLBERTON, R.C.Sales Manager, Laser Electronics Pty., Ltd. P.O. Box 359, Southport, Queensland, 4215, Australia. Telephone: 61 75 53 2066. FAX: 61 7553 3090.

HOLLINSWORTH, T.R. Contact for Scope Optics Ltd., Noon Lane, Barnet, Hertfordshire EN5 SST, England, United Kingdom. Telephone: (44)(81) 4412283.

HOLMES, Burton. Contact for Burton Holmes International, 1004 Larrabee Street, West Hollywood, CA 90069 USA. Telephone: (213) 652 0970.

HOPE, C.L. Sales Manager, Amblehurst Ltd., 52 Invincible Road, Farnborough, Hants GU14 7QU, England, United Kingdom. Telephone: (44) 0252520052. FAX: (44) 0252 373871.

HOPKINS, Anthony. Contact for Light Engineering; New Clear Imports. 12 New St. Johns Road, St. Helier, Jersey, Channel Islands, England, United Kingdom. Telephone: (44)(534) 30614.

HORGAN, M.F. Contact for Wentworth Laboratories, Ltd. Sunderland Road, Sandy, Bedfordshire, SG19 lRB, England, United Kingdom. Telephone: (44) (0767) 81221. FAX: (44) (0767) 291951.

HOSKINS, Greg. Vice-President, Robert Sherwood Holographic Design, Inc. 400 West Erie Street, Chicago, IL 60610 USA. Telephone: (312) 944 3200.

HOUDE-WALTER, Susan. Contact, University of Rochester, Institute of Optics, Rochester, NY 14627 USA.

HOWELLS, Malcolm. Contact for Lawrence Berkeley Laboratory. University of California. Building 80-101, Berkeley, CA 94720 USA. Telephone: (415) 486 4000.

HOWLETT, Glenn. Rr Box 2873, Warren, VT 05674 USA. Telephone: (802) 496 2576. FAX: (802) 496 6488. Description: Holographic artist; reflection holograms.

HUBEL, Paul. Contact for University of Oxford-Holography Group, Department of Engineering Science, Parks Road, Oxford OX1 3PJ, England, United Kingdom

HUDSON, Philip M.G. Director, Amblehurst Ltd., 52 Invincible Road, Farnborough, Hants GU14 7QU, England, United Kingdom. Telephone: (44) 0252520052. FAX: (44) 0252373871.

HUFF, Lloyd. Contact for University of Dayton, Research Institute. 300 College Park, Dayton, OH 45469 USA. Telephone: (513) 229 3221.

HUG, Bill. President, Omnichrome. 13580 Fifl.h Street, Chino, CA 91710 USA. Telephone: (714) 627-1594. FAX: (714) 591-8340.

HYDE, D.J. Vice-President, Kendall Hyde Ltd. Kingsland Industrial Park, Stroudley Road, Basingstoke, Hants., RG24 OUG, England, United Kingdom. Telephone: (44) 0256 840 830. FAX: (44) 0256 840 443.

I

IAN-FREESE, Richard. Contact for Advanced Environmental Research. Route 2, Box 2948, Davis, CA 95616, USA.

IBSEN, Per. Vice-President, Superior Technology Implementation. Hjortekaersvej 99B, 2800 Lyngby, Denmark. Telephone: (45)(45) 933358. FAX: (45)(45) 930 353.

INEICHEN, Beat. Contact for Stoltz AG, Tafernstrasse 15, CH-5405 Baden Dattwil, Switzerland. Telephone: (41)(056) 840151.

IOVINE, Laura. Contact for Images Company South, P.O. Box 2251, Inverness, Florida, 32651 USA. Telephone (904) 344-8540.

ISHII, Setsuko. 1-23-26-404 Kohinata, BunkyoKu, Tokyo, Japan. Telephone: (81)(03) 945 9017. Description: Holographic artist; making DCG, Transmiasion, Embossed holograms.

IVES, Janet. Contact for Holotec Pic. 7 Cameron Road, Seven Kings, Easex, IG1 3DF, England, United Kingdom. Telephone; (44X01) 597 8004.

J

JACKSON-SMITH, Roaemary. P.O.Box 2850, Key Largo, Florida 33037 USA. Telephone: (516) 3243000. Description: Artist.

JACOBSON, Dr. Alex. Contact: CVI Laser Corporation, CVI West. 470 Lindbergh Avenue, Livermore, CA 94550 USA.

JAEGER, M. President, Jaeger Graphic Technology, J.G.T.-Holofoil S.A. 20 Avenue des Dllsirs,B-1140 Bruaels, Belgium Telephone: 00-322-7359551. FAX: 733 1035.

JAIN, Dr. Anil. Contact for APA Optics, Inc. 2950 Northeast 84th Lane, Blaine, MN 55432 USA. Telephone: (612) 7844995. FAX: (612) 784 2038.

JAMES, D.J. President, Lumonics Inc. 105 Schneider Road,Kanata, Ottawa, Ontario K2K 1Y3, Canada.

JAMES, Randy. Owner, Randy JameslHolography. P.O. Box 305, Santa Cruz, CA 95061 USA. Telephone: (408) 458 4213.

JANNSON, Dr. Joanna. President, Physical Optics Corporation. 2545 W 237th Street, Suite B, Torrance, CA 90505 USA. Telephone: (213) 530-1416. FAX: (213) 530-4577.

JAOUDE, Majdeline A. Contact for Aasociates of Science and Technology (AST) Inc., 2450 Lancaster Road, Suite 36, Ottawa K1B 5N3, Canada.

JEONG, Tung H. President, Integraf; Director of Lake Forest Holography Workshops. P.O. Box 586, Lake Forest, IL 60045 USA. Telephone: (708) 234 3756. FAX: (708) 615 0835.

JOHANN, Larry. President, Southern Indiana Holographics. 6841 Newburgh Rd, Evansville, IN, 47751. Telephone: (812) 474-0604.

JOHNSON, D.A. President, Lambda!fen Optics, Division of Optical Corp. of America. One Lyberty Way, Westford, MA 01886 USA. Telephone: (508) 692-8140. FAX: (508) 6929416.

JONSSON,' E.A. Contact for Laserlabbet, Box 521, SE 581 06 Linkoping, Sweden.

JORDAN, Carmen. Mng. Marketing Services, LiCONix, 3281 Scott Boulevard, Santa Clara, CA, 95054 USA.

JOSHI, Dr. S.B. Contact for York University, Department of Physics, 4700 Keele St., North York, Ontario, Canada M3J 1P3.

JOYEux. D. Contact for LURE. Institut d'Optique, BP 147, F-91403 Orsay, Cedex, France.

JUNG, Dieter. Viuonvillestrasse 11, D 10000 Berlin 41, Federal Republic of Germany. Telephone: (49)(30) 7718 431. Description: Holographic artist.

JUPTNER, Werner. Contact for Bremer Institute fur Angewandte. Strahltechnik, BIAS, Ermlandstraße 59, D-2820 Bremen 71, Federal Republic of Germany.

JUREWICZ, Arlene. Contact for Synchronicity Holograms, Box 4235, Lincolnville, ME 04849 USA. Telephone: (207) 763 3182.

K

KAC, Eduardo. Rua Hilario de Gouveia, 1101 1002, Rio de Janeiro RJ Cep 22040, Brazil. Telephone: (55)(021) 237 6012. Description: The basis of my work is verbal syntax. I make holopoems; poems conceived in space and time, that change their features according to the beholder's viewpoint.

KAPONIS, Notis. Contact for Laser Graphics. AG. Dimitriou 150,546 35 Theasaloniki, Greece.

KASPRZAK, Henryk. Contact, Technical University of Wroclaw. Institute of Physics, Wybrvzeze Wyspianskiego 27, PL-50-370 Wroclaw, Poland

KATSUMA, Hidetoshi. Contact for Tokai University, Department of Electro Photo Optics, 1117 Kitakaname Hiratsuka City, Kanagawa 259-12, Japan.

KAUFFMAN, R.E. Sales Manager, Amblehurst

Ltd., 52 Invincible Road, Farnborough, Hants GU14 7QU, England, United Kingdom.

KAUFMAN, Andreas. Graf von Galen Strasse 5, D-4800 Bielefeld 1, Federal Republic of Germany. Telephone: (49)(521) 102 269. Description: Holographic artist.

KAWAHARA, Takao. Marketing Director, Holomedia Inc. 3-15-22, Takaban, Meguro-ku, Tbkyo 152 Japan. Telephone: (81) (03) 793 2321. FAX: (81)(03) 793 2322.

KELLISON, Kirt. Contsct for Robert Sherwood Holographic Design, Inc. 400 West Erie Street, Chicago, IL 60610 USA. Telephone: (312) 944 3200.

KELLY, Ed. President, Keystone Scientific Co. P.O. Box 22, Thorndale, PA 19372 USA. Telephone: (215) 384-8092, FAX: (215) 384 8093.

KELLY, Jim. Contsct for The Whole Picture, A Gallery of Holography. 634 Parkway, Gatlinburg, TN 37738 USA. Telephone: (615) 436 3650.

KELLY, Mary. Customer Service, Third Dimension Arts, Inc. 1241 Andersen Drive, Suites C & D, San Rafael, CA 94901 USA. Telephone: (415) 485 1730. FAX: (415) 485 0435.

KENDALL, M.D. President, Kendall Hyde Ltd. Kingsland Industrial Park, Stroudley Road, Basingstoke, Hants., RG24 OUG, England, United Kingdom. Telephone: (44) 0256 840 830. FAX: (44) 0256 840 443.

KENRICK, Paul. Sales Administrator, Melles Griot,1770 Kettering Street, Irvine, CA 92714 USA Telephone: (714) 261-5600. FAX: (714) 261-7589.

KHANNA, Shyam. Contact for Columbia University, Deptartment of Otolaryngology, 630 West 168th Street, New York, NY 10032 USA. Telephone: (212) 305 3993.

KILPATRICK, Jack. System Sales, Laser Resale Inc. 54 Balcom Road, Sudbury, MA 01776 USA. Telephone: (508) 443 8484. FAX: (508) 443 7620.

KIMBALL, Randy. Sales Manager, LiCONix, 3281 Scott Boulevard, Santa Clara, CA, 95054 USA.

KINCEL, David. Vice-President, Aerotech Inc., Electro Optical Division, 101 Zeta Drive, Pittsburgh, PA 15238 USA.

KLEMPNER, Jonathan. Contact for Light Harmonies Inc. 93 Lake Shore Drive, Oakland, NJ 07436 USA.

KLINE, Richard. Contact for Central Michigan University, Art Department. Mt Pleasant, MI 48859 USA. Telephone: (517) 774 3025.

KLUEPFEL, Brian. Contact for Ross Books, P.O. Box 4340, Berkeley, CA 94704 USA. Telephone: (415) 8412474. FAX: (415) 8412695.

KLUTE, Jeff. Research, Holographies North Inc. 444 South Union Street, Burlington, VT 05401 USA. Telephone: (802) 658-2275. FAX: (802) 862-6510.

KNIGHT, Roger C. Marketing Director, Light Fantastic PLC. 4EIF Gelders Hall Road, Shepshed, Leicestershire LE12 9NH, England,

United Kingdom.

KOCHER, Clive. Contact for Royal Photographic Society, Salisbury College of Art, Southampton Road, Salisbury, Wiltshire, England, United Kingdom.

KODERA, Tbkio. Contact for DAI Nippon Printing Co. Ltd.Cental Research Institute 12, 1- Chome Ichigsya-Kagscho, Shinjuku-ku, Tbkyo 162, Japan. Telephone: (81X03) 266 2310.

KONTNIK, Lewis. Publisher, Holography News.3932 McKinley Street N.W., Washington, D.C. 20015 USA. Telephone or FAX:(202) 986-3464.

KOOI, A. Manging Director, Optilas B.V. P.O. Box 222, 2400 AE Alphen, NO Rijn, The Netherlands. Telephone: (31) 01720 31234. FAX: (31) 0172043414.

KORIL, Jerry. Contact for Creative Label, 2450 Estes Drive, Dept. M, Elk Grove Village, IL 60007 USA. Telephone: (708) 956 6960.

KORTZ, Dr. Hans Peter. Vice-President, Adlas GmbH & Company, Seeland Strasse 67, D -2400 Luebeck 14, Federal Republic of Germany.

KOST, Ann. Contact for Burleigh Instruments, Burleigh Park, Fishers, NY 14453, USA. Telephone: (716) 924-9355. FAX: (716) 924-9072.

KRAMER, Charles. Contact for Holotek Ltd., 300 East River Road, Rochester, NY 14623 USA. Telephone: (716) 424 4996. FAX: (716) 424 4958.

KREISCHER, Cody. Contact for Kreischer Optics, Ltd. 906 N. Draper Rd., McHenry, IL 60050 USA. Telephone: (815) 344-4220. FAX: (815) 344-4221.

KREMER, Peter. Vice-President, AKS Holographie-Galerie GmbH. Potsdamer StraBe 10, 4300 Essen 1, Federal Republic of Germany

KRIEG, Christine. Marketing Assistant. 3210 Porter Drive, Palo Alto, CA 94306 USA.

KUBOTA, Thshihiro. Contact for Kyoto Technical University. Dept. of Photographic Technology, Matsugasaki, Sakyo-ku, Kyoto 606, Japan.

KURZEN, Aaron. P.O. Box 3233, Stony Creek, CT 06405 USA. Telephone: (203) 488 4711. Description: Fine art holograms.

L

LACEY, Lee. Contact for Holo/Source Corporation. 21800 Melrose Avenue, Suite 7, Southfield, MI 48075 USA. Telephone: (313) 355 0412. FAX: (313) 355 0437.

LACHAUD, Denis. Contact for Hologram. Industries, 42-44, rue de Trucy, F-94120 Fontenay sous bois, France.

LACOSTE, Russell. Vice-President, Sales and Marketing. American Bank Note Holographies, Inc. 500 Executive Boulevard, Elmsford, New York 10523 USA. Telephone: (914) 592-2355. FAX: (914) 5923248.

LACZYNSKI, Andrew. Contact for The Holography Development Group. 23 Neville Park Blvd, Tbronto, Ontario M4E 3P5 Canada. Telephone: (416)-691-9381. FAX: (416)-691-0407.

LA DUCA, Tim. Sales Manager, Third Dimension Arts, Inc. 1241 Andersen Drive, Suites C & D, San Rafael, CA 94901 USA. Telephone: (415) 485 1730. FAX: (415) 485 0435.

LAFGREN, Darlene. Customer Service, General Holographies, Inc. P.O. Box 82247, Bumaby B.C., V5C 5P7 Canada. Telephone: (604) 4356654. FAX: (604) 4327326.

LAFRENIERE, Anik. 4060 Boulevard St. Laurent, Montreal, Qullbec, H2W 1Y9 Canada. Telephone: (514) 286 9619. Description: artist.

LAGANIS, Evan D. Senior Product Specialist, E.I. Dupont De Nemours & Co. Inc Optical Element Venture, Experimental Station, P.O. Box 80352, Wilmington, DE 19880-0352 USA.

LAMBERT, Chris & Carole.Contacts for Laza Holograms. 47 Alpine Street, Reading, Berkshire RG1 2PY, England, United Kingdom. Telephone: (44) 0734 589 026. FAX: (44) 0734 571974.

LANCASTER, Ian. Contact for Holography News; Ian Lancaster Holographies Consultancy, 1 Erica Court, Wych Hill Park, Woking, Surrey GU22 OJB England, United Kingdom. Telephone or FAX: (44) 483 740 689.

LANDRY, John. Contact for University of California. Advanced Imaging Center, College of Engineering, Santa Barbara, CA 93106 USA. Telephone: (805) 892-4341. FAX (805) 893-3262.

LANG, Tim. Contact for JK Lasers Ltd. Cosford Lane, Swift Valley, Rugby, Warwickshire CV21 1QN, England, United Kingdom.

LANGENBECK, P. Contact for Ing.-Agentur rur neue Technologie in Optick und Precision Engineering. D-7771 Frickingen 2, Federal Republic of Germany.

LANTZ, Fran. Sales Representative for Applied Holographies, 1721 Fiske Place, Oxnard, CA 93033 USA. Telephone: (805) 385-5670. FAX: (805) 385-5671.

LARKIN, Alexander. Contact for Moscow Physical Engineering Institute. Kashirskoe Shosse 1, Moscow, 115409 USSR.

LARSON, Steve, Roger, Anne. Contacts for Portson, Inc. (Laser Images). 9201 Quivira, Overland Park, KS 66215-3905 USA. Telephone: (913) 4927010. FAX: (913) 492 7099.

LAUK, Matthias. Contact for Museum fUr Holographie & neue visuelle Medien. Pletschmiihelenweg 7, D -5024 Pulheim 1, Federal Republic of Germany. Telephone: (49)(02) 233385 1053. FAX: (49)(02) 23852158.

LAW, Linda. Owner, Linda Law Holographies. 8 Crescent Drive, Huntington, NY 11743 USA.

LAWRENCE, Gary. Contact for North American Holographics Inc. P.O. Box 451, 103 East Scranton Avenue, Lake Bluff, IL 60044 USA. Telephone: (708) 234 4244.

LEAFLOOR, Stephen. President, Starlight Holographic Inc. 73 Stable Way, Kanata, Ontario K2M 1A8, Canada. Telephone: (613) 2350440. FAX: (613) 5927647.

LEBINE, Pauline. Contact, Ealing Electro-Optics Inc. New Englander Industrial Park,

Holliston, MA 01746 USA Telephone: (508) 429 8370.

LEE, Sing. Contact for University of California. San Diego. Dept. Electrical and Computer Engineering, La Jolla, CA 92093 USA.

LEECE, Larry. Contact for Myers Printing, Inc. 914 East Gier Street, Lansing, MI, 48908, USA. Telephone: (1) (BOO) 333-3645. FAX: (517) 482-0550.

LEFLOCH, H.C. Contact for Aerospatiale-Ets D'Aquitaine, Saint-Medard-en-Jalles, F-33165 Bordeaux, France.

LEIS, H.G. Contact for Daimler Benz AG. D-7000 Stuttgart 60. Gennany.

LEITH, Emmett. Contact for University of Michigan, Department of Electrical and Computer Engineering; EECS Building, Ann Arbor, MI 48109-2122 USA. Telephone: (313) 764 9545.

LEKKI, WalterJ. President, Corion Corp. 73 Jeffrey Ave., Holliston, MA 01746 USA Telephone: (508) 429-5065. FAX: (508) 429-8983.

LESAR, Christopher J. 100 Hickory Lane, Lancaster, OH 43130 USA. Telephone: (614) 654 0862. Description: Holographic artist.

LESEBERG, Detlef. Contact for University Essen, Fachbereich 71PIPhysik, Universitstsstrasse 2, D-4300, Essen I, Federal Republic of Germany. Telephone: (49)(0201) 183 3019.

LESSARD, Roger A. Professor, Universite Laval. Dept. Physique-COPL, Pavilion Vachon, SteFoy, Quebec, G1K 7P4 Canada.

LESSING, Rainer. Contact: Spindeler & Hoyer GmbH. Postfach 3353, Koenigaallee 23, D-3400 Goettingen, Federal Republic of Germany.

LESTER, J .L. President, Metaplast Electrochemicals Corp. 67 Whitson Street, Hempstead, NY 11550 USA. Telephone: (516) 481 4530. FAX: (516) 481 7320.

LEV, Steven. Vice-President, Chromagem. 1871 Selma, Youngatown, OH 44503 USA. Telephone: (216) 799 0323. FAX: (216) 747 9371.

LEVINE, Chris. 32 Lexington Street, London WlR 3HR, England, United Kingdom. Telephone: (44)(1) 437 8992. Description: Artist

LEVINE, Jeffrey. Contact for Another Dimension/AD 2000, 948 State Street, New Haven, CT 06511 USA ..

LEVY, Larry. President, American Laser Corporation, 1832 South 3850 West, Salt Lake City, Utah 64104 USA. Telephone: (801) 972-1311. FAX: (801) 972-5251.

LEVY, Robert. Vice-President, HololSource Corporation. 21800 Melrose Avenue, Southfield, MI 48075 USA .. Telephone: (313) 355 0412. FAX: (313) 355 0437.

LEVY, Uri. Vice-President, Holo-Or Ltd. P.O. Box 1051, Rehovot 76110, Israel..

LIEBERMAN, Dan. Contact for Hologramas De Mexico. PINO 343, Local 3, Col. Sta. Ma La Ribera, 06400 Mexico, D.F. Mexico. Telephone: (525) 547-2033. (525) 547-1893.

LIEBERMAN, Larry . Contact for Holographic Images, Inc., 1301 Dade Boulevard, Miami

Beach, FL 33139 USA. Telephone: (305) 531 5465. FAX: (305) 531-3029.

LIEGEOIS, Dr. Christian D. Contact for X-IAL "Les Algorithmes", Parc d'Innovation, F-67400 TIlkirch, France, and Arbeitskreis Holografie B.V. (Federal Republic of Germany);Telephone: (33) 88 67 44 90. FAX: (33) 88 67 40 66. German telephone:(49)(2831) 3034.

LIGHT, Gail. 1521 Revere Circle, Schaumburg, IL 60193 USA. Telephone: (312) 351 9545. Description: Artist; Fine art originals.

LIJN, Liliane. 99 Camden Mews, London EC2A 2AA, England, United Kingdom. Telephone: (44)(1) 485 8524. Description: artist.

LIN, Simon. Sales Engineer, Superbin Ltd., 5F-3, 792, Tun Hua South Road, P.O. Box 59555, Taipei, Taiwan 10662., Republic of China. Telephone: (886)(02) 733 3920. FAX: (886) (02) 732 5442.

LINCOLN, Margaret. Contact for Light Angels, The Corridor, High Street, Bath Spa, Avon BA1 5AJ, England, United Kingdom.

LIPKA, Gerold. Vice President, Topac GmBH, AuJ'm Eickholt 47, 4830 Gutersloh I, West Germany. Telephone: 05241 580192. FAX: 052 41 58549.

LISSACK, Selwin. Contact for Laserworks. P.O. Box 2408, Orange, CA 92669 USA. Telephone: (714) 832 2686.

LIVNEH, David. Contact, Holography Ltd. 21 Hakomemiut Str., Herzlia Pituah, Israel.

LLOYD, S. 655 Sixth Street, Trafford, PA 15085 USA. Description: Consultant.

LOHMANN, Adolf. Contact for Universitat Erlangen-Nurnberg, Physikalisches Institute, Erwin-Rommel Strasse I, D·8520 Erlangen, Federal Republic of Germany. Telephone: (49) (09131) 857408.

LOKBERG, Ole. Contact for Norges Tekniske Hogskole, Institute for Almen Fysikk, Sem Saelands v 7, N-7034 Trondheim-NTH, Norway.

LOMAX, Jeff. President, Markem Systems Ltd Ladywell Trading Estste, Eccles New Road, Salford, M5 2DA England, United Kingdom.

LONG, Kathryn. Contact for Light Impressions, Inc. 149-B Josephine Street, Santa Cruz, CA 95060 USA.

LOUlE, Stephanie. Customer Service, Smith & McKay, 96 North Almaden Boulevard, San Jose, CA 95110-2490 USA. Telephone: (408) 292 8901 FAX: (408) 2920417.

LOUlS, Henry. Sales Manager, Laser Optics, Inc. 111 Wooster St., Bethel, CT 06801. (mailing address) P.O. Box 127, Danbury, CT 06813 USA. Telephone: (203) 744-4160. FAX: (203) 798-7941.

LOUSHIN, Sharilyn. ContactiMarketing for Advanced Optics, Inc. 5 E. Old Shakopee Rd., Minneapolis, MN 55420 USA.

LOVE, V. Contact for Op-Graphics (holography) Ltd. Unit 4: Tech north , 7 Harrogate Road, Leeds LS7 3NB, England, United Kingdom. Telephone: (44)(0532) 628 687. FAX: (44) (532) 374 182,.

LOWENTHAL, Serge. Contact for Universite

de Paris-Sud, Institute d'Optique, F-91405 Drsay, France. Telephone: (33)(9) 416750.

LUBETSKY, Neal. Owner, Point of View Dimensions, LTD. 45-2903 River Drive South, Jersey City, NJ 07310 USA. Telephone: (201) 626 8844.

LUNGERSHAUSEN, Arnold. Contact for Rochester Institute of Technology. One Lomb Memorial Drive, P.O. Box 9887, Rochester, NY 14623 USA. Telephone: (716) 475 2770. FAX: (716) 475-5804.

LUSIGEN, R. Customer Service, Laser Applications, Inc. 12722 Research Parkway, Orlando, FL 32826 USA. Telephone: (407) 380 3200. FAX: (407) 3819020.

LUTON, Chris. President, Holocrafts of Europe, Barton Mill House. Barton Mill Road. Canterbury, Kent, CT1 1BY England. Telephone: 0227 463223 FAX: 0227 450399.

LYSOGORSKI, Charles. 271 Keyes Road, Honeoye Falls, NY 14472 USA. Telephone: (716) 533 1258. Description: Technical and artistic holographer; Fabrication and consulting. Telephone: (313) 426-0452.

M

MACGREGOR, A.E. Contact for Newcastle Upon Tyne Polytechnic, Department of Physics, Ellison Building, Newcastle upon Tyne, NE1 SST. England, United Kingdom. Telephone: (44)(091) 2356453. FAX: (44)(09l) 235 8017.

MACSHANE, Jim & Elaine. Contact for MacShane HolographylLaser Arts Programs. 512 West Braeside Drive, Arlington Heights, IL 60004 USA.

MACY, Preston. Sales Manager, Evergreen Laser Corporation, Unit B, Commerce Circle, Durham, CT 06422 USA. Telephone: (203) 349-1797 (413) 731-8813.

MADDUX, Gene. Contact for Wright Petterson Air Force Base, Structure Division AFSC, Dayton, OH 45433 USA. Telephone: (513) 255-5159.

MADER, Bob. Owner, Bob Mader Photography. P.O. Box 796728, Richardson, TX 75080 USA. Telephone: (214) 690 5511.

MAGGI, Ruggero. Contact for Amazon, C. So Sempione 67, 20149-Milano, Italy.

MAJEAU, Celine. Customer Service, Holocor I.B.F. Printing Inc. 95 des Sulpiciens, L'Epiphanie, Quebec JOK 1JO, Canada.

MAJERI, H. Customer Service, Jaeger Graphic Technology, J.G.T--Holofoil S.A. 20 Avenue des Desirs,B-1l4O Brussels, Belgium Telephone: 00, 322-7359551. FAX: 733 1035.

MALINA, Dr. Roger F. President, ISAST/ Leonardo, P.O. Box 75, 1442A Walnut Street, Berkeley, CA 94709 USA. Telephone: (415) 845 8306. FAX: (415) 8416311.

MALMQVIST, Sven. Contact for Saab-Scania, S- 581 88 LINKOPING, Sweden. Telephone: (46)(013) 129020.

MALLOTT, Michael. Contact for Laser Light Designs,2412 Kennedy Way, Antioch, CA 94509 USA. Telephone: (415) 7543144.

MANDERSCHEID, Thierry. Contact for Citrüen Industrie. 35, rue Grange Dame Rose, F-92360 Meudon-Ia-Foret, France.

MARCHAND, Jean-Pierre. Contact for IBOU Inc. CP 214, Cap-de-Ia-Madeleine, Quebec, G8T 7W2 Canada. Telephone: (819) 295 5229. FAX: (819) 295 5229.

MARCINOWSKI, Gerhard. Product Manager, Adlas GmbH & Co., Seeland Strasse 67, D-2400 Luebeck 14, Federal Republic of Germany.

MARGOLIS, Mark S. President, Rainbow Symphony Inc. 6860 Canby Road 11120, Reseda, CA 91335 USA. Telephone: (818) 708-8400. FAX: (818) 708-8470.

MARHIC, Michael. Vice-President, Holicon Corporation. 906 University Place, Evanston, IL 60201 USA; Northwestern University, Dept. of Electrical Engineering, Technological Institute, Evanston, IL 60208 USA. Telephone: (708) 4914310. FAX: (708) 491 7955.

MARKOV, Vladimir. Contact, Institute ofPhysics. Ukrainian Academy of Sciences, Prospect Nauki 46, 252 650 Kiev 28, USSR Telephone: (7) 222158.

MARKS, Gerald. Contact for Gerald Marks Studio, 29 West 26th Street, New York, NY 10010 USA. Telephone: (212) 889 5994.

MARSHALL, John. Contact for Institute of Opthalomology, Jud Street, London WC1, England, United Kingdom. Telephone: (44)(071) 3879621.

MARTRE, Mr. President, Aerospatiale, Ets D'Aquitaine, Sanit-Medard-en-Jalles, F-33165 Bordeaux, France. Telephone: (33) 5657 3480. FAX: (33) 5657 3548.

MASCIS, Michael. Sales Manager, Corion Corp. 73 Jeffrey Ave., Holliston, MA 01746 USA Telephone: (508) 429-5065. FAX: (508) 429-8983.

MASSENBURG, Henry. General Manager, Spectra-Physics Inc., Laser Products Division. 1250 West Middlefield Road, Box 7013, Mountain View, CA 94039-7013 USA Telephone: (415) 9612550. FAX: (415) 9694084.

MATHIEU, Marie-Christiane. Contact for Lea Productions Hololabl 3970, Boulevarde St. Laurent, Montreal, Qu~ H2W 1Y3, Canada. Telephone: (514) 849 4325.

MATTHIESEN, Johannes. Contact for Holocom. Lange Stra88e 51, D-2117 Kakenstorf, Federal Republic of Germany. Telephone: (49) (04186) 6510.

McCAFFREY, Avi. Vice-President, 3D Gallery, 207 Queen's Quay West, Thronto, Ontario, M6J 1A7 Canada. Telephone: (416) 359 0417.

McCAIN, Richard & Clare. President, Vice-President, McCain Marketing. 10962 North Wauwatosa Road 76W, Mequon, WI 53092, USA.

McCLEAN, James. 1 Employee at this addre88. 809 Marquette NW, ALbuquerque, NM 87102 USA. Telephone: (505) 243 8400. Description: Artist

McCORMACK, Sharon. Contact for School of Holography, 263 Montego Key, Novato, CA 94949 USA. Telephone: (415) 824 3769.

McGANN, Donna. Customer Service, Excitek, Inc. 277 Coit Street, Irvington, NJ 07111 USA. Telephone: (201) 3721669. FAX: (201) 372 8551.

McGEORGE, Neville. Contact for Rosewell Ltd. Blacknest Estate, Bentley, Alton, Hants., England, United Kingdom. Telephone: (44) (44) 42023605. FAX: (44)(44) 420 22517.

McGOWAN, Dr. J . William. President, AB-Bociates of Science and Technology (AST) Inc., 2450 Lancaster Road, Suite 36, Ottawa KIB 5N3, Canada.

McGREW, Steve. Contact for Light Impressions, Inc. 149-B Josephine Street, Santa Cruz, CA 95060 USA.

McINTYRE, Jim. Royal College of Art, Holography Unit. 175 Kings Cross Road, Kings Cross, London WCIX 9B2, England United Kingdom. Telephone: 44 01 584 5020, 44 01 837 9363. Description: Artist.

McKAY, Dave. President, Smith & McKay, 96 North Almaden Boulevard, San Jose, CA 95110-2490 USA. Telephone: (408) 292 8901 FAX: (408) 2920417.

McMAHAN, Robert, Sr. Contact for McMahan Electro-Optics, 2160 Park Avenue, (Orlando Division), Winter Park, FL 32789 USA.

McNAMARA, Maureen. Contact for The Hologram Schoppe. P.O. Box 318, 591 Tonawanda, Buffalo, NY 14202 USA.

McNAUGHTON, Jim. Marketing Director, Kaiser Optical Systems, 371 Parkland Plaza, 'Ann Amor, MI 48106. Telephone: (313) 665-8083

MEANS, Marcia M. 86 Hungry Hollow Road, Spring Valley, NY 10977 USA. Telephone: (914) 356 5203. Description: M.A. Art History, Columbia University 1979 - thesis on holography; Curator "The holographic instant" Museum of Holography, NYC 1987; Contributing editor Holosphere 1979-1988; Coauthor "Holography: Memories in light," documentary videotape; author exhibition catalogs.

MEDORA, Michael. Contact for Medora Waves, Studio 8, 1 Ranelagh Gardens, London SW6 3PA, England, United Kingdom. Technical Director, Raven Holographies, Telephone: 0730- 821612. FAX: 0730-821260.

MENNING, Melinda.171 Hopetown Avenue, Vaucluse 2030, Sydney N.S.W. Australia. Telephone: (61)(2) 337 1916. Description: Artistic holographer.

MEREDITH, Dennis. President, Meredith Instruments. 6403 North 59th Avenue, Glendale, AZ 85301 USA. Telephone: (602) 934 9387.

MERLIN. Achim Konz, Wassenberger Strasse 47, 5138 Heinsberg, Federal Republic of Germany. Telephone: (49)(2452) 6072. Company description: Artistic holography.

MERRICK, Michael G.,1002 Meadowview Drive, Mendota, IL 61342-1444 USA. Description: Artistic holographer. Description: Fine art originals

MERRILL, Mark C. 410 Riverdale-Studio B, Glendale, CA 91204 USA. Telephone: (818) 247 6458. Description: Artistic holograms; fine art originals.

MESTANCIK, Phillip E. Contact for Polaroid Corporation, 730 Main Street-lA, Cambridge, MA 02139 USA. Telephone: (BOO) 237 5519 or (617) 577-4307. FAX: (617) 577 5434.

METZ, MICHAEL. P.O. Box 332, Yorktown Heights, NY 10598 USA. Telephone: (914) 245-0997. Consultant specializing in holography, optical system analysis and design, laser scanning systems.

METZLER, Elaine. Contact for Asociacion Espanola de Holografia, Avda. Filipinas 3S-1A, Madrid 3, Spain.

MEULIEN, Odile. Contact for Holography World, (Branch of Art Science and Technology Institute). 800 K Street N.W., Techworld Plaza, Suite 54, Washington, D.C. 20001 USA. Telephone (202) 408-1833.

MEUSE, Ron. President, Mu's Laser Works. 1328 Dunsterville Avenue, Victoria, B.C., Canada V8Z 2X1. Telephone: (604) 479-4357.

MEYRUEIS, P. Contact for Universite Louis Pasteur, Photonics Systems Laboratory. 7 rue de L'Universite, 67000 Strasbourg, France. Telephone: (88) 355750.

MICHELITSCH, Barry. Vice-President, Holographic Developments LTD. Box 1035, Delta, B.C., V4M 3T2 Canada.

MIKHAYLOV, V.P. Contact for A.N. Sevchenko Research Institute of Applied Physical Problems. 220106 Minsk, USSR.

MILLER, D. Contact for Holographic Design, Inc. 1084 North Delaware Avenue, Philadelphia, PA 19125 USA. Telephone: (215) 425 9220. FAX: (215) 425 9221.

MILLER, Irvin. Contact for Modern Optics, 270 East Bonita Avenue, Pomona, CA 91767, Telephone: (714) 596-7741. FAX: (714) 596-9033.

MILLER, Mr. Peter. 2 Foxes Lane, MousehoIe, Cornwall, TR19 6QQ England, United Kingdom. Description: Holographic artist, consultant.

MILLMAN, Frank. CEO, Holographic Images Inc. 1301 Dade Boulevard, Miami Beach, FL 33139, USA Telephone: (305) 531 5465. FAX: (305) 531-3029.

MILLS, Karl. Contact for Alpha Photo Products, Inc., 985 Third Street, P.O. Box 23955, Oakland, CA 94623 USA. Telephone: (415) 893 1436 (ext. 1). FAX: (415) 834-8107.

MISTRY, Rohit. President, Jayco Holographics. 29/43 Sydney Road, Watford, Herta, WDI 7PY England, United Kingdom. Telephone: (44) 923246760. FAX: (44) 923 247 769.

MITAMURA, Shunsuke. Professor: Institute of Art & Design, University ofTsukuba. I-I, Tennodai, Tsukuba, Japan 305. Telephone: (81) (0298) 53 2833. FAX: (81)(0298) 53 6508.

MOLIN, Nils-Erik. Contact for Lulea University of Technology, Dept. of Mechanical Engineering, S-951 87 Lules, Sweden.

MOLTENI, William. 43 Hilltop Road, West Long Branch, New Jersey 07764 USA. Telephone: (617) 484 3592. Description: Marketing consultant.

MOORE, Duncan. Contact for University of Rochester, Institute of Optics, Rochester, NY 14627. Telephone: (716) 275 5248.

MORAINE, Mary. Sales Manager, Meredith Instruments. 6403 North 59th Avenue, Glendale, AZ 85301 USA. Telephone: (602) 934 9387.

MOREE, Samuel. Contact for New York Holographic Labs. P.O. Box 20391, 'lbmpkins Square Station, 176 East 3rd Street, New York, NY 10009 USA. Telephone: (212) 254 9774. or (212) 6741007.

MORETERUD, Alan. President, Imagen Holography, 135 Brush Creek Road, Suite B44, Snowmass Village, CO 81615 USA. Telephone: (1) 800- 43-PULSE and (303) 923-2905.

MORGUN, Yuri. Contact for Institute of Electronics.. BSSR Academy of Sciences-Minsk, 22 Logoiski Trakt, 220841 Minsk 90, USSR. Telephone: (7) Minsk 65-35-14.

MORI, John L. Contact for Illinois Valley Magnetic Resonanoe. 4005 Progress Boulevard, Peru, IL 61354 USA. Telephone: (815) 2238674.

MORRIS, G.M. President, Rochester Photonies Corporation. 67 Nettlecreek Street, Fairport, NY 14450 USA. Telephone: (716) 377-7990. FAX; (716) 377-7913.

MULHEM, Dominique, I, Rllsidence les Camélias, F-92600 Asnires, France. Telephone: (33) (1) 47 94 82 42. Description: I make fine art holography mixed with painting. This is called Holopainting. I exibit in art galeries and museums throughout the world. I have edited a monograph.

MULVANEY, Mark. President, Letterhead Press Inc., 155 North 120th Street, Wauwatosa, WI 53226 USA. Telephone: (414) 258 1717.

MUNDAY, Rob. Contact for Munday Spatial Imaging. 39 Pyrcron Road, Chertsey, Surrey KT16 9HT, England, United Kingdom. Telephone: (44)(0932) 564 899.

MURATA, M. Contact for Mitsubishi Heavy Industries Ltd., Nagasaki Technical Institute, II Akunoura-machi, Nagasaki 850-91 Japan. Description: Holographic non-destructive testing; industrial research.

MURRAY, Lisa. Director, Atlanta Gallery of Holography, 75 Bennett Street, Suite E-2, Atlanta, GA 30309 USA. Telephone: (404) 352-3412.

MURRAY, Jeffrey. President, Holography Institute, P.O. Box 446, Petaluma, CA 94953 USA. Telephone: (707) 778 1497.

MURRAY, Maria. Manager, Inrad Inc., 181 Legrand Avenue, Northvale, NJ 07647 USA. Telephone: (201) 767-1910. FAX; (201) 767-9644.

MURRAY, Rod. Course Leader, Royal College of Art, Holography Unit, Darwin Building, Kensington Gore, London SW7 2EU, England, United Kingdom. Telephone: (44)(01) 584 5020.

MUSKOVITZ, Aaron. P.O. Box 1022, South Lake Tahoe, CA 95705 USA. Telephone: (916) 544 5989. Description: Artistic holographer.

MUTH, August. Partner: Lasart Ltd. P.O. Box 703, Norwood, CO 81423 USA.

MYRE, Robert. 6090 Waverly, Montreal, Quebec, H2X 2A3 Canada. Telephone: (416) 533 4692. Description: artist.

N

NAEVE, Ambjorn. Contact for Dialectica AB, Skanegatan 87, 6tr, 8-116 37 Stockholm, Sweden.

NAIMARK, Michael. 216 Filbert Street, San Francisco, CA 94133 USA. Telephone: (415) 391 4817. Description: Holographic artist; fine art holograms.

NAKAJIMA, Masato. President, Holomedia Inc. 3-15-22, Takaban, Meguro-ku, 'lbkyo 152 Japan. Telephone: (81) (03) 793 2321. FAX; (81) (03) 793 2322.

NAKAMURA, Ikuo. Artistic holographer. 321 West 47th Street, N.Y., New York 10036. Telephone: (212) 315-2346. FAX: (212) 586-1181.

NALIMOV, Igor. Contact for Cinema & Photo Research Institute. NIKFI, Leningradsky, Prospect 47, Moscow USSR. Telephone: (7) (1570) 2923

NEKRASOVA, Larisa. Contact for Scientific Council on Exhibitions of the USSR Academy of Sciences, 30, Varilov Street, Moscow USSR. Telephone: (7)(135) 64 86.

NELSON, Drew. Business Development Manager, Laser Ionics Inc. 701 South Kirkman Road, Orlando, FL 32811 USA. Telephone: (407) 298 1561. FAX; (407) 297 4167.

NEMTZOW, Scott. 242 East Highland Avenue, Philadelphia, PA 19118. Telephone: (215) 242 2848. Description: Holographic artist.

NEUMAN, John. President, Laser Technology, Inc. 1055 West Germantown Pike, Norristown, PA 19403 USA

NEWELL, William. Vice-President, Laser Ionics Inc. 701 South Kirkman Road, Orlando, FL 32811 USA. Telephone: (407) 298 1561. FAX: (407) 297 4167.

NEWMAN, Paul. Sheerwater Lodge, Sheerwater Road, Woodham, Weybridge, Surrey KT15 3QL, England, United Kingdom. Telephone: (44)(09323)42396. Description: Personal artworks for exhibition and sale; tuition at all levels from beginners to M.A.; research into architectural and interior design applications; design for commercial display; technical consultancy.

NIEDZIALKOWSKI, George. Contact for Holographic Shop of Milwaukee, 5644 Parking Street, Greendale, WI 53129 USA.

NIGGEBRüGGE, Jorg. Customer Service, KolbeDruck, Coloco GmbH & Co. KG. 1m. Industrigellinde 50, Postfach 1103, D-4804 Versmold, Federal Republic of Germany. Telephone: (05423 2431 (-5). FAX: 05423 41230.

NORAIS, Filipe Vallada P. President, Ibero Gestäo Integrada e Tecnologica LDA. Apartado 1267, 4104 Porto Codex, Portugal. Telephone: (351-2) 301276.

O

OBZANSKY, J.J. & M.A. President, Contact, 'lbwne Laboratories, Inc.P.O.Box 460-HM, One U.S. Highway 206, Somerville, NJ 08876-0460 USA. Telephone: (908) 722 9500 FAX; (908) 722-8394.

ODHNER, Jeff. President, Odhner Holographies. 833 Laurel Avenue, Orlando, FL 32803 USA. Telephone: (407) 894 7966.

OHNUMA, K. Contact for 'lbppan Printing Co. Ltd., Central Research Institute. 5, 1-chome 'raito, Taito-ku, 'lbkyo 110, Japan.

OHZU, Hitoshi. Contact for Waseda University. Dept. of Applied Physics, School of Science and Engineering, 3-4-1, Ohkubo, Shinjuku-Ku, 'lbkyo 160, Japan.

OLIVER, Jane. Contact: Madem Systems Ltd.. Ladywell Trading Estate, Eccles New Road, Salford, M5 2DA England, United Kingdom.

OLLIER, Bernard. Contact for ERBA. 12, rue Libergier, F-51loo Reims, France. Telephone: (33)(26) 664 452.

OLMSTED, George. Contact for Lambda!l'en Optics, Division of Optical Corp. of America. One Lyberty Way, Westford, MA 01666 USA. Telephone: (508) 692-8140. FAX; (508) 6929416.

ONDREJIK, Charles. Sales Manager, 'lbwne Laboratories, Inc. P.O. Box 460-HM, One U.S. Highway 206, Somerville, NJ 06676-0460 USA. Telephone: (908) 722 9500 FAX; (908) 722-8394.

ORAZEM, Vito. Contact for Holo GmbH Holografielabor Oanabriick. MindemerStr. 205, D-4500 Osnabruck, Federal Republic of Germany. Telephone: (49)(0)(541) 7102 173. FAX; (49)(541) 7102 176.

ORR, Edwina. President, Richmond Holographic Studios. 6 Marlborough Road, Richmond, Surrey, England, TWIO 6JR United Kingdom. Telephone: (44)(1) 940 5525. FAX: (44)(1) 948 6214.

OSADA, Richard B. Contact for Wonderous Light Galleries, Tabor Center, Bridge Mart, 1201 16th Street, Denver, CO 80202 USA. Telephone: (303)399 2603.

OSADA, Richard M. Contact for Lazer Wizardry, Light Fantastic, 11022 West Oregon Place, Lakewood, CO 80226 USA; 2026 S. High Street, Denver, CO 80210 USA. Telephone: (303) 987 9438

OSE, Teruji. Contact for Kyoto Technical University. Dept. of Photographic Technology, Matsugasaki, Sakyo-ku, Kyoto 606, Japan.

OSTLER, Kevin D. Vice-President, Ion Laser Technology Inc. 263 Jimmy DolittIe Road, Salt Lake City, UT 84116 USA. Telephone: (801) 537 1587. FAX; (801) 537 1590.

OSTROVSKY, Juri. Contact for Leningrad Subsidiary in Machinery Science. Academy of Sciences of the USSR, Bolshoi Av. 61, 199178 Leningrad, USSR. Telephone: (7X247) 9185.

OTT, Hans-Peter. Curator Museum rur Holographie & neue visuelle Medien. Pletschmiihelenweg 7, D-5024 Pulheim I, Federal Republic of Germany. Telephone: (49)(02) 233385 1053. FAX. (49)(02) 238 52158.

OUTWATER, Chris. Vice-President, Applied Holographies Corp, 1721 Fiske Place, Oxnard, CA 93033 USA. Telephone: (805) 385-5670. FAX: (805) 385-5671.

OWEN, Harry. Contact for Pilkington PE Ltd. Glascoed Road, St. Asaph, Clwyd LL17 OLL, England, United Kingdom.

OWEN, Robert. Contact for NASA Marshall Space Flight Center. Space Sciences Laboratory, ES 73, Huntsville, AL 35812 USA.

OZAWA, Mr. S. Contact for Toppan Printing Co. Ltd. (Branch office). 1 Embarcadero Center, Suite 2106-MP, San Francisco, CA 94111 USA. Telephone: (415) 9827733. FAX: (415) 956 2551.

P

PADNOS, W.R. 2019 North Damen Avenue, Chicago, IL 60647 USA. Telephone: (213)384 2647. Description: Holographic artist; make fine art holograms, silver halide transmission and reflection; and presentation support.

PAGE, Michael. Contact for Ontario College of Art, 100 McCaul Street, Toronto, Ontario M9W 1W1, Canada. Telephone: (416) 977-5311-

PALAMAS, Yannis. Contact for: Laser Graphics. AG. Dimitriou 150, 546 35 Thessaloniki, Greece. Telephone: (30)(031) 908 087. FAX(30) (031) 234 173.

PALETZ, Jim. Contact for Palco Marketing, Inc. 11110 48th Avenue North, Plymouth, MN 55442. Telephone: (612) 559-5539.

PALMER, Caroline. 49 Upper Woodford, Salisbury, Wiltshire SP4 6NU, England, United Kingdom. Telephone: (44)(72) 273 471. Description: Holographic artisUconsultant on applications in architecture and museums.

PANICO, John. President, Images Company. P.O. Box 313, Jamaica, NY 11419 USA. Telephone: (718) 706-5003.

PARKER, Richard. Contact for Rolls-Royce PLC--Advanced Research Laboratory, P.O.Box 31, Derby DE2 8BJ, England, United .Kingdom. Telephone: (332) 242424. FAX: (332) 249936.

PARKER, S.C.J. Contact for British Aerospace PLC. Sowerby Research Centre, FPC: 267, P.O. Box 5, Filton, Bristol BS12 7QW, England, United Kingdom. Telephone: (44)(0272) 366 842. FAX: (44)(0272) 36 3733.

PATTERSON, Rich. Contact for Reynolds Metals Co. Flexible Packing Division, 6603 West Broad Street, Richmond, VA 23230 USA Telephone: (804) 2813969.

PAWLUCZYK, Romuald. Contact for Litton Systems Canada Ltd, 25 Cityview Drive, Rexdale, Ontario M9W 5A 7 Canada.

PEPPER, Andrew. 22 Haldane Road, London E6 3JJ, England, United Kingdom. Telephone: (44)(1) 471 1609. FAX: (44) (8l) 318 1439. Description: Fine art holograms; holography education.

PERIASAMY, Ammasi. Contact for Medical University of South Carolina, Dept. of Anatomy & Cell Biology, 171 Ashley Avenue, Charleston, SC 29425-2203 USA. Telephone: (919) 966-

2941.

PERKINS, K. Customer Service, Lumonies Inc. 105 Schneider Road, Kanata, Ottawa, Ontario K2K 1Y3, Canada.

PERRY, Dr. John & Barbara D. President, VicePresident, Holographies North Inc. 444 South Union Street, Burlington, VT 05401 USA. Telephone: (802) 658-2275. FAX: (802) 862-6510.

PERSCH, Ellen. Customer Service, Images Company. P.O. Box 313, Jamaica, NY 11419 USA. Telephone: (718) 706-5003.

PETERSEN, Joel. 7343 Adams Street, Paramount, CA 90723 USA. Telephone: (213) 634 0434. Description: Holographic artist; fine art holograms.

PETICOV, Antonio, 712 Broadway, New York, NY 10003 USA. Telephone: (212) 529 0465. Description: Makes fine art originals.

PETTERSSON, Sven-Goran. Contact for Lund Institute of Technology, Department of Physies, Box 118, S-221 00 LUND, Sweden.

PHILLIPS, Nicholas. Contact for Loughborough University of Technology, Dept. ofPhysies, Loughborough, Leicesterchire LE11 3TU, England, United Kingdom.

PINK, P. Contact for Holography Institute, P.O. Box 446, Petaluma, CA 94953 USA. Telephone: (707) 778 1497.]

PIZZANELLI, David. 24 Thirsk Road, London, SW11 5SX, England, United Kingdom. Telephone: (44) 071-350 0702. Description: Holographic artist.

PLUMB, Chris. Sales Manager, Electro Optics Developments Ltd., Howards Chase, Pipps Hill Industrial Estate, Basildon, Essex SS14 3BE, England, UK. Telephone: (44)(268) 531344. POE, Nelson. Contact for Active Image. P.O. Box 97, Boulder Creek, CA 95006 USA.

POLENTA, Manuela. Sales Manager, Metamorphosi Olografia Italia SRL. Via Lecco 6, 20124 Milano, Italy. Telephone: (39)(2) 204 9943. FAX: (39)(2) 204 1625.

PONZO, Sam. Contact for Spectrogon, 550 County Avenue, Secaucus, NJ 07904 USA. Telephone: (201) 867 4888. FAX: (201) 8672191.

POSTNIKOFF, Brian. President, 3D Gallery, 207 Queen's Quay West, Toronto, Ontario, MBJ 1A7 Canada. Telephone: (416) 359 0417.

PRICE, Kim. Customer Service, The Diffraction Company, Inc. P.O. Box 151, Riderwood, MD, 21152 USA. Telephone: (301) 666 1144. FAX: (301) 472 4911.

PRICE, Stu. Contact for Shipley Chemical Co. 1457 McArthur, Whitehall, PA 18052 USA. Telephone: (215) 820 9777.

PRICONE, Robert. President, Holographic Industries, Inc. 3 Warwick Lane, Lincolnshire, IL 60069 USA. Telephone: (708) 680 1884. FAX: (708) 680 0505.

PROSTEV, A. Contact for S.1. Vavilov, State Optics Insti tute, 199064 Leningrad, USSR.

PROVENCE, Steve. Contact for Steve Provence Holography. 15220 Fern Avenue, Boulder

Creek, CA 95006 USA. Telephone: (408) 338 2051.

PRYPUTNIEWICZ, Ryszard. Contact for Worcester Polytechnic Institute. Mechanical Engineering Department, 100 Institute Road, Worcester, MA 01609-2280 USA. Telephone: (508) 8315536. FAX: (508) 8315483.

PUMMER, Herbert. Customer Service, Coherent Laser Group, 3210 Porter, Palo Alto, CA 94304.

PYE, Tim. 19 Wonderland Avenue, Tamarama, Sydney, 2026 Australia. Telephone: (61X2) 306 611. Description: Artistic holographer.

Q

QUALLS, Steve. Contact for Apollo Lasers, Inc. 12351 Research Parkway, Orlando, FL 32826 USA. Telephone (407) 281-4103

QUINLAN, Denis. 48 Clifton Street, North Balwyn, Victoria 33104 Australia. Telephone: (61)(3) 857 8655. Description: Marketing consultant.

R

RACEY, Carl. Sales Manager, Amazing World of Holograms. Corrigan's Arcade, Foreshore Road, South Bay, Scarborough, North Yorkshire YO11 1PB, England, United Kingdom; Contact for Laser Light Image. 101 Spring Bank, Hull, HU3 1BH, England, United Kingdom. Telephone: (44)(0482) 26744 . FAX: (44) (0482) 492 286.

RALLISON, Richard. President, Ralcon. Box 142,8501 South 400 West, Paradise, UT 84328 USA. Telephone: (801) 245 4623. FAX: (801) 245 6672.

RAMUNKE, Thomas. Contact for Dream Images. Telephone: (49) 02236 43138 (gallery: 02644-3147). FAX: (49) 02236-82369.

RANKIN, Kevin. Application Engineer, Physical Optics Corporation, 2545 W 237th Street, Suite B, Torrance, CA 90505 USA. Telephone: (213) 530-1416. FAX: (213) 530-4577.

RAPKE, Marc. Contact for Holographic Marketing, Inc., 9250 S.W. First Street, Plantation, Florida 33324, USA. Telephone: (305) 474-9965. FAX: (305) 472-0101.

RATHJE, Barbel. Contact for Holographic Art. Werderstralle #73, Bremen 2800, Federal Republic of Germany. Telephone: (49)(421) 555 690. FAX: (49)(421) 556 202.

RAVE RAT DE BOISHEU, J-C. Contact for Holographic Creations. 26- Rue Daniel Stern, Paris 75015, France. Telephone: (33)(1) 45 78 8742.

RAYFIELD, Dave & Holly. President, Vice-President, Holographic Products, Inc. 755 South 200 West, Richmond, UT 84333 USA. Telephone: (801) 752-4225. FAX: (80l) 752-4467.

RAYMAN, R. Sales Manager, Lumonies Inc. 105 Schneider Road,Kanata, Ottawa, Ontario K2K 1Y3, Canada.

REALS, Bob. Contact for State Education Department of New York, Bureau of Arta and Music Education. "Holography in the Classroom" Project, Room 681, Albany, NY 12234. Telephone: (518) 474-8773. FAX: (518) 473-3072.

REDZIKOWSKl, Mark. Product Manager, Agfa Gevaert Inc.-Industrial Division, 100 Challenger Road, Ridgefield Park, NJ 07660 USA. Telephone: (201) 440 2500 ext. 4226. FAX: (201) 440 5733.

REED, Judy. President, JR Holographies, Suite 1660, 100 Wilshire Blvd., Santa Monica, CA 90401-1135 USA. Telephone (213) 393-2388. FAX: (213) 393-8611.

REED, Terry. Contact for Newport Corporation. 18235 Mt. Baldy Circle, P.O. Box 8020, Fountain Valley, CA 92708-8020 USA. Telephone: (714) 963 9811. FAX: (714) 9632015.

REID, Ray. Contact for Omnichrome. 13580 Fifth Street, Chino, CA 91710 USA. Telephone: (714) 627-1594. FAX: (714) 591-8340.

REILLY, John. 238a Gloucester Terrace, London W2, England, United Kingdom. Telephone: (44)(01) 243 0601. Description: Artist.

REINGRUBER, Juergen. Sales Manager, Adlas GmbH & Co., Seeland Strasse 67, D-24OO Luebeck 14, Federal Republic of Germany.

RELANZON, Santiago. Contact for Lasergrafies, Pens y Valero 130, 2a Valencia, Spain. Telephone: (34)(96) 333 3013.

RENVALL, Ms. Mikaela. Customer Service, Starcke KY. P.O. Box 22, SF-32811, Peipohja, Finland. Telephone: +(358) (39) 360700. FAX: +(358) (39) 367230.

REUTERSWARD, Carl Fredrick. 6 Rue Montolieu, 1030 BussignylLausanne, Switzerland. Telephone: (41)(021) 701 0514. Description: Artistic holography.

REZNY, Abe. Owner, Laser Light Ltd. 57 Grand Street, New York, NY 10013 USA. Telephone: (212) 226 7747.

RHODY, Alan. Sales Manager, A.H. Prismaticl Holos Gallery. 1792 Haight Street, San Francisco, CA 94117 USA Telephone: (415) 2214717. FAX: (415) 2214815.

RICE, Graham. Contact for Newcastle Upon Tyne Polytechnic, Department ofPhysies, Ellison Building, Newcastle upon Tyne, NE1 8ST. England, United Kingdom. Telephone: (44) (091) 2358453. FAX: (44)(091) 2358017.

RICHARDS, Beverly. Contact for Laser Institute of America, Education Division, 12424 Research Parkway 11130, Orlando, FL 32826 USA. Telephone: (407) 380-1553. FAX: (407) 380-5588.

RICHTER, Thomas. President, Richter Enterprises. 640 19th Street, Manhattan Beach, CA 90266-2509 USA. Telephone: (213) 546-5107. FAX (213) 545-6757.

RICKERT, Susan. Owner, Hologram Land, 43 Main Street SE, 11131 Riverplace, Menneapolis, MN, 55414, USA.

RITCHER, Michael. Contact for Ritcher Holograms. Adolf Kolping Strasse 16, D-4050 Monchenglabbach, Federal Republic of Germany.

RICHTER, Thomas. President, Richter Enterprises. 640 19th Street, Manhattan Beach, CA 90266-2509 USA. Telephone: (213) 546-5107. FAX (213) 545-6757.

RITSCHER, Eve. Contact for Eve Ritscher Associates Ltd., 73 Allfarthing Lane, Wands worth, London SW18, England, United Kingdom.

RIZZI, Mrs. M. Luciana, Eng., CISE SpA Technologie Innovative. via Reggio Emilia, 39, 20090 Segrate, Milano, Italy. Mailing address: P.O.Box 12081, 1-20134 Milano, Italy. Telephone: (39)(2) 21672634. FAX: (39)(2) 2167 2620.

ROA, Warna J. 2450 Lancaster Road-Unit 36, Ottawa, Ontario, Canada, KIB 5N3. Telephone: (613) 521 2557. Description: Marketing consultant.

ROBINSON, Anthony. Contact for Robinson Hologram Lighting Systems. 5 Hillside Cottages, Owismoor Road, Camberley, Surrey GU15 4SU, England, United Kingdom. Telephone: (44)(0344) 762739.

ROBINSON, D.W. Contact for National Physical Laboratories, Queens Road, Teddington, Middlesex TW11 OLW, England, United Kingdom. Telephone: (44)(01) 977 3222.

ROBINSON, George. Contact, Holos, 3001 Hennepin Avenue, Space E202, Minneapolis, MN 55408 USA. Telephone: (612) 824-1404.

ROBITAILLE, Gaetan. Vice-President, Lasiris Inc. 3549 Ashby, Ville St. Laurent, Que, Canada H4R 2K3.

ROGERS, Grahame. Sales Manager, Cambridge Lasers, Inc., 43216 Christy Street, Fremont, CA 94538 USA. Telephone: (415) 651-0110. FAX: (415) 651-1690.

ROGERS, John. Sales Manager, Smith & McKay, 96 North Almaden Boulevard, San Jose, CA 95110-2490 USA. Telephone: (408) 292 8901 FAX: (408) 2920417.

ROHLER, R. Contact for University of Munich, Institute of Medical Optics, Thresienstrasse 37, D-8000 Munich 2, Federal Republic of Germany.

ROOSE, Stephan. Contact, Free University of Brussels, Department of Applied Physics (ALNA), Faculty of Applied Sciences, Brussels, Belgium.

ROSEWELL, Michael P. 2302 South Damen Avenue, Chicago, IL 60608 USA. Telephone: (312)254 2577. Description: Fine art originals, represent Micraudel(Schiltigheim, France) and their thermoplastic camera.

ROSS, Franz. President, Ross Books. P.O. Box 4340, Berkeley, CA 94704 USA. Telephone: (415) 8412474. FAX: (415) 8412695.

ROSS, Jonathan. 4 Macaulay Road, London SW4 OQX, England, United Kingdom. Telephone: (44)(01) 622 7729. FAX: (44)(01) 622 5308. Description: Holography consultant and collector.

ROSSIGNOL, Yves. Contact for Holograma Laboratoire Holographique, 41 rue Mariziano, CH-1227 Geneva, Switzerland.

ROSSING, Thomas. Contact: Northern Illinois University. Department of Physies, DeKalb, IL 60115-2854 USA. Telephone: (815) 7531772.

ROTH, Stephen. Contact for Mind's Eye: Holographic Consultants, 17329 Zola Street, Granada Hills, CA 91344 USA. Telephone: (818) 360 6023.

ROTHAUS, Valli. Contact for Space Age Designs Inc, P.O. Box 72, Carversville, PA 18913 USA. Telephone: (215) 297 8490.

ROTTENKOLBER, H. Contact for Rottenkolber Holo-System GmbH, Henschelring 15, 0-8011 Kirchheim / Munich, Federal Republic of Germany.

ROY, Joanne. 3575 Rue de Bullion, Montreal, Quebec, H2X 3Al Canada. Telephone: (514) 845 4419. Description: Fine art holography

RUDERMAN, Warren. President, Inrad, Inc., 181 Legrand Avenue, Northvale, NJ 07647 USA. Telephone: (201) 767-1910. FAX: (201) 767-9644.

RUSCHMAN, Henry. Contact for Alpha Foils, Inc. P.O. Box 152, Bernardsville, NJ 07924 USA. Telephone: (908) 766-1500. FAX: (908) 766-1501.

RYDEN, Hans. Contact for Karolinska Institutet, School of Dentistry, Box 4064, 8-141 04 HUDDINGE, Sweden. Telephone: (46X08) 774 0080.

S

SACHTLER, Ek. Contact for Ilford Inc. West 70 Century Road, Paramus, New Jersey 07653 USA. Telephone: (201) 265 6000. FAX: (201) 265-0894.

SADDEQ, Walled. Customer Service, Wonders of Holography, P.O. box 1244 Jeddah, Saudi Arabia. Telephone: (966)(2) 652 0052. FAX: (966)(2) 651 1325.

SAITO, Takayuki. Contact for Fuji Photo Optical Co. , Ltd. No. 324, 1-Chome, Uetake-Machi, Omiya, Japan. Telephone: (81)(04) 866 30111. FAX: (81)(04) 86510

SALAM, M. A. Sales Manager, Wonders of-Holography Gallery. P.O. Box 1244, Jeddah 21431, Saudi Arabia. Telephone: (966)(2) 652 0052. FAX: (966)(2) 651 1325.

SALG, Joseph. General Manager, Laser Applications, Inc. 12722 Research Parkway, Orlando, FL 32826 USA. Telephone: (407) 380 3200. FAX: (407) 381 9020.

SAMSON, Kevin. Sales Manager. Light Impressions, Inc. 149-B Josephine Street, Santa Cruz, CA 95060 USA.

SANDWELL, R.S. Vice-President, Lumonies Inc. 105 Schneider Road,Kanata, Ottawa, Ontario K2K lY3, Canada.

SANTOS, Antonio A. Rua Da Quintanda, 194-Sala 404, CEP 20091 , Rio de Janeiro, Brazil. Telephone: (55)(021) 233 5590. Branch office of Newport Corporation, Fountain Valley, CA USA.

SAPAN, Jason. Contact for Holographic Studios. 240 East 26th Street, New York, NY 10010 USA. Telephone: (212) 686 9397. FAX: (212) 4818645.

SAXBY, Graham. Contact for Wolverhampton Polytechnic, Holography Unit. Holography consulting, editor/writer of books & articles on holography and photography. 3 Honor Avenue, Goldthorn Park, Wolverhampton, West Midlands WV4 5HF, United Kingdom. Telephone: (44) 902 313004. FAX: (44) 90225015.

SCHAFFNER, David. Contact for Newport Holograms. 3412 Via Oporto, Suite 2, Newport Beach, CA 92663 USA. Telephone: (714) 675 1337.

SCHEIR, Peter. Contact for Another Dimension! AD 2000, 948 State Street, New Haven, CT 06511 USA.

SCHICKER, Scott. Vice-President, Ralcon. Box 142, 8501 South 400 West, Paradise, UT 84328 USA. Telephone: (801) 245 4623. FAX: (801) 245 6672.

SCHMELZER, Dr. Carlo. Contact for Studio fur Holographie. Moosstral3e 27, D·8031 Eichenau, Federal Republic of Germany (FRG). Telephone: 08141 70831.

SCHMIDT, David. Owner, David Schmidt Holography. 23962 Craftsman Road, Calabasas, CA 91302 USA. Telephone: (818) 992 1541. FAX: (818) 703 1182.

SCHOENBECK, Gerhard. Contact for Kranwerk Union AG. D·4330 MulheimIRubr, Federal Republic of Germany.

SCHORR, Alan and Allison. Contacts for Visions of the Future, 1040 N. Kings Highway, Suite 717, Cheny Hill, N.J. 08034 USA. Telephone: (609) 667·5622. FAX: (609) 667·6289.

SCHULTZ, Lynn & Bob. Contacts for E.C. Schultz & Company. 333 Crossen, Elk Grove Village, IL 60007 USA. Telephone: (312) 640 1190. FAX: (312) 640 1198.

SCHUMACHER, Dr. Werner. Contact for'Ibpac GmbH. Aufm Eickholt 47, 4830 Gutersloh I, West Germany. Telephone: 0 52 41 580192. FAX: 0 52 4158549.

SCHUMANN, Walter. Contact for Swiss Federal Institute of Technology, Laboratory of Photoelasticity, Ramistrasse 101, CH·8092 Zurich, Switzerland. Telephone: (41)(0380 246 000.

SCHVARTZMAN, Frederic. General Manager, The Foreign Dimension, Suite B, 8JFl., Central Mansion, 270-276 Queen's Road Central, Hong Kong. Telephone: (852)(542·0282. FAX: (852) 541·6011.

SCHWEITZER, Dan. Contact for New York Holographic Labs. P.O.Box 20391, Tompkins Square Station, 176 East 3rd Street, New York, NY 10009 USA. Telephone: (212) 254 9774. or (212) 6741007.

SCIAMMARELLA, Cesar. Contact for Illinois Institute of Technology, Mechanical & Aerospace Engineering. Engineering Building IH, Room 2460, Chicago, IL 60616 USA. Telephone: (312) 567 3249.

SCRIFF, Joe. Contact for Mitutoyo Measuring Instruments. 18 Essex Road, Paramus, NJ 07852 USA. 'Thlephone: (201) 368 0525.

SEKULIN, Robert. Contact for Rutherford and Appleton Laboratories, Chilton, Didcot, Oxon 0X11 0QX, England, United Kingdom. Telephone: (44)(0235) 21900.

SENECHAL, Jacques. 90 Temple Fortune Lane, London NW11 7TX, England, United Kingdom. Telephone: (44)(1) 458 5825. Description: Artistic holographer, large format holograms.

SEROV, O.B. Contact for VNIKTK Kultura, ul. Intusiastov 34,1051118 Moskow, USSR.

SEURAY, Dominique. Customer Service, Atelier Holographique de Paris, 13, Passage Courtois, F·75011 Paris, France.

SEYDEL, Bill. Sales Manager, HololSource Corporation. 21800 Melrose Avenue, Southfield, MI 48075 USA. Thlephone: (313) 355 0412. FAX: (313) 355 0437.

SHANNON Robert. Contact for The University of Arizona, Optical Science Center, Tucson, Arizona 85721 USA. Telephone: (602) 6216997.

SHAPIRO, B.I. Contact for State Research and Prtoject Institute of Chemica-Photographic Industry, 125167 Moskow, USSR.

SHERMAN, H.T. President, Advanced Optics, Inc. 5 E. Old Shakopee Rd., Minneapolis, MN 55420 USA.

SHERWOOD, Robert. President, Robert Sherwood Holographic Design, Inc. 400 West Erie Street, Chicago, IL 60610 USA. 'Thlephone: (312) 944 3200.

SHETKA, Stanley. Contact for World Art Prtoject, 10247 40th Street West, Webster, MN 55088 USA.

SHEW, Ellen M. President, Chimeric Images, Inc. 713 112 Mai n Street, Lafayette, IN 47901 USA. Telephone: (317) 742 0586.

SHIOZAKI, Yumiko. Contact for Holo Arp; Independent artist. 7·5·18·Ryoke Urawa, Saitama, Japan. 'Thlephone: (81X0488) 317 723. Description: Artistic holographer. Telephone: (81) (03) 574 8307. FAX: (81)(03) 574 8377.

SHOLLY, Kate. Customer Service, ISAsrl Leonardo, P.O. Box 75, 1442A Walnut Street, Berkeley, CA 94709 USA. Telephone: (415) 845 8306. FAX: (415) 8416311.

SHREVE, Loy. Contact for T.A.I. Incorporated. 12021 South Memorial Parkway, Huntsville, AL 35803 USA. Telephone: (205) 881 4999. FAX: (205) 880 8041.

SHVARTS, K.K. Contact for Physics Institute. Latvian SSR Academy of Sciences, 229021 Riga-Salaspils, USSR. Telephone: 007·01321 947642.

SIA, Teck·Mong. President, Electech Distribution Systems, 605·A, MacPherson Road, #03·03, Citimac Industrial Complex, Singapore 1336. Telephone: (65)286 9933. FAX: (65) 284 3256.

SIEGEL, Steven. Partner: Lasart Ltd. P.O.Box 703, Norwood, CO 81423 USA.

SIKORSKY, Zbigniew. Contact, Institute of Plasma Physics and Laser Microfusion, P.O. Box 49, 00·908 Wroclaw, Poland.

SIMMS, Lloyd. Sales Liaison, Electro Optical Industries, Inc. 859 Ward Drive, Santa Barbara, CA 93111 USA.

SIMSON, Bernd & Paula. Contact, General Holographics, Inc. P.O. Box 82247, Burnaby B.C., V5C 5P7 Canada Telephone: (604) 435 6654. FAX: (604) 4327326.

SINCERBOX, Glenn. Contact for IBM Almaden Research Center. K69/803, 650 Hany Road, San Jose, CA 95120 USA. Telephone:

(408) 927 1937.

SITCH, B.J. Vice-President, George M. Whiley Limited. Firth Road, Houston Industrial Estate, Livingston, West Lothian EH54 5DJ, Scottland, United Kingdom. Telephone: (0506) 38611 FAX: (0506) 38262.

SIVY, George. President, Gray Scale Studios Ltd. 4500 19th Street, #588, Boulder, CO 80304 USA. Telephone: (303) 442 5889. FAX: (303) 442 5889.

SIXT, W. Customer Service, Steinbichler Optotechnik GmbH. Am Baubof 4, D·8201 Neubeuern, FRG Federal Republic of Germany. Telephone: (0049) 80351017. FAX: (0049) 8035 1010.

SJOLINDER, Sven. Contact, Royal Institute of Technology, Institute of Photography, 8-100 44 Stockholm, Sweden.

SKANDE, Per. Contact for Martinsson Elektronik AB, Instrumentvagen 16, Box 9060, 8-126 09 HAGERSTEN, Sweden.

SKIPNES, Olav and Bodiil. Contacts for Interferens Holografie, Halvor Hoels Gt. 6, N·23oo Hamar, Norway. Telephone: (065) 25050 or (065) 30659.

SKRYPNYK, Susan. Vice-President, Starlight Holographic Inc. 73 Stable Way, Kanata, Ontario K2M 1A8, Canada. Telephone: (613) 2350440. FAX: (613) 5927647.

SMELZER, Geert T. A. Contact for Technical University of Eindhoven. Faculty of Architec·ture, Calibre Institute, P.O. Bax 513, Eindhoven, NI.r5800MB, The Netherlands.

SMIGIELSKI, Paul. President, Holo 3. rue de l'Industrie, 66300 Saint·Louis, France. 'Thlephone: (33)(89) 69 82 08.

SMITH, Prof. Contact, British Aerospace PLC. Sowerby Research Centre, FPC: 267, P.O. Box 5, Filton, Bristol BS12 7QW, England, United Kingdom.

SMITH, Steve. Contact for Beddis Kenley Engineering Ltd., Unit 3, Elland 'Thrrace, Holbeck, Leeds LS11 9NW England, UK Telephone; (44) (0532) 465 979.

SMITH, Steven. Contact for The Lasersmith, Inc. 1000 West Monroe Street, Chicago, IL 60607 USA. Telephone: (312) 733 5462.

SNYDER, Bruce. Contact for Holomart Pic. Hamilton House, 1 Temple Avenue, London EC4Y OHA, England, United Kingdom. Telephone: (44)(01) 353 4212. FAX: (44)(071) 353 3325.

SOARES, Oliverio. Contact for Universidade do Porto, Laboratorio de Fisica, Praca Gomes'Thixeira, P·4000 Porto, Portugal.

SOBOLEV, G.A. Contact for FTI Joffe, Politechnicheskaya 26, Academy of Sciences of the USSR, 194021 Leningrad, USSR.

SOTT, Gudrun. Sales Manager, AKS Holographie·Galerie GmbH. Potsdamer Straße 10,4300 Essen I, Federal Republic of Germany. Telephone: (49)(020l) 704 562.

SOUPARIS, Hugues. President, Hologram. Industries, 42-44, rue de Trucy, 94120 Fontenay sous bois, France. Telephone: 1 4394 1919. FAX: 143940032.

SOWDON, Michael. Contact for Fringe Research Holographies Inc.1l79A IGng Street West, Suite 008, Toronto, Ontairo M6K 3C5, Canada. Telephone: (416) 535 2323.

SPALDING, Jean. Contact for Norland Producta, Inc. 695 Joyce Kilmer Avenue, P.O. Box 145, North Brunswick, NJ 08902 USA. Telephone: (908) 545 7828. FAX: (908) 545 9542.

SPIERINGS, Walter. Contact, Dutch Holographic Laboratory. Kanaal dyk Noord 61, 5642 JA Eindhoven, The Netherlands. Telephone: (31)(40) 817 250. FAX: (31)(40) 814 865.

SPIRO, Shari. Consultant, Sillcocks Plastics International. 310 Snyder Avenue, Berkeley Heights, NJ 07922 USA. Telephone: (800) 922 0958. FAX: (201) 665 1856.

SPREER, Elmar. Contact for Phoenix Holograms, Traubenstr. 41, D-2913 - Apen, Federal Republic of Gennany. Telephone: (49)044 89 15198

SPRINGER, Greg. Customer Service, LiCONix, 3281 Scott Boulevard, Santa Clara, CA, 95054 USA.

STALLARD, PENN. P.O. Box 4851, Chicago, IL 60680-4851. FAX: (clo Marble Supply International) (312) 278-8989. Description: Sculptor/Holographer. I work in bronze, stone (marble and granite) and holography, and combine these mediums/material both sculpturally and architecturally. Also, design and fabricate furniture combining stone and holography.

STARCKE, Mr. Ari-Veli & Mrs. Milla-Ritta. President, Vice-President, Starcke KY. P.O. Box 22, SF-32811, Peipohja, Finland. Telephone: +(358) (39) 360700. FAX: +(358) (39) 367230.

STASELKO, F.I. Contact for S.l. Vavilov, State Opties Institute, 199064 Leningrad, USSR.

ST. CYR, Suzanne. President, Holographic Applications. 21 Woodland Way, Greenbelt, MD, 20770 USA. Telephone: (301) 345 4652. FAX: (301) 345 4853.

STECKER, Debbie. Contact for Metrologic Instruments, Inc. Coles Road at Route 42, P.O. Box 1458 USABIackwood, New Jersey 08012. Telephone: (609) 228 8100. FAX: (609) 228 6673.

STEINFELD, Belle. Contact for Dell Optics Company, 25 Bergen Blvd., Fairview, NJ 07022, USA. Telephone: (201) 941-1010. FAX: (201) 941-9524.

STEINBICHLER, Dr. H. President, Steinbichler Optotechnik GmbH. Am Bauhof 4, D-8201 Neubeuern, FRG Federal Republic of Germany. Telephone: (0049) 80351017. FAX: (0049) 80351010.

STENSBORG, Jan. Customer Service, Superior Technology Implementation. Hjortekaersvej 99B, 2800 Lyngby, Denmark. Telephone: (45) (45) 933358. FAX: (45)(45) 930353.

STERN, Rudy. Contact for Let There Be Neon. P.O. Box 337, Canal Street Station, New York, NY 10013 USA.

STETSON, Karl. Contact for United Technologies Research Center. Optics and Acoustics 86, Silver Lane, East Hartford, CT 06108 USA. Telephone: (203) 727 7060. FAX: (203) 7277852.

STEYER, Dr. Bernhard. President, Adlas GmbH & Co. Seeland Strasse 67, D-2400 Luebeck 14, Federal Republic of Germany.

STIEF, G. Vice-President, Labor Dr. Steinbichler, Am Bauhof 4, D-8201, Neubeuern, Federal Republic of Gennany. Telephone: (0049) 8035 1018. FAX: (0049) 80351010.

STOHL, Mr. Robert G.' Re:Laser Regulations, Electro Optical Specialist, S.J. Fed.Bldg--US Court House, 280 South 1st Street, Rm 2062, San Jose, CA 95113 USA.

STOCKLER, Len. 7227 Eastwood Street, Philadelphia, PA 19149 USA. Telephone: (215) 331 5067. Description: Clearing house for practical holography. Classes and individualized workshops. Reference library. Prototype Designs for optical hardware, vibration isolation tables, sensitized materials processing,lighting and display. Educational consulting and project management.

STRAZDS, Glenn. Contact for Holo Laser Tech Ltd, 7 Fraser Ave, #16, Thronto, Ontario M6K 1 Y7 Canada. Telephone: (416) 534-7419.

STREATOR, Stacey, Customer Service, Holographics North, 444 South Union Street, Burlington, VT 05401 USA. Telephone: (802) 658-2275. FAX: (802)862-6510.

STRIEBIG, Jocelyne. Deputy Director. Holo 3. rue de I'Industrie, 68300 Saint-Louis, France. Telephone: (33)(89) 69 82 08.

STROKE, George. Contact for Messerschmitt-Boelkow-Blohm. Zentrale Entwicklung MBB, Postfach 801109, D-8000 Munich 2, Federal Republic of Germany.

STURGESS, Rosemary. Marketing Manager, Laser Light Expressions Pty. Ltd. Holoptics. 3 Gibbons Street, Telopea, New South Wales, Australia 2117. Telephone: (612) 890 1233. FAX: (612) 890 1243.

SUGARMAN, Steven. Contact for Holographic Concepts, Inc. 1711 St. Clair Avenue, St. Paul, MN 55115. Telephone: (612) 698-6893.

SVENNSON, Lennart. Contact for Royal Institute of Technology. Department of Industrial Metrology, S-10044 Stockholm, Sweden. Telephone: (48)(08) 790 7823. FAX: (48)(08) 790 8219.

SWEENEY, Donald. Contact for Sandia National Laboratories. Combustion Research Facility. Livermore, CA 94550 USA. Telephone: (415) 294 3138.

SYNOWIEC, George. Contact for JK Lasers Ltd. Cosford Lane, Swill; Valley, Rugby, Warwickshire CV211QN, England, United IGngdom.

SYNOWIEC, J .A. Contact for Lumonies, Ltd.. Cosford Lane, Swill Valley, Rugby, Warwickshire, CV21 1QN, United Kingdom.

SZANTO, David. Contact for Ontario Science Centre, 770 Don Mills Road, Don Mills, Ontario, M3C 1T3 Canada. Telephone: (416) 429 4100.FAX: (416) 429 5961.

SZeKELY, Zsolt. Sales Manager, Artplay Holographic Studio. H-1191 Budapest, Ady Endre ut. 8, Hungary.

SZPAKOWSKI, Brian. Sales Manager. A.H. Prismatic, Inc. 285 West Broadway, New York, NY 10013. USA. Telephone: (212) 219 0440. FAX: (212) 219 0443.

T

TALBaIT, M. Contact for Laser Arts, 1712 Cathedral Street, Plano, TX 75023 USA. Telephone: (214) 423 0158

TAYLOR, Wes. Custonner Service, Aerotech Inc., Electro Optical Division, 101 Zeta Drive, Pittsburgh, PA 15238 USA.

TEITEL, Michael. Contact for Massachusetts Institute of Technology, M.l.T. Media Laboratory, Spatiat Imaging Group, 20 Ames Street, El5- 416, Cambridge, MA 02139 USA.

THALEMANN, Udo. President .. Thpac GmBH. Auf'm Eickholt 47, 4830 Gutersloh I, West Germany. Telephone: 0 52 41 580192. FAX: 0 52 41 58549.

THIELKER, Klaus. Contact, Dream Images. Postfach 1602, Venneerweg 15, D-5047 Wesseling, Federal Republic of Germany Telephone: (49) 02236 43138 (gallery: 02644-3147). FAX: (49) 02236 82369.

THOMPSON, Bare. President, A.H. Prismatic, LTD. New England House, New England Street, Brighton, BN1 4GH England, United Kingdom. Telephone: (44) 0273 686 966. FAX: (44) 0273676 692.

THORNTON, Gary. Customer Service, Lenox Laser. 1 Green Glade Court, Phoenix, MD 21131 USA. Telephone: (301) 592-3106. FAX: (301) 592-3362.

TIZIANI , Hans. Contact for University of Stuttgart. Institute of Applied Optics, Pfaffenwaldring 9, D-7000 Stuttgart 80, Federal Republic of Gennany. Telephone: (49X0711) 685 6075.

TOBIN, John. Contact for Laser Light Expressions Pty. Ltd. Holopties. 3 Gibbons Street, Telopea, New South Wales, Australia 2117. Telephone: (612) 890 1233. FAX: (612) 890 1243.

TOMKINS, Don. Contact for Advanced Holographies, LTD. 243 Lower Mortlake Rd.,Unit 11, Richmond, Surrey TW9 2LL, England, United Kingdom.

TOMKINS, J on. President, Spectra-Physics, Boundary Way, Hemel Hempstead, Herts. HP2 7SH, England, United Kingdom. Telephone: 0442 232 322. FAX: 0442 68538.

TOMKO, Martha. Director, Museum of Holography, 11 Mercer St., NY, NY 10013 USA. Telephone: (212) 925 0581 or 925-0526. FAX: (212) 334-8039.

TORCHIA, John. Contact for Bobst Group, 148-T Harrison Avenue, Roseland, NJ 07068 USA. Telephone: (201) 226 8000.

TRAYNOR, David. Vice-President, Richmond Holographic Studios. 6 Marlborough Road; Richmond, Surrey, England, TWIO 6JR United Kingdom. Telephone: (44)(1) 940 5525. FAX: (44)(1) 948 6214.

TRIBILLON, Gilbert. Contact for Universite de Franche-Comte, Laboratoire d'Optique, L.A. 214, UFR Sciences et des Techniques, F-25030 Besancon Cedex, France.

TRIBILLON, Dr. Jean Louis, H., Contact for Holo-Laser. 6, rue de la Mission, Ecole, 25480 Miserey, France. Telephone: (33)(1) 45 315 275. FAX: (33Xl) 48 331702.

TROLINGER James. Partner, Metrolaser. 18006 Skypark Circle, Suite 108, Irvine, CA 92714-6428 USA. Telephone: (714) 553-0688. FAX: (714) 553-0495.

TSUFURA, Lisa. Technical Manager, Melles Griot,1770 Kettering Street, Irvine, CA 92714 USA Telephone: (714) 261-5600. FAX: (714) 261-7589.

TSUGE, Hibiki . Customer service, Holomedia Inc. 3-15-22, Takaban, Meguro-ku, Tokyo 152 Japan.

TSUJIUCHI, Jumpei. Contact for Chiba University. Faculty of Engineering, 1-33 Yayoicho, Chiba 260, Japan. Telephone: (81X0472) 511 III ext. 2874.

TUNNADINE, Graham .46 Calthorpe Street, London, WCl England, United Kingdom. Telephone: (44)(1) 278 1572. Description: Holographic artist.

TURNER, Brian. President, Excitek, Inc. 277 Coit Street, Irvington, NJ 07111 USA. Telephone: (201) 372 1669. FAX: (201) 3728551.

TURUKANO, Boris. Contact for Institute of Nuclear Physies, Leningradska obI., 188350 Gatchina, USSR.

TYLER, Douglas E. 1450 Broadway Street, Niles, MI 49120-2123USA. Telephone: (616) 683 0934. Description: holographic consulting.

TYRER, John. Contact for Loughborough University, Dept. Mechanical Engineering, Loughborough, Leicestershire LE11 3TU England, United Kingdom.

TZONG, Mr. Tang Yaw. Contact for Institute of Optical Science/Central University. Chung-Li 32054, Taiwan, RO.C. Telephone: (886)(3) 425 7681. FAX: (886)(3) 425 8816.

U

UMEKI, Spencer J. Customer Service, Alpha Photo Products, Inc. 985 Third Street, P.O. Box 23955, Oakland, CA 94623 USA Telephone: (415) 893 1436 (ext. 1). FAX: (415) 834-8107.

UNBEHAUN, Klaus. President, Holopublic. Hirschstrasse 84, D-56oo Wuppert&I-2,Federal Republic of Germany (FRG). Telephone: (49) 020284118.

UNTERSEHER, Fred. 3463 State Street, Suite 304, Santa Barbara, CA 93105 USA. Telephone: (805) 967-9727 FAX: (818) 549-0534 Description: Artist & Technical Consultant: (Education + Training Courses. HOLOGRA-PHY HANDBOOK Coauthor), (Laboratory ConstructionDichromate, Silver, Photoresist), Pulse Systems + Imaging Design), (Dichromate Mastering + Production), Artist Portfolio-Installation, Sculptural & Wall Mounted pieces incorporating holograms).

UPATNIEKS, Juris. Contact for Environmental Research Institute of Michigan, Optical Science Lab., ACD, P.O. Box 8618, Ann Amor, MI 48107 USA. Telephone: (313) 9941220.

VRAM, Amy & Marvin. Contact for U.K. Gold Purchasers, Inc. dba Holograms Unlimited. 5858 S.P.LD., Sunrise Mall, Corpus Christi, TX 78412 USA. Telephone: (512) 993 0211. FAX: (512) 993-0467.

UYEMURA, T. Contact for University of Tokyo. Faculty of Engineering, Hongo 7-3-1, Bunkyoku, Toyko, Japan.

V

VAN DER HAUTE, J . President, Center for Applied Research in Art and Technology, 72 Lange Boomgaardstraat, B 9000 Gent, Belgium. Telephone: (32) (91) 626384. FAX: (32) (91) 237326.

VAN DER MOLEN, Mandy. Contact for Holotec CC. P.O. Box 5144, Brackengardens, 1452 Transvaal, South Africa. Telephone: (27)(011) 8641292.

VAN DUYNE, Tom. Contact for Laser Resale, Inc. 54 Balcolm Road, Sudbury, MA 01776 USA.

VAN HAMERSVELD, Eric. Contact for Pacific Holographies, Inc. 1245 Stone Drive, San Marcos, CA 92069 USA. Telephone: (619) 4719044.

VANIN, Valery A. Dep't Head. Soviskusstva vlo mexhdunarodnaya Kniga. Art holography department, 141120 Ttyazino, Moscow region, USSR. Telephone: (7) 238 4600. FAX: (7) 230 2117.

VAN OORSCHOT, R Sales Manager, Dutch Holographic Laboratory. Kanaal dyk Noord 61, 5642 JA Eindhoven, The Netherlands. Telephone: (31)(40) 817 250. FAX: (31)(40) 814 865.

VAN RENESSE, Ruud. Contact for TNO Institute of Applied Physics. Department of Optics, P.O. Box 155, NL-2600 AD Delft, The Netherlands.

VASILLIEVA, N.V., Contact for Cinema & Photo Research Institute. NIKFI, Leningradsky, Prospect 47, Moscow USSR.

VELONA, F. President, CISE SpA Technologie Innovative. via Reggio Emilia, 39, 20090 Segrate, Milano, Italy. Mailing address: P.O.Box 12081,1-20134 Milano, Italy.

VEST, Charles. Contact for The University of Michigan. College of Engineering, Chrysler Center, Ann Amor, MI 48109-2092 USA.

VIENOT, Jean-Charles. Contact, Ecole Nationale Superieure d'l ngenieurs. clo Groupe de Laboratoires, C.M.RS., 5, Av. D'Edimbourg, BLD Marechal.Jwn, F-14032 Caen-Cedex, France.

VIKRAM, C.S. Contact for Pennsylvania State University. Applied Research Laboratory, P.O.Box 30, State College, PA 16804 USA

VILA, Doris. 157 East 33rd Street, New York, NY 10016 USA. Telephone or FAX: (212) 686 5387. Description: Artistic holographer; holog-

raphy education.

VILLANI, S. Vice-President, CISE SpA Technologie Innovative. via Reggio Emilia, 39, 20090 Segrate, Milano, Italy. Mailing address: P.O.Box 12081,1-20134 Milano, Italy.

VINCENT, Ken. Coordinator, Photon League. 110 Sudbury Street-Unit B, Toronto, Ontario M6J lA7 Canada. Telephone: (416) 531 7087. FAX: (416) 536-4291

VISSIOS, Xanthippos. Contact for Laser Graphics. AG. Dimitriou 150, 546 35 Thessaloniki, Greece.

VOGD, H.T. Contact for Que Sera Sera, P.O. Box 29, 9700 AA, Groningen, The Netherlands. Telephone: (31)(050) 140417. FAX: (31)(050) 144142.

VOGEL, Jon. Contact for Holographics (UK) Ltd. 32 Lexington Street, London WIR 3HR, England, United Kingdom. Telephone: (44)(071) 4378992. FAX: (44) (01) 494 0386.

VOLKER, Mirau. Schosserstrasse 93, 46 Dortmund I, Federal Republic of Gennany. Telephone: (49)(8035) 1017. Description: Artistic holographer.

VON BALLY, Gert. Contact for University of Munster. Ear, Nose and Throat Clinic, Kardinal von Galen Ring 10, D-4400 Munster, Federal Republic of Germany. Telephone: (49)(0251) 836 861. FAX: (49)(0251) 836

VUKICEVIC, Dalibor. Contact for Trend, Miramarska 85, 41000 Zagreb, Yugoslavia. Telephone: (38)(041) 511 426.

W

WADDELL, P. Contact for University of Stratchclyde. Mechanical Engineering Group, Glasgow, Scotland, United Kingdom.

WAHLBERG, Bjorn. Contact for Swede Holoprint. Duvhoksgatan 6A, Malmo 21460, Sweden. Telephone: (46) (040) 89821.

WAITTS, Robert, James. President, Contact for Crown Roll Leaf, Inc. 91 Illinois Ave., Paterson, NJ 07503 USA. Telephone: (201) 742-4000. FAX: (201) 742-0219.

WALDEN, N. Managing Director, Laser Electronics Pty., Ltd. P.O. Box 359, Southport, Queensland, 4215, Australia. Telephone: 61 75 53 2066. FAX: 61 75 53 3090.

WALE, RD. Contact for Galvoptics Ltd. Harvey Road, Basildon, Essex, SS13 lES, England, United Kingdom. Telephone: (44)(0268) 728 077. FAX: (44)(0268) 590 445.

WALKER, Julie. mContact for Massachusetts Institute of Technology. M.LT. Arts and Media Technology, E15-416, 20 Ames Street, Cambridge, MA 02139 USA.

WALKER, Neil. Contact, White Tiger Holograms. Johannes Verhulststraat 45, 1071 MS Amsterdam, The Netherlands. Telephone: (31) (20) 797270. FAX: (31)(20) 790 896.

WALTER, Eve. Contsct for Three-D Light Gallery, 107 The Commons, ltacha, NY 14850 USA.

WANLASS, Mike. Contact for Spectratek Corporation, 1510 Cotner Avenue, Los Angeles, CA 90025 USA. Telephone (213) 473-4966. FAX: (213) 477-6710.

WANYUN, Huang. Contact, Beijing Nonnal University. Analysis and Testing Centre, Beijing 100875, Peoples Republic of China.

WARD, A. A. Contact for Oxford University. Department of Engineering Science, Parks Road, Oxford OX1 3PJ England, United Kingdom. Telephone: (44) (0865) 273 805.

WATERMAN, Tracy. Sales Manager, ISASTI Leonardo, P.O. Box 75, 1442A Walnut Street, Berkeley, CA 94709 USA. Telephone: (415) 845 8306. FAX: (415)8416311.

WATSON, John. Contact for University of Aberdeen, Dept. of Engineering, Kings College, Aberdeen AB9 2UE, Scotland, United Kingdom.

WATTS, Mike. Contact for Light Angels, The Conidor, High Street, Bath Spa, Avon BA1 5AJ, England, United Kingdom.

WEBER, Sally. 1240 N. Ventura Avenue, Ventura, CA 93001 USA. Telephone: (805) 648 6419. Description: Sally Weber is an independent artist producing Sculptural and architectural installations using holographic optical elements (HOEs). Weber's projects integrate environmental design with elements including water and sunlight.

WEBER, Mr. A. Vice-President, Holo 3. rue de l'Industrie, 68300 Saint-Louis, France. Telephone: (33)(89) 69 82 08.

WEHMEIER, Helge. President, Agfa Gevaert Inc.--Industrial Division, 100 Challenger Road, Ridgefield Park, NJ 07660 USA. Telephone: (201) 440 2500. FAX: (201) 440 5733.

WEIL-ALVARON, Hans. Established in 1933. bstra Tullgatan 8, 8-211 28 Malmo, Sweden. Telephone: (46)(40) 129956. Contact: Lektor H.

Herman Weil. Description: Hans Weil's inventions were made in the period 1933-1937, while Gabor invented holography in 1948, the laser was invented 1962 and the first laser-illuminated hologram was exposed as late as 1964.

WEISS, Daniel. Contact for Tridimensionale Hologramas. Alberto, Alcocer, 38-2D, 28016 Madrid, Spain. Telephone: (34)(481) 290 745.

WELBY, Mr. Dominic. 2 Foxes Lane, Mousehole, Cornwall, TR19 6QQ England, United Kingdom. Description: Holographic artist, consultant.

WELFORD, W. T. Contact for Imperial College of Science. Optics Section, Blackett Laboratory, London SW7 2BZ, England, United Kingdom.

WELLS, Karan. Contact for Deep Space Holographies, 1328 Dunsterville Avenue, Victoria, British Columbia, Canada V8Z 2X1. Telephone: (1)(604) 384-3927.

WELLS, Sandrajean , P.O. Box 927, Federal Station, Worcester, MA 01601 USA. Telephone: Unlisted. Description: Holographic artist; fine art holograms.

WESLY, Ed. Contact for The School of The Art Institute of Chicago. 5331 North Kenmore Avenue, Chicago, IL 60640 USA Telephone: (312) 443-3883 (Holography lab). WESTPHAL, Carlo. President, die dritte dimension. Frankfurter

Straße 132 -134, D-6078 Neu Isenburg, Federal Republic of Germany. Telephone: (49)(06102) 33367. FAX: (49)(06102) 36709.

WESTPHAL, Paul. Contact for Amity Photonies Co.,26 Gibbs Road, Amity Harbor, NY 11701 USA. Telephone: (516) 789-1099. FAX: (516) 226-8701.

WEY, Marc. Contact for Dennison Manufacturing, 300 Howard Street, Framingham, MA 01710 USA. Telephone: (508) 879-0511.

WHITE, John. Contact for Coburn Corporation. 1650 Corporate Road West, Lakewood, NJ 08701 USA. Telephone: (908) 367-5511. FAX: (908) 367-2908.

WILBUR, Fred. Contact: Elusive Image. 135 West Palace Avenue, Suite 102, Santa Fe, NM 87501 USA. Telephone: (214) 720 6060.

WILLENBORG, George. Contact for Holographic Concepts. 14 Cove Road, Forestdale, MA 02644 USA. Telephone: (508) 477 2488.

WILLIAMS, Chas . . Contact for Laser Image Design. 3031-K Nihi Street, Honolulu, Hawaii 96819 USA

WILLIAMS, H.M. Contact for Brandtjen & Kluge, Inc, 539 Blanding Woods Road, St. Croix Falls, WI 54024 USA. Telephone: (715) 483 3265. FAX: (715) 483-1640.

WILLIAMSON, Ken. Vice-President, Markem Systems Ltd .. Ladywell Trading Estate, Eccles New Road, Salford, M5 2DA England, United Kingdom

WILSON, Brett and Roslyn. Lazart Holographies. 22 Erina Valley Road, Erina, New South Wales 2250, Australia. Telephone: (61X043) 676 245. FAX: (61)(043) 652 306.

WILSON, Richard. Contact for Three D Gallery. Queen's West, 207 Quay, Toronto, Ontario, MSJ 1A4Canada.

WISE, Peter. Contact for Wise Instruments. Unit 9, Hollins Business Centre, Marsh Street, Stafford, ST16 3BG, England, United Kingdom. Telephone: (44)(0785) 223535

WITTIG, Siegmar & Rita. Contact for Wittig Fachbuchverlag, 10 Chemnitzer Strasse, D-5142 Huckelhoven I, Federal Republic of Germany. Telephone: (49)(2433) 84412. FAX: (49)(2433) 86356.

WöBER, Irmfried. Contact for . Wober Design Hololab Austria. Kahlenbergstra6e 6, A-3042 Würmla, Austria. Telephone: (43) 02275 8210. FAX: (43) 227582105.

WOODD, Glenn P. Business Development Manager, IIford Limited. Mobberley, Knutsford, Cheshire WA15 7HA, England, United Kingdom. Telephone: (44)(565) 50000. FAX: (44)565 872 734.

WOODD, Peter H.L. Managing Director, Light Fantastic PLC. 4EIF Gelders Hall Road, Shepshed, Leicestershire LE12 9NH, England, United Kingdom.

WOODD, Terry. Sales Director, Raven Holographies Ltd., Old Saw Mills, Nyewood, Near Petersfield, Hampshire, GU3 15HX, England, United Kingdom. Telephone: 0730-821612. FAX:

0730-821260.

WUTHNOW, Alan W. President, Alpha Photo Products, Inc. 985 Third Street, P.O. Box 23955, Oakland, CA 94623 USA Telephone: (415) 893 1436 (ext. 1). FAX: (415) 834-8107.

WYANT, James. Contact for WYKO Corporation. 1955 East Sixth Street, Tucson, AZ 85719 USA. Telephone: (602) 741-1044. FAX: (602) 294-1799.

WYND, Hugh C. , Christoper W. President, VicePresident, The Diffraction Company, Inc., P.O. Box 151, Riderwood, MD 21152 USA. Telephone: (301) 666 1144. FAX: (301) 472 4911.

Y

YANNUZZI, Paul. President, Archeozoic Incorporated, 777 Gravel Hill Rd., Southampton, PA 18966 USA. Telephone: (215) 322-4915.

YEAGER, Carol. RD 4 Box 335, Catskills, NY 12414 USA. Telephone: (508) 943 2007. Description: Teaches holography.

YERKES, Elizabeth T. East Coast Editor, Ross Books. P.O. Box 4340, Berkeley, CA 94704 USA.

YETKA, Charlie. Contact for Transfer Print Foils Inc. P.O. Box 518, 9 Cotters Lane, East Brunswick, NJ 08816 USA. Telephone: (908) 722 9500 FAX: (908) 722-8394.

YU, Francis. Contact for Pennsylvania State University. Dept. of Electrical Engineering, 121 East University Park, PA 16802 USA.

Z

ZANELLI, Don. Customer Service, Ion Laser Technology Inc. 263 Jimmy Dolittle Road, Salt Lake City, UT 84116 USA. Telephone: (801) 537 1587. FAX: (801) 537 1590.

ZEC, Dr. Peter. President, Deutsche Gesellschaft rur Holografie e.V. Lerchenstra6e 142 a, D-45oo Osnabrock, Federal Republic of Gennany. Telephone: (49)(0541) 7102 199. FAX: (49) (0541) 74297.

ZELLERBACH, Gary. 18 Bonview Street, San Francisco, CA 94110. Telephone: (415) 282-3646. FAX: (415) 282-4013. Holographic consultant; services provided regarding retail and custommade holograms.

ZEMAN, Lee. 55 Ann Street, New York, NY 10038 USA. Telephone: (212) 732 1854. Description: Holographic artist.

ZILEBA, Edward. Contact for Aptec Engineering Limited. 4251 Steeles Avenue West, Downsview, Ontario, M3N 1 V7 Canada.

ZUCKER, Rich. Contact, Bridgestone Graphic Technologies, Inc. 375 Howard Avenue, Bridgeport, CT 06605. USA. Telephone: (203) 366-1595. FAX: (203)366-1667

DID YOU BORROW THIS COPY?

If so, now is the time to order your own personal copy of the Holography Market Place. This international directory for the holography industry is the first and only resource of its kind. You will refer to the HMP day after day, so don't you want one of your own?

Standing Order Plan:

If you would like to receive the *Holography Market Place* each year, ask to be put on our Standing Order Plan, where we will ship you one of the first copies each year.

COST:

For USA delivery:
US$38.00 (Includes UPS shipment)

For delivery outside USA:
We will accept your check drawn on any bank in the world for any one of the following amounts:
US$51.00 (Includes Airmail Shipment) or
£26.00 (Includes Airmail Shipment) or
DM76.00 (Includes Airmail Shipment)

TO ORDER:

Credit card customers can order by phone. Call: (800) 367 0930 or (415) 841 2474
OR FAX your order to: (415) 841 2695.

For orders by mail, fill out the information below and enclose check or credit card information.

NAME: _____

TITLE:_____

COMPANY:_____

ADDRESS: _____

CITY/STATE (PROVINCE)/: _____

COUNTRY/ZIP CODE: _____

PHONE/FAX:_____

FOR CREDIT CARD ORDER:

CHECK ONE: ☐ MASTER CARD - or ☐ VISA - or ☐ AMERICAN EXPRESS

CREDIT CARD NUMBER: _____

CREDIT CARD EXPIRATION DATE: _____

SINGATURE: _____

ADDRESS ALL CORRESPONDENCE TO:

HOLOGRAPHY MARKET PLACE
ROSS BOOKS
P.O. BOX 4340
BERKELEY, CA LIFORNIA 94704
USA

14

Bibliography

For further reading on holography and its many applications, we provide this bibliography. It has been divided into ten categories: artistic holography; general holographic applications; exhibition catalogues; general dissertations; industrial holography; industrial dissertations; holography marketing; periodicals,; non-English holography material, and laser-related titles.

If you are aware of any other sources which should be included, please forward all information to Holography Market Place.

Artistic Holography

Allen, Judy. Lasers and Holograms. England: Puffin Books, UK, 1985.

Anderson, John. Holography. Tempe: Art Dept., Arizona State University, 1979. Series title: Northlight; no_11.

Barrett, N. S. Lasers & Holograms. Boston: Watts,1985. Series Title: Picture Library.

Berley, Lawrence F. Holographic mind, holographic vision: a new theory of vision in art and physics. 1st ed. Bensalem, PA: Lakstun Press, 1980.

Burkig, Valerie. Photonics: The New Science of Light. [n.l.l: Enslow Publishers,1986.

Caulfield, H. John; (et al). Holography Works. New York: Museum of Holography (NYC),1984.

Centerbeam. Otto Piene and Elizabeth Goldring, eds. Introduction by Lawrence Alloway.

Cambridge, MA: Center for Advanced Visual Studies, Massachusetts Institute of Technology, 1980.

Dowbenko, George. Homegrown holography. Garden City, N.Y.: Amphoto, 1978.

Easy Way to Make Reflection Holograms (no author). Embee Press, 1986.

Falk, David R. (et alia). Seeing the Light: Optics in Nature, Photography Color, Vision, & Holography. New York:Wiley,1985.

Finch College, New York. Museum of Art. Contemporary Study Wing. "N dimensional space".

Prepared by Ted McBurnett. Introd. by Elayne H. Varian. New York: 1970.

Furst, Anton. et alia. Light fantastic. 2d ed. London: Bergstrom & Boyle Books, 1977.

Graphics in motion: from the special effects film to holographics. John Halas, ed. New York: Van Nostrand Reinhold, 1984.

Griffiths, John. Lasers & Holograms. New York: Silver Publishers, 1983. Series Title: Exploration & Discovery Series.

Heckman, Philip. The Magic of Holography. New York, Macmillan & Co.,1986.

The Holographic paradigm and other paradoxes: exploring the leading edge of science. Ken Wilber, ed. 1st ed. Boulder: Shambhala, 1982.

Holographic recording materials. H. M. Smith,ed. Contributions by R. A. Bartolini... [et aLl. Berlin: Springer-Verlag, 1977. Series title: Topics in applied physics; v. 20.

Holography Redefined: Thresholds. Barilleaux, Rene P., Editor. New York: Museum of

Holography,I984_

International Exhibition of Holography. Jeong, Tung and Michael Croydon, Editors. Lake Forest, IL: Lake Forest College Press,I982.

(Second) International Exhibition of Holography. Jeong, Tung, Editor.Lake Forest, IL: Lake Forest College Press,1985.

Jeong, Tung H. Display Holography: Proceedings of the International Symposium,1982; Vol. I. Lake Forest, IL: Lake Forest College Press,I983.

_ Display Holography: Proceedings of the International Symposium,1982; Vol. II. Lake Forest, IL: Lake Forest College Press,1986.

Kallard, Thomas. Laser Art & Optical Transforms. New York: Optosonic Press,1979.

Kasper, Joseph Emil and Steven Feller. The complete book of holograms: how they work and how to make them. New York: Wiley, 1987. Series title: The Wiley science editions.

___ . The hologram book. Englewood Cliffs, N.J .: Prentice-Hall,1985.

Lancaster, Ian M. (et alia). The Holographic Instant: Pulsed Laser Holography. New York: Museum of Holography, 1987.

Light vistas light visions. Sponsored by the Department of Art, St. Mary's College. Notre Dame, IN: St. Mary's College, 1983.

Lucie-Smith, Edward. Art In The Seventies.Ithaca, NY: Cornell University Press, 1980. Contrib: R. Berkhout, H. Casdin-Silver, P.Claudius.

Neumann, Don Barker. "l'he effect of scene motion on holography". Columbus, OH: dissertation, 1967.

Palin, Michael. The Mirrorstone: A Ghost Story with Holograms. New York: Knopf,1986.

Walton, Paul. Space-light: a holography and laser spectacular. London: Routledge & Kegsn Paul,1982.

Artistic Holography: Exhibit Catalogues And Miscellany

Bienal Internacional de sao Paulo, Catalogo Geral. sao Paulo, Brasil: Fundacao Bienal de sao Paulo, 1985. See ·Entre a Ciencia EA Ficcao", pp167-197, re: holographers M. Baumstein, H. Casdin-Silver, J.W.Garcia.

Alice in the Light World. 'Ibkyo, Japan: The Ashai Shimbun, 1978.

Art Transition:. Cambridge, MA: MIT Center for Advanced Visual StudieslUniversity Film Study Center, 1975. See H. Casdin-Silver Holography.... pp30-32.

Critic's Choice: The Craft of Art; Peter Moore's Liverpool Project 5. Liverpool, England: November 3, 1979. See E. Lucie-Smith, "New Attitudes, New Materials, New Techniques".

____.. Liverpool, England:Walker Art Gallery, 1979. No. 84-110. See W.D.L. Scobie, ·Arts Review".

Electra 83. Paris: Les Amis du Musee d'Art de la Ville de Paris, 1983. See F.Popper, ·Electricity and Electronics in the Art of the 20th century", pp 46-50.

Expansion. Internationale Biennale fur Graphik und Visuelle Kunst. Horst Gerhard Haberl, Generalsekretar. Wien, Austria: Internationale Biennale flir Graphik und Visuelle Kunst, 1979. See O.Peine "MIT-Center for Advanced Visual Studies", ·Sky Events" pp 232- 239; E. Goldring ·Documentation room", ·Centerbeam" p. 232; H.Casdin-Silver ·Holography, a holographic environment" p. 234; G. Kepes ·Art of the Environment" p. 99.

Fantasy of Holography. 'Ibkyo, Japan: Seibu Museum of Art, 1976. Itsuo Sakane, ed. Contr: Shuntoro Tankawa, Junpei Tsujiuchi.

Harriet Casdin-Silver Holography. New York: Museum of Holography, 1977. First one person exhibition at the Museum.

High Technology and Art 1986. 'Ibkyo, Japan: Tokyo Shimbun & Nagoya Shimbun, with Associates of Art and Technology, Japan.

'Holography redefined', 'Thresholds'. Harriet Casdin-Silver with Dov. Eylath. New York: Museum of Holography, 1984. Group exhibition.

Inter. Quebec, Canada: Les Editons Intervention, Printemps 86, no.31, 1986. See E. Shapiro, ·Art, Perception et Holographie" p. 32; L.Heaton, "Tire D'une Entrevue avec Harriet Casdin-Silver" pp32-3.

Images in Time and Space. Ottawa, Ontario, Canada: Association of Science and Technology Inc, 1987. Travelling exhibit.

International Holography. London, England: The Photographers' Gallery, 1980.

Light and Substance. New Mexico: University of New Mexico Art Museum, 1973-75. History of photography, holography by S. Benton, H.Casdin-Silver. Van deren Coke, org.

MultiMedia Exhibition. Kansas City, MO: Nelson Gallery at Atkins Museum of Fine Arts, 1970. See ·Holography by H.Casdin-Silver".

Otto Peine und CAVS: 20th Anniversary CAVS. Karlsruhe, West Germany: Badischer Kunstverein, 1988. See H. Casdin-Silver, A. Cheji, D. Jung, J.Powell.

Sky art conference-'81. Cambridge, MA: MIT Center for Advanced Visual Studies, 1981. L. Burgess, E. Goldring, B. Kracke, O. Piene, eds. See H.Casdin-Silver, ·Sky work: solar-tracked hologram series" p.49.

Sky art conference '83. Cambridge, MA: MIT Center for Advanced Visual Studies with der Landeshaupstadt Munchen der BMW AG und der Digital Equipment GmbH, 1983.

General Holography Information

Abramson, Nils H. The making and evaluation of holograms. London: Academic Press, 1981.

Advances in Holography. Farhat, Nabil H.-Editor. New York: Marcel Dekker,1975.Vol. 1.

Advances in Holography. Farhat, Nabil H.-Editor. New York: Marcel Dekker,1976.Vol. 2.

Advances in Holography. Farhat, Nabil H.-Editor. New York: Marcel Dekker,1976.Vol. 3.

Applications of Holography: January 21-23, 1985, Los Angeles, California. Lloyd Huff, chair, ed. Bellingham, WA: SPIE- the International Society for Optical Engineering, 1985. Series title: Proceedings of SPIE--the International Society for Optical Engineering; v. 523.

Berner, Jeff. The holography book. New York: Avon Books,1980.

Berry, Michael V. Diffraction of Light by Ultrasound. New York: Academic Press,I967.

Cadig Liaison Centre. Reference Library. A compendium of Cadig bibliographies: Metrication, Fluidics, Explosive techniques in engineering, Holography, Carbon fibres. Coventry (Warwickshire): Cadig Liaison Centre, 1970.

Cathey, W. Thomas. Optical information processing and holography. New York:Wiley, 1974. Series title: Wiley series in pure and applied optics.

Caulfield, H. J . and Sun Lu. The applications of holography. New York, Wiley-Interscience, 1970. Series title: Wiley series in pure and applied optics.

Chambers, R. P. and J .S. Courtney-Pratt. Bibliography on Holograms. New York: Bell Telephone Laboratories,I966.

Chuguy, Yu. V. and N. T. Kolesova. Bibliraphyon holography, 1971-1972. Translated from Russian. London: Scientific Information Consultants Ltd, 1976.

Coherent optical processing: seminar, August 21-22, 1974, San Diego, CA. Palos Verdes Estates, CA: Society of Photo-optical Instrumentation Engineers, 1975. Series title: Society of Photo-optical Instrumentation Engineers Seminar proceedings; v. 52.

Coherent optics in mapping: tutorial seminar and technology utilization program, March 27- 29, 1974, Rochester, N.Y. N. Balasubramanian, Robert D. Leighty, eds. Jointly sponsored by American Society of...Palos Verdes Estates, CA: SPIE,1974. Series titl e: Society of Photo-optical Instrumentation Engineers Proceedings ; v. 45.

Collier, Robert Jacob, C. B. Burckhardt and L.H. Lin. Optical holography. New York: Academic Press, 1971.

Collings, Neil. Optical pattern recognition using holographic techniques. Wokingham, England; Reading, MA. : Addison-Wesley Pub. Co., 1988. Series title: Electronic systems engineering series.

Collings, Neil. Optical Pattern Proceedings. Symposium on High-Power Lasers and Optical Computing, OEILASE '90, 14-19 January 1990, Los Angeles, California. SPIE-the International Society of Engineering; v. 12ll.

Computer-generated holography II : ll-12 January 1988, Los Angeles, CA. Sing H. Lee, chair/

editor; the International Society for Optical Engineering; cooperating ... Bellingham, WA., USA: SPIE--the International Society for Optical Engineering, 1988.Series title: Proceedings of SPIE-the International Society for Optical Engineering; v. 884.

Conference on Fourier Optica, Lasers and Holography, Mysore, 1971. Proceedings of the Conference on Fourier Optics, Lasers and Holography, Mysore, November 11-15, 1971. Madras, India: Institute of Mathematical Sciences, 1971. Series title: Matscience report ; 77.

Conference on Holography and Optical Filtering, Marshall Space Flight Center, 1971: Holography and optical filtering; proceedings. Washington, Scientific and Technical Information Office, National Aeronautics and Space Administration.

Conference on Holography and Optical Filtering, 1972: Marshall Space Flight Center. Holography and optical filtering; proceedings. Washington, Scientific and Technical Information Office, National Aeronautics and Space Administration, 1973. Series title: United States. National Aeronautics and Space Administration NASA SP -299.

Defense Documentation Center (U.S.) Holography: a DDC bibliography, January 1970-September 1972. Alexandria, VA: Defense Documentation Center, Defense Supply Agency, 1973.

Denisiuk, IU. N. Fundamentals of holography. Translated from the Russian by Alexander Chubarov. Rev. from the 1978 Russian ed. Moscow: Mir, 1984.

DeVelis, John B. and George O. Reynolds. Theory and applications of holography. Reading, MA: Addison-Wesley Pub. Co., 1967.

Developments in holography: seminar-in-depth; proceedings. Brian J. Thompson and John B. DeVelis,eda. Redondo Beach, CA: Society of Photo-optical Instrumentation Engineers, 1971. Series title: Society of Photo-optical Instrumentation Engineers S.P.I.E.seminar proceedings, v. 25.

Dudley, David D. Holography; a survey. Washington, DC: Technology Utilization Office, National Aeronautics and Space Administration, 1973. Series title: NASA SP-5118.

Eichert, Edwin S. and Alan H. Frey; Randomline, Inc. Holography in driver education, training, testing, and research. Washington, DC: National Highway Traffic Safety Administration,1978. Series title: United States. National Highway Traffic Safety Administration Report; no. DOT HS-803 035.

Engineering Applications of Holography Symposium, 1972: Los Angeles: Proceedings. Redondo Beach, CA: Society ofPhoto-optical Instrumentation Engineers, 1973.

The Engineering uses of coherent optics: proceedings and edited discussion of a conference held at the University of Strathclyde, Glasgow, 8-11 April,1975 . Organised by the University, in association with the ... Cambridge, Eng. : Cambridge University Press, 1976.

The Engineering uses of holography. Robertson,

Elliot R. and James M. Harvey, eds. Cambridge, England: Cambridge University Press, 1970.

European Hybrid Spectrometer Workshop on Holography and High-Resolution Techniques, 1981: Strasbourg, France. Photonics applied to nuclear physics, 1. European Hybrid Spectrometer Workshop on Holography and High-Resolution Techniques, Strasbourg, Council of Europe, 9-12 November 1981. Geneva: European Organization for Nuclear Research, 1982. Series title: CERN (Series) ; 82-01.

An External Interface for Processing 3-D Holographic and X-ray Images: Werner Juptner, Thomas Kreis (eds.). Berlin: Springer-Verlag,1989.

Fiber Diffraction Methods. French, Alfred D. and Kenn Gardner, Editors. Series Title: ACS Symposium Ser.,; No. 141. New York: American Chemical Society,1980.

Firth, Ian Mason. Holography and computer generated holograms. London, Mills and Boon, 1972. Series title: M & B monograph EElll.

Francon, M. Holography. Expanded and revised from the French edition. Translated by Grace Marmor Spruch. New York: Academic Press, 1974.

Handbook of Optical holography. H. J. Caulfield, ed . Contributors, Gilbert April ... ret al.l. New York: Academic Press, 1979.

Hariharan, P. Optical holography : principles, techniques, and applications. Cambridge: Cambridge University Press,1983. Series title: Cambridge monographs on physics.

Hildebrand, B. P. and B. B. Brenden. An introduction to acoustical holography. New York: Plenum Press, 1972.

"Holographic detection of intraocular pathology in the presence of cataracts: final report". By George o. Reynolds ... ret al.l. Burlington, MA: Technical Operations, 1974.

Holographic optics: design and applications : 13- 14 January 1988, Los Angeles, CA. Ivan Cindrich, ChairlEditor ; SPIE--The International Society for Optical Engineering; cooperating ... Bellingham, WA., USA: SPIE--The International Society for Optical Engineering, 1988.Series title: Proceedings of SPIE--The International Society for Optical Engineering ; v. 883.

Holographic optics: optically and computer generated: 19-20 January 1989, Los Angeles, CA . Ivan N. Cindrich, Sing H. Lee chairs/editors; sponsored by SPIE--The International Society for Optical. .. Bellingham, WA., USA : SPIE--The International Society for Optical Engineering, c1989. Series title: Proceedings of SPIE--the International Society for Optical Engineering ; v. 1052.

Holography. Redondo Beach, CA: Society of Photo-optical Instrumentation Engineers, 1968. Series title: Society ofPhoto-opticalInstrumentation Engineers S.P.I.E. seminar proceedings, v. 15.

Holography Applications. Wang. Editor. Bellingham, WA: SPIE,1986.

Holography: Critical Reviews. Huff, L., Editor. Bellingham, WA: SPIE, 1985.

Holography : January 24-25, 1985, Los Angeles, CA. Lloyd HulT, chair,ed. Bellingham, WA: SPIE-the International Society for Optical Engineering, 1985. Series title: Proceedings of SPIE-the International Society for Optical Engineering; v. 532. Series title: SPIE critical reviews oftechnology series; 12th.

Holography; seminar-in-depth, May, 1968, San Francisco CA. B.G.Ponseggi and Brian J . Thompson, eds. Redondo Beach, CA: Society of Photooptical Instrumentation Engineers, 1972. Series title: Society of Photo-optical Instrumentation Engineers Proceedings, v.15.

Holography techniques and applications: ECOl, 19-21 September 1988, Hamburg, Federal Republic of Germany / Werner P.O. Juptner, chair/editor; EPS--European Physical Society, Europtica--the... Bellingham, WA. : SPIE--the International Society for Optical Engineering, 1989. Series title: Proceedings ofSPIE--the International Society for Optical Engineering ; v. 1026.

I. Aroslavskii, L. P. and N. S. Merzlyakov. Methods of digital holography. Translated from Russian by Dave Parsons. New York: Consultants Bureau, 1980.

International Commission for Optics. Congress, 10th : 1975 : Prague, Czechoslovakia. Recent advances in optical physicsz: proceedings of the Tenth Congress of the International Commission for Optics, August 25-29, 1975, Prague, Czechoslovakia. Bedrich Havelka and Jan Blabla, eds. Olomouc: Palacky University; Prague: Society of Czechoslovak Mathematicians and Physicists, 1976.

International Conference on Applications of Holography and Optical Data Processing,1976: Jerusalem, Israel. Applications of holography and optical data processing: proceedings of the international conference, Jerusalem, August 23- 26, 1976. E. Marom, A. A. Friesem, and E. Wiener-Avnear, eds. 1st ed. Oxford: Pergamon Press, 1977.

International Conference on Computer-generated Holography, 1983: San Diego, CA. International Conference on Computer-generated Holography, August 25-26,1983, San Diego, CA: proceedings. Sing H. Lee, chair, ed. Bellingham, WA: SPIE--The International Society for Optical Engineering,1983. Series title: Proceedings of SPIE--the International Society for Optical Engineering; v. 437.

International Conference on Computer-generated Holography: 2nd: 1988: Los Angeles, CA. Computer-generated Holography II: 11-12 January 1988, Los Angeles,California, [proceedingsl . Sing H. Lee, chair/ed. Sponsored by SPIE--The International Society for Optical Engineering; ... Bellingham, WA: SPIE--The International Society for Optical Engineering, 1988. Series title: Proceedings of SPIE--the International Society for Optical Engineering; v. 884.

International Conference on Holographic Systems, Components and Applications,1987: Churchill College. Holographic systems, components and applications: Churchill College, Cambridge, 10th-12th September, 1987. London: Institution of Electronic and Radio Engineers, 1987. Series title: Publication / Institution of

Electronic and Radio Engineers; no_ 76.

International Conference on Holographic Systems, Components, and Applications. 2nd:1989: University of Bath.

International Conference on Holography Applications, l986: Peking, China. International Conference on Holography Applications: 2-4 July, 1986, Beijing, China . Dahang Wang, chair. Jingtang Ke, Ryszard J . Pryputniewicz, cds. Sponsored by COS-Chinese Optical Society- Bellingham, WA: SPIE,l987. Series title: Proceedings of SPIE--the International Society for Optical Engineering; v. 673.

International Conference on Holography, Optical Recording, and Processing of Information; 21-24 May 1989, Varna, Bulgaria; Y.N. Denisyuk, T.H. Jeong, chair/cochair; sponsored by The Bulgarian Academy of Sciences, ... ret al.). Bellingham, WA_, USA: SPIE--the International Society for Optical Engineering, 1990.

International Congress on High-Speed Photography, 11th:1974: Imperial College, London. High speed photography : proceedings of the eleventh International Congress on High Speed Photography, Imperial College, University of London, September 1974. P. J . Rolls, ed. London: Chapman & Hall: distributed in the USA by the Society of Photo-Optical Instrumentation Engineers, 1975.

International Congress on High Speed Photography, 12th:1976: .Toronto, Canada. Proceedings of the 12th International Congress on High Speed Photography (Photonics), Toronto, Canada, 1-7 August 1976. Martin C. Richardson. Bellingham, WA: Society of Photo-Optical Instrumentation Engineers, 1977. Series title: SPIEv.97.

International Congress on High Speed Photography and Photonics, 13th: 1978: Tokyo. Proceedings of the 13th International Congress on High Speed Photography and Photonics-Tokyo, 20-25 August 1978. Shin-ichi Hyodo, ed. Tokyo : Japan Society of Precision Engineering; [New York): distributed (outside Japan) by Society of Photo-Optical Instrumentation E'Igineers, 1979. Series title: SPIE v. 189.

International Optical Computing Con ference, 1974: Zurich. Digest of papers. New York, Institute of Electrical and Electronics Engineers, 1974.

International Optical Computing Con ference, 1975: Washington, D.C. Digest of papers: International Optical Computing Conference, April 23-25, 1975, Washington, D.C. Sponsored by the Computer Society of the Institute of Electrical and Electronic Engineers, in cooperation... New York: Institute of Electrical and Electronics Engineers, 1975.

International Symposium on Acoustical Holography: Acoustical holography. New York: Plenum Press, 1967.

International Symposium on Acoustical Holography. 1st: 1967: Huntington Beach, CA. 'Acoustical holography; proceedings. New York: Plenum Press, 1969.

International Symposium on Acoustical Holography and Imaging,7th: 1976: Chicago, IL. Recent advances in ultrasonic visuslization. Lawrence W. Kessler,ed. New York: Plenum Press, 1977. Series title: International Symposium on Acoustical Holography and Imaging Acoustical holography; v. 7.

International Symposium on Acoustical Holography and Imaging, 8th, Key Biscayne, FL, 1978. Ultrasonic visualization and characterization. A. F. Metherell, ed. New York: Plenum Press, 1980. Series title: International Symposium on Acoustical Holography and Imaging Acoustical imaging; v. 8.

Klein, H. Arthur. Holography. With an introd. to the optics of diffraction, interference, and phase differences 1st ed. Philadelphia: Lippincott, 1970.Series title: Introducing modem science.

Kock, Winston E. Radar, sonar, and holography: an introduction. New York: Academic Press, 1973.

Kostelanetz, Richard. On Holography. RK Editions, 1979.

Kock, Winston E. Lasers and holography; an introduction to coherent optics 1st ed.Garden City, N.Y., Doubleday, 1969.

Kurtz, Maurice K. Study of potential application of holographic techniques to mapping; final technical report. Lafayette, IN: Purdue Research Foundation, Purdue University, 1971.

Laser Holography in Geophysics. Takemoto, Shuzo, Editor. New York, Wiley,1989.

Lehman, Edward J . Applications of Holography: a bibliography with abstracts. Springfield, VA.: NTIS, 1975.

Lehmann, Matt. Holography; technique and practice. London: Focal Press, 1970. Series title: The Focal library.

Light and its uses: making and using lasers, holograms, interferometers,and instruments of dispersion: readings from Scientific American. San Francisco: W. H. Freeman, 1980.

Lingenfelder, P. G. Holography manual; a compilation of laboratory techniques commonly used in the construction of holograms including refinements developed at NELC .. San Diego, CA: Naval Electronics Laboratory Center, 1969. Series title: NELC Technical document 47.

McNair, Don. How to make holograms. 1st ed. Blue Ridge Summit, PA: Tab Books,1983.

"A Multi-frequency synthetic detecting holography with high depth resolution". Peking, China: The Research Group of Holography, Chinese Academy of Geological Sciences, [s.n.), 1976.

NATO Advanced Study Institute on Optical and Acoustical Holography, 1971: Milan. Optical and acous tical hologra phy; proceedi ngs of the NATO Advanced Study Institute on Optical and Acoustical Holography, Milan, Italy, May 24- June 4, 1971. Ezio Camatini,ed. New York: Plenum Press, 1972.

Okoshi, Takanori. Three-dimensional imaging techniques. New York: Academic Press, 1976.

Optical & Acoustical Holography. Camatini, E.Editor. New York: Plenum Publishing,1972.

Optical Computing Symposium, Darien, Conn.,

1972. Digest of papers presented at the 1972 one-day-in-depth Optical Computing Symposium, April 12, 1972 at the Noroton School, Darien, Connecticut. Naval Underwater Systems Center and IEEE Computer Society, Eastern... [s.I.) Institute of Electrical Engineers, 1972.

Optical Information Processing and Holography, Cathey,W.Thomas.,Editor. Series Title: Pure & Applied Optics Series. New York: Wiley, 1974.

Optics and photonics applied to three-dimensional imagery (Image 3-0): presented as part of the Optics, Phototonics, and lconics Engineering Meeting (OPIEM), November 26-30, 1979, Stiasbourg, France. Bellingham, WA: Society of Photo-Optical Instrumentation Engineers, l980. Series title: Society of Photo-optical Instrumentation Engineers Seminar proceedings; v. 212.

Optics in engineering measurement: 3-6 December 1985, Cannes, France. William F. Fagan, chair,ed. Organized by SPIE--the International Society for Optical Engineering, ANRT--Association Nationale de ... Bellingham, WA: SPIE--the International Society for Optical Engineering, 1986. Series title: Proceedings of SPIE--the International Society for Optical Engineering; v. 599.

Optics in entertainment: January 20-21, 1983, Los Angeles, California. Chris Outwater, chair,ed. Bellingham, WA: SPIE--the International Society for Optical Engineering,1983. Series title: Proceedings of SPIE--the International Society for Optical Engineering; v. 391.

Optics in entertainment: January 26-27, 1984, Los Angeles, CA. Chris Outwater, chair, ed. Bellingham, WA: SPIE--the International Society for Optical Engineering,1984.Series title: Proceedings of SPIE--the International Society for Optical Engineering; v. 462.

Optics, Photonics, and Iconics Engineering Meeting,1979: Strasbourg, France. Optics and photonics applied to three-dimensional imagery (IMAGE 3-0): presented as part of the Optics, Photonics, and Iconics Engineering Meeting (OPIEM), November 26-30, 1979, Strasbourg, France. Bellingham, WA: Society of Phota-optical Instrumentation Engineers, 1980. Series title: Society of Photo-optical Instrumentation Engineers Proceedings; v. 212.

Optics Today. John N. Howard, ed. New York, N.Y: American Institute of Physics, 1986. Series title: Readings from Physics today; no. 3.

Ostrovskii , IU. I. Holography and its application. Translated from the Russian by G. Leib. Moscow: Mir, 1977.

Outwater, Chris. and Eric Van Hamersveld. Guide to practical holography. Beverly Hills, CA: Pentangle Press,1974.

Pattern recognition studies: seminar-in-depth, proceedings / Society of Photo-Optical Instrumentation Engineers. [Redondo Beach, Calif.) : the Society, [cl969). Series title: S.P.I.E. seminar proceedings; v. 18.

Pattern Recognition & Acoustical Imaging. Ferrari, Editor. Bellingham, WA: SPIE,I987.

Periodic structures. gratings. moire patterns. and diffraction phenomena: July 29-August 1. 1980. San Diego. CA. C.H. Chi. E.G. Loewen. C.L. O'Bryan III. ads. Bellingham. WA: Society of Photo-optical Instrumentation Engineers.1981. Series title: Proceedings of the Society of Photo-optical Instrumentation Engineers; v. 240.

Pethick. J. On holography and a way to make holograms. Ontario: Belltower Enterprises. 1971.

Photonics applied to nuclear physics. 2: proceedings; Strasbourg. Council of Europe. 5-7 December 1984. Geneva: CERN. 1985. Series title: Nucleophot.

Photopolymer device physics. chemistry. and applications: 17- 19 January 1990. Los Angeles. California. Roger A. Leasard. chair/editor; sponsored by SPIE--the International Society for Optical Engineering. Bellingham. WA . • USA : SPIE--the International Society for Optical Engineering. 1990.

Photoreactive materials and their applications. P. Gunter. J.-P Huignard. eds. Contributions by A.M. Glass ... let al.l. Berlin: Springer-Verlag. 1988. Series title: Topics in applied physics; v. 61. ete.

Pietsch. Paul. Shumebrain. Boston: Houghton Mimin. 1981.

Pisa. Edward J .• S. Spinak & A.F.Metherell. Color acoustical holography. Huntington Beach. CA: Douglas Advanced Research Laboratories. 1969. Series title: Douglas Advanced Research Laboratories. Research communication 109.

Practical holography: 21-22 January 1986. Los Angeles. CA. Tung H. Jeong. Jacques E. Ludman chair.eds. Presented in cooperation with American Association of Physicists in Medicine ... let al.l. Bellingham. WA: SPIE-The International Society for Optical Engineering. 1986. Series title: Proceedings of SPIE--the International Society for Optical Engineering; v. 615.

Practical holography II: 13-14 January 1987. Los Angeles. CA. Tung H. Jeong. chair/ed. Sponsored by SPIE--the International Society for Optical Engineering. in coopera tion wi th Center for .. Bellingham.WA: SPIE--the International Society for Optical Engineering. 1987. Series title: Proceedings of SPIE--the International Society for Optical Engineering; v. 747.

Practical holography III : 17-18 January 1989. Los Angeles. CA. Stephen A. Benton. chair/editor; sponsored by SPIE--the International Society for Optical Engineering; cooperating organizations. Applied .. Bellingham. WA . • USA : SPIE. 1989. Series title: Proceedings of SPIE--the International Society for Optical Engineering; v. 1051.

Proceedings of the information processing and holography symposium ICALEO 83. Symposium heads: David Casasent. Milton T. Chang. Organized with American Society of Metals ... let al.l. Sponsored by Laser Institute ... Toledo. OH: The Institute. 1984. Series title: LIA (Series) ; v. 41.

Proceedings of the Inspection. Measurment [sicl and Control and Laser Diagnostics and Photochemistry. ICALEO '84. Donald Sweeney. Robert Lucht. eds. Organized in cooperation with ... The American ...Toledo. OH: LIA-Laser Institute of America. 1985. Series title: LIA (Series) ; v. 45. 47.

Processing and display of three-dimensional data: August 26-27. 1982. San Diego. CA. James J. Pearson. chair.ed. Bellingham. WA: SPIE-The International Society for Optical Engineering. 1983. Series title:Proceedings of SPIE--the International Society for Optical Engineering; v. 367.

Processing and display of three-dimensional data II : August 23-24. 1984. San Diego. CA. James J. Pearson. chair.ed. Cooperating organizations. Optical Sciences Center. University of Arizona.... - Bellingham. WA: SPIE--the International Society for Optical Engineering.1984. Series title: Proceedings of SPIE--the International Society for Optical Engineering; v. 507.

Progress in holographic applications: 5-6 December 1985. Cannes. France. Jean Ebbeni. chair. ed. Organized by SPIE--the International Society for Optical Engineering. ANRT- Association Nationale de la ... Bellingham. WA: SPIE--the International Society for Optical Engineering.1986. Series title: Prooeedings of SPIE--the International Society for Optical Engineering; v. 600.

Progress in holography: 31 March ·2 April 1987. The Hague. The Netherlands. Jean Ebbeni. chair/ed. Organized by ANRT--Association nationale de la recherche technique. SPIE--The International Society for ... Bellingham.WA: SPIE. 1987. Series title: Proceedings of SPIE--The International Society for Optical Engineering; v. 812.

"Project Search". Subcommittee on Feasibility of Automated Fingerprint IdentificationlVerification. An experiment to determine the feasibility of holograph assistance to fingerprint identification. Sacramento. CA: 1972. Series title: Project Search Technical report. no. 6.

Recent advances in holography III: February 4-5. 1980. Los Angeles. CA. Tzuo-Chang Lee. Poohsan N. Tamura. eds. Bellingham. WA: Society of Photo-optical Instrumentation Engineers. 1980. Series title: Society of Photo-optical Instrumentation Engineers Proceedings; v.215.

Saxby. Graham. Practical holography. New York. N.Y.: Prentice-Hall Intern~tional. 1988.

___ . Holograms: How to Make & Use Them. Masson. France: Focal Press. 1980.

Saxby. John. Holograms. New York: Focal Press. 1980.

Schultz. Jerold M. DiffTaction for Materials Scientists. New York: Prentice-Hall. 1982.

Schumann. Walter and J .-P. Zurcher. D.Cuche. Holography and deformation analysis. Berlin: Springer-Verlag. 1985. Series title: Springer Series in Optical Sciences; v. 46.

Smith. Howard Michael. Principles of holography. New York. Wiley-Interscience.1969.

___ . Principles of holography. 2d ed. New York: Wiley.1975.

Solem. Johndale C. High-intensity X-ray holography: an approach to high-resolution snapshot imaging of biological specimens. Los Alamos. N.M.: Los Alamos National Laboratory.1982.

Solymar. L. and D.J . Cooke. Volume holography and volume gratings. London: Academic Press. 1981.

Soroko. Lev Markovich. Holography and coherent optics. Translated from Russian by Albin Tybulewicz; with a foreword by George W. Stroke. New York: Plenum Press. 1980.

Sources of Physics Teaching: Atomic Energy. Holography. Electrostatics; Vol. 4. Noakes. G. REditor. New York: Coronet Books.1970.

Spencer John R Holographic Infonnation Storage and Retrieval. England: National Reprographic. 1975.

Stroke. George W. An introduction to coherent optics and holography. New York: Academic Press. 1966.

___ . An introduction to coherent optics and holography. 2d ed. New York: Academic Press. 1969.

Symposium on Applications of Holography in Mechanics.1971: University of Southern California. Symposium on Applications of Holography in Mechanics. W. G. Gottenberg. ed. New York: American Society of Mechanical Engineers. 1971.

Three-dimensional imaging; April 21-22. 1983. Geneva. Switzerland. Jean Ebbeni. Andre Monfils. chairmen-editors. Bellingham. WA: SPIEthe International Society for Optical Engineering. 1983. Series title: Proceedings of SPIEthe International Society for Optical Engineering; v. 402.

Three-dimensional imaging: August 25-26. 1977. San Diego. CA. Stephen A. Benton. ed. Presented by the Society of Photo-optical Instrumentation Engineers. in conjunction with the IEEE Computer... Bellingham. WA: SPIE. 1977. Series title: Society of Photo-optical Instrumentation Engineers Proceedings ; v. 20.

Ultrasonic Imaging & Holography: Medical. Sonar. & Optical Applications. Stroke. George W., and Jumpei Tsujiuchi. Editors. New York: Plenum Publishing.1974.

United States-Japan Science Cooperation Seminar on Pattern Information Processing in Ultrasonic Imaging. 3rd: 1973: University of Hawaii. Ultrasonic imaging and holography: medical. sonar. and optical applications: [proceedingsl. George W. Stroke ... let al.l.ed. New York: Plenum Press. 1974.

United States.Japan Seminar on Information Processing by Holography. 2nd :1969 : Washington. D.C. Applications of holography; proceedings. Euval S. Barrekette. ed. New York: Plenum Press. 1971.

Unterseher. Fred. Jeannene Hansen and Bob Schlesinger. Holography handbook: making holograms the easy way. Berkeley. CA: Ross Books. 1982.

Vasilenko. G. I. (Georgii Ivanovich) and L.M. Tsibul'kin. Image recognition by holography. Translated from Russian by Albin Tybulewicz. New York : Plenum PresslConsultants Bureau. 1989.

Weinstein. L. Albertovich. Theory of Diffraction & the Factorization Method: Generalized Wiener Hopf Technique.Golem Publications. 1969. Electromagnetics Series, Vol. 3.

Wenyon, Michael. Understanding holography. New York: Areo Pub. Co., 1978.

___ . Understanding holography. Newton Abbot, Eng.: David & Charles,1978.

___ . Understanding holography. 2nd Arco ed. New York: Areo Pub., 1985.

Wolff J. (et alia). Light Fantastic; Lasers and Holography Explained. England: Gordon Fraserl Bergstrom & Boyle, 1977

Yaroslavskii, L. P., and N. S. Merzlyakov. Methods of Digital Holography. New York: Plenum Publishing,1980.

Yu, Francis T.S. Introduction to diffraction, information processing, and holography. Cambridge, MA: MIT Press , 1973.

General Holography: Information --dissertations

Coello-Vera, Agustin Elias. "Scanned acoustic imaging in the ocean: a study of holographiclike systems and their limitations". 1978.

Elliott, John Douglas. "Computer simulation of the holographic image degradation due to transmission of the signal through a random noise media".1971.

Eu, James Kim-Tzong. "Studies in spatial tiltering". 1974.

Fischer, Wolfgang Klaus. Methods for acoustic holography and acoustic measurements. Newark, N.J. : [s.n.], 1972.

George, Daweel Joseph. "Holography as applied to jet breakup and an analytical method for reducing holographic droplet data". 1972.

Kurtz, Maurice K. "Potential uses of holography in photogrammetric mapping". 1971.

Landry, Caliste John. "tntrasonic imaging by Brillouin-Bragg diffraction: development of an operational system with prospective applications in medical diagnosis and material testing". 1972.

Lee, Hua. "Development and analysis of the back-projection method for acoustical imaging". 1980.

Liu, Charles Yau-chi. "Some topics in holographic image formation".1974.

Mensa, Dean L."Techniques for microwave imaging". 1980.

Powers, John Patrick. "Some aspects of the application of Bragg diffraction oflaser light to the imaging and probing of acoustic fields". 1970.

Ramos, George Urban. "I. On the fast fourier transform; II. On the computations in digital holography". 1970.

Schlusaler, Larry. "Improvement of the horizontal resolution of a Bragg-diffraction imaging system and motion limitations of a holographic system". 1978.

Schueler, Carl Frederick. "Development and applications of computer-assisted acoustic holography". 1980.

Schwank, James Ralph. "Refractive holography". 1974.

Sherman, George Charles. "Wavefront reconstruction and its application to the study of the optical properties of atmospheric aerosols". 1969.

Shuman, Curtis Alan. "Holographic imaging through moving diffusive media". 1973.

___ . "Holographic imaging through moving scatterers".1972.

Stone, William Ross. "The concept, design, and operation of a demonstration holographic radio camera". 1978.

Strand, Timothy C. "Comparison of analog and binary holographic data storage". 1973.

Sutton, Jerry Lee. "Broadband acoustic imaging". 1974.

Tricoles, Gus Peter. "Some topics in microwave holography". 1971.

Tse, Nie But. "Digital reconstruction of acoustic holograms". 1979.

Vourgourakis, Emmanuel John's. "Coherence limitations on holographic systems". 1967.

Wang, Keith Yu-Chih. "Threshold contrasts for various acoustic imaging systems". 1972.

Wollman, Michael Thomas. "An experimental acoustical holographic system for eventual use in the ocean". 1975.

Industrial Holography

Acoustical Holography; Vol. 1. Metherell, A.F., Editor. New York: Plenum Publishing,1969.

Acoustical Holography; Vol. 2. Metherell, A.F., Editor. New York: Plenum Publishing, 1970.

Acoustical Holography; Vol. 3. Metherell, A.??F., Editor. New York: Plenum Publishing,1971.

Acoustical Holography; Vol. 4. Wade, Glen, Editor. New York: Plenum Publishing, 1972.

Acoustical Holography; Vol. 5. Green, Philip S.,Editor. Plenum Publishing,1974.

Acoustical Holography; Vol. 6. Booth, N., Editor. New York: Plenum Publishing,1975.

Acoustical Holography; Vol. 7. Kessler, L.W., Editor. New York: Plenum Publishing,1977.

Acoustical Imaging; Vol. 8. Metherell, A.F., Editor. New York: Plenum Publishing,1980.

Acoustical Imaging; Vol. 9. Wang, Keith, Editor. New York: Plenum Publishing, 1980.

Acoustical Imaging; Vol. II. Powers, John P.,Editor. New York: Plenum Publishing, 1982.

Acoustical Imaging; Vol. 11. Ash, Eric A. and C.R. HiII ,Editors. New York: Plenum Publishing, 1983.

Acoustical Imaging, Vol. 15. Jones, Hugh W., Editor. Plenum Publishing, 1987.

Acoustic Imaging: cameras, microscopes, phased arrays, and holographic systems. Glen Wade, ed. New York: Plenum Press, 1976.

Acoustic Surface Wave & Acousto-Optic Devices. Kallard, Thomas-Editor. Series Title: State of the Art Review Series; Vol. 4. New York: Optosonic Press,1971.

Aldridge, Edward E. Acoustical holography. Watford: Merrow Publishing Co. Ltd., 1971. Series title: Merrow monographs, practical science series 1.

Applications of Holography. Barrekette, E. S., Editor. New York: Plenum Publishing,1971.

Applications of Holography in Mechanics: Symposium, University of Southern California, 1971. Symposium Staff; Gottenberg, W. G., Editors. Books on Demand UMI, Reprint of 1971 edition.

Applications of Holography & Optical Data Precessing: Proceedings of an International Conference, Jerusalem, 1976. Marom, E.; Avnear Wiener and A.A. Friesem, Editors. London: Pergamon Press,1977.

Applications oflasers to photography and information handling; proceedings, two-day seminar. Richard D. Murray, ed.Washington, DC: Society of Photographic Scien tists and Engineers, 1968.

Basov, N. G. Lasers & Holographic Data Precessing. USSR: Mir Publications, 1985.

___ . Lasers & Holographic Data Processing. England, Colletts: State Mutual Books,1984.

Beiser, Leo. Holographic scanning. New York: Wiley, 1988. Series title: Wiley series in pur and applied optics.

Brcic, Vlatko. Application of holography and hologram interferometry to photoelasticity: lectures held at the Department for Mechanics of Deformable Bodies. 2d ed. Wien: Springer-Verlag, 1974. Series title: Courses and lectures ; no. 7.

Bristol University Electron Microscopy Group. Convergent Beam Electron Diffraction of Alloy Phases. Mansfield, J ., Editor. A Hilger UK: Taylor & Francis, 1984.

Business Communications Staff. Holography: New Commercial Opportunities. [s.l]: BCC,1986.

Butters, John N. Holography and its technology. London: P. Peregrinus,1971; Published on behalf of the Institution of Electrical Engineers. Series title: Institution of Electrical Engineers I.E.E. monograph, series, 8.

Ceccon, Harry L. Holographic techniques for nondestructive testing of tires. Washington, D.C: National Highway Traffic Safety Administration, 1972.

Conference on Holographic Instrumentation Applications,1970: Ames Research Center. Holographic instrumentation applications. Prepared by NASA Ames Research Center. Boris Ragent and Richard M. Brown, eds. Washington, DC: Scientific and Technical Information Division, National Aeronautics and Space Administration,1970. Series title: NASA SP ; 248.

Dirtoft, Ingegard. Holography: A New Method for Defonnation Analysis of Upper Complete Dentures in Vitro & in Vivo. New York: Coronet Books,l985.

An External Interface for Processing 3-0 Holographic & X-Ray Images. Juptner, W., Editor New York: Springer-Verlag!Research Reports,1989.

Flow visualization and aero-optics in simulated environmenta:21-22 May 1987, Orlando, Florida. H. Thomas Bentley III, chair/ed. Sponsored by SPIE--the International Society for Optical Engineering. Bellingham, WA: SPIE--the International Society for Optical Engineering,1987. Series title: Proceedings of SPIE--the International Society for Optical Engineering; v. 788.

Hartman, W. F. Acoustic Emission: Advances in Acoustic Emission. American Society for Nondestructive Testing,1981.

Holographic data nondestructive testing: October 4-8, 1982, Croatia Hotel de Luxe, Dubrovnik, Yugoslavia. Dalibor Vukicevic, chair,ed. Sponsored by the International Commission for Optics (I CO) [andl ... Bellingham, WA: SPIE--the International Society for Optical Engineering, 1983. Series title: Proceedings of SPIE--the International Society for Optical Engineering; v. 370.

Holographic nondestructive testing. Robert K. Erf, ed. New York: Academic Press, 1974.

Holographic nondestructive testing: status and comparison with conventional methods: 23-24 January 1986, Los Angeles, California. Charles M. Vest, chair, ed. Presented in cooperation with American Association .. _Bellingham, WA: SPI-Ethe International Society for Optical Engineering, 1986. Series title: Proceedings of SPIE-the International Society for Optical Engineering; v. 604.8eries title: SPIE critical reviews of technology series; 15th.

Holographic Nondestructive Testing. Err, Robert K, Editor. New York: Academic Press,1974.

Holographic Nondestructive Testing: Critical Review of Technology. Vest, Charles., Editor. Bellingham, WA: SPIE,l986.

Holography in Medicine: International Symposium Proceedings. Greguss, Pal. Editor.I.P.C.Sci.& Technology, 1976.

Holography in Medicine & Biology. Von Bally, G., Editor. New York: Springer-Verlag,1979. Series Title: Springer Series in Optical Sciences,; Vol. 18.

Industrial applications of holographic nondestructive testing: May 3-5, 1982, Brussels. J . Ebbeni, chair, ed. Sponsored by SPIE--the International Society for Optical Engineering; with the support ... Bellingham, WA: SPIE--the International Society for Optical Engineering, 1982. Series title: Proceedings of SPIE--the International Society for Optical Engineering; v. 349.

Industrial Applications of Holography. Robillard, Jean, Editor. Oxford, UK: Oxford University Press, 1989 and 1990.

Industrial Radiography Holography. American Society for Nondestructive Testing, 1983.

International Symposium on Holography in Bio-medical Sciences,1973: New York. Holography in medicine: proceedings of the International Symposium on Holography in Biomedical Sciences, New York, 1973. Pal Greguss, ed. Guildford, Eng: IPC Science and Technology Press, 1975.

International Workshop on Holography in Medicine and Biology,1979: Munster, Germany. Holography in medicine and biology: proceedings of the International Workshop, Munster, Fed. Rep. of Genna ny, March 14-15, 1979. G. von Bally, ed. Berlin: Springer-Verlag, 1979. Series title: Springer Series in Optical Sciences; v 18.

Jones, Robert; and Catherine M. Wykes. Holographic and Speckle Interferometry. Cambridge: Cambridge University Press, 1983.

Jones, Robert; and Catherine M. Wykes. Holographic & Speckle Interferometry. Cambridge University Press,1989. Series title: Cambridge Studies in Modern Optics,; No.6.

Nondestructive holographic techniques for structures inspection. R. K. Erf. .. [et aLl. WrightPatterson Air Force Base, OH: Air Force Materials Laboratory, Air Force Systems Command, 1972.

Ostrovsky, Y. I. and M.M. Butusov. Interferometry by Holography. New York: Springer-Verlag, 1980. Series Title: Springer Series in Optical Sciences,; Vol. 20.

Schumann, Walter, and M. Dubas. Holographic Interferometry: From the Scope of Deformation Analysis of Opaque Bodies. Series Title: Springer Series in Optical Sciences,; Vol. 16. New York: Springer-Verlag,1979.

Spanner, Jack C. Acoustic Emission Testing: Acoustic Emission: Techniques & Applications. American Society for Nondestructive Testing, 1974.

Tropical Meeting on Hologram Interferometry and Speckle Metrology,1980, June 2-4 : North Falmouth, MA. A digest of technical papers presented at the Tropical Meeting on Hologram Interferometry and Speckle Metrology, June 2-4, 1980, Sea Crest Hotel, North Falmouth, Cape Cod, MA. [s.1.1: Optical Society of America,1980.

Vest, Charles M. Holographic interferometry. New York: Wiley, 1979. Series title: Wiley series in pure and applied optics.

Industrial Holography Dissertations

Dallas, William John. "Computer holograms: improving the breed". 1971.

Dzekov, Tomislav Angel. "Microwave holographic imaging of aircraft with spacebome illuminating source ·. 1976.

Matthews, Barbara Kubitz. "Application of holographic methods to the analysis of flexural vibrations of annular sector plates ·. 1976.

Su, Kung-Yen. "The fabrication of an opto-acoustic transducer for real-time diagnostic imaging systems ·, 1982.

Holography Marketing

Holography, 1971-72. Kallard, Thomas-Editor.

Series Title: State of the Art Review Series, Vol. 5. New York: Optosonic Press,1972.

Holography: Exploiting the Leading-Edge Developments. Chicago: Technical Insights,1987.

Holography Marketplace. Ross, Franz and Elizabeth Yerkes, Eds. Berkeley , CA: Ross Books, 1989.

Holography Marketplace, 2nd Edition. RoBS, Franz and Elizabeth Yerkes, Eds. Berkeley, CA: Ross Books,1990.

Holography Marketa. International Resource Developments,1984.

Industrial and commercial applications of holography: August 24-25,1982, San Diego, CA. Milton Chang, chair,editor. Bellingham, WA: SPIE--the International Society for Optical Engineering, 1983. Series title: Proceedings of SPIE-the International Society for Optical Engineering; v. 353.

Kallard, Thomas. Holography; state of the art review, 1969. New York: Optosonic Press,l969. Series title: State of the art review, 1.

____ . Holography; state of the art review ... 1970. Holography in1970: an overview by Dr. Dennis Gabor. New York: Optosonic Press,1970. Series title: State of the art review, no. 3.

Miller, Richard K. (et alia).Holography. New York: Future Tech Surveys, 1989. Series title: A Survey on Technology & Markets Ser.,; No. 51

Periodicals

Acoustical Holography. International Symposium on Acoustical Holography. New York: Plenum Press. v.1-7, (1969-1977).

Acoustical holography, proceedings. International Symposium on Acoustical Holography and Imaging. (-1973).

Acoustical imaging and holography. New York : Crane, Russak, 1978-1979.

Advances in holography. New York, M. Dekker, 1975-76.

Acoustical holography; [proceedingsl: Acoustical imaging, 1978. International Symposium on

Acoustical Holography and Imaging. New York: Plenum Press, 1978.

Afterimage. Rochester, NY: Visual Studies Workshop, V.12, no.7, Feb. 1985. See: A. SargentWooster, "Manhattan shortcuts, Harriet CasdinSilver's Thresholds' · p 19.

Fundamentals and applications of optical data processing and holography. Ann Arbor: University Michigan Engineering Summer Conferences.

Holography News. (newsletter). Washington, DC.: Louis Kontnick, since 1987. (see listing for address).

The Holo-gram. (newsletter). Allentown, PA: Frank Defreitas, since 1983. (See listing for address).

New Scientist. London, England: 1977. See: R. Weale "Art: Holography by H.Casdin-Silver" (June 29).

REFERENCES IN LANGUAGESOTHERTHAN ENGLISH

French Titles

Caussignac, Jean Marie. Visualisation d'ecoulements aerodynamiques dans les compresseul'8 par interferometrie holographique_ Chatillon, France: Office national d'etudes et de recherches aerospatiales, 1972. Series title: France. Office national d'etudes et de recherches aerospatiales Note technique, 190.

Francon, M. Holographie. Paris: Ma88on, 1969. Series title: Recherche appliquee.

International Symposium of Holography, Besancon, France, 1970. Applications de l'holographie; comptes rendus du Symposium international d'holographie. Applications of holography; proceedings of the International Symposium of Holography. Besancon 6-11 juillet 1970. Besancon: Laboratoire de physique generale et optique, Univel'8ite de Besancon, 1970.

Pinson, G., A. Demailly, and D. Favre.La Pensee: approche holographique. Lyon: Pre88es universitaires de Lyon, 1985. Series title: Collection Science des systemes. Serie theorie des systemes.

Voropaiev, N. Dictionnaire d'Electronique Quantique, Holographie et Optoelectronique. France: French & European, 1983

German Titles

Claus, Jurgen. ChippppKunst: Computer, Holographie, Kybernetik, Laser. Originalausg. FrankfurtlM.: Ullstein,I985. Series title: Ullstein Materialien.

Kiemle, Horst, [undl Dieter Ross. Einfuhrung in die Technik der Holographie. Frankfurt am Main, Akademische Verlagsanstalt, 1969. Series title: Technisch-physikalische Sammlung Bd.8.

Laserbeugung an elektronenmikroskopischen Aufnahmen. Ludwig Reimer ... ret al.l. Opladen: Westdeutscher Verlag, 1973. Series title: Forschungsberichte des Landes Nordrhein- Westfalen; Nr. 2314.

Licht-Blicke: Holographie, die 3. Dimension fur Technik und kunst : [Ausstellungl 7. Juni-30. September 1984, Deutsches Filmmuseum Frankfurt am Main. Schirmherr, Bundesprasident a. D. Walter Scheel; ... Frankfurt am Maain: [Deutsches Filmmuseuml,1984. Interviews with: S.Benton, M.Benyon, R.Berkhout, H.Casdin- Silver, F.Mazzero, S.Moree, N.Phillips, G. Schneider-Siemssen, D.Schweitzer. Articles by: M. Schneckenburger, et alia.

Mehr Licht: Kunstlerhologramme und Lichtobjekte = More light : [artists's [sicl holograms and light objectsl. herausgegeben von Achim Lipp und Peter Zec. Hamburg: Fielmann im E. Kabel Verlag, 1985.

Menzel, Eric, W. Mirande [undl I.Weingartner. Fourier-Optik und Holographie Wien: Springer-Verlag, 1973.

Optoelectronik in der Technik : Vortrage des 6. Internationalen Kongre88es Laser 83 Optoelektronik = Optoelectronics in engineering : proceedings of the 6th International Congress, Laser 83 Optoelektronik I herausgegeben. Berlin: Springer-Verlag, 1984.

Schreier, Dietmar unter Mitarbeit von W. Hase.. [et al.1 Synthetische Holografie. Weinheim: Physik-Verlag, 1984.

Universitatsbibliothek Jena. Zusammenstellung in- und auslandischer Patentschriiten auf dem Gebiet der Holographie. Berichtszeit: 1948-1970. Gesamtleitung: Konrad Marwinski, Informationsabt., 1971. Series title: Universitatsbibliothek Jena Bibliographische Mitteilungen, Nr.12.

Voropaev, N. D. Woerterbuch der Quantenelektronik, Holographie und Optoelektronik French & European,I983.

Zec, Peter. Holographie: Geschichte, Technik, Kunst. Koln: DuMont,1987.

Portuguese (Articles)

Catalogo of VII Salilo Nacional de Artes PlAstiCBS, "As t~ dimensOOs do signo verbal", Eduardo Kac, Museu de Arte Moderna, Rio de Janeiro, pp. 43-44, 1984.

Folha de SIlo Paulo, "Na holografia, 0 cinema tridimensional do futuro", Eduardo Kac, SIlo Paulo, August 21, 1985.

Folha de Silo Paulo, "Harriet, pioneira na arte do laser, expOe na Bienal", Eduardo Kac, Silo Paulo, October 2, 1985.

Folha de SIlo Paulo, "Holografia impressa sera prduzida comercialmente", Eduardo Kac, SIlo Paulo, May 20, 1987.

Folha de SIlo Paulo, "Ingl~s mostra seu percurso na holografia", Marion Strecker, Silo Paulo, June 20, 1989.

Jornal do Brasil, "AArte da Sfntese nos holopoemas", Reynaldo Roels Jr., Rio de Janeiro, Septemebr 29, 1985.

Jornal do Brasil, "A holografia da urn passo A frente", Eduardo Kac, Rio de Janeiro, September 29,1985.

Jornal do Brasil, "Intelig~ncia e high tech", Reynaldo Roels Jr., Rio de Janeiro, December I, 1986.

Jornal do Brasil, "Urn unic6rnio na matematica", Reynaldo Roels Jr., Rio de Janeiro, November 29, 1988.

Jornal do Brasil, "Holografia gerada por computador", Marcelo Tognozzi, Rio de Janeiro, July 2, 1989.

O Estado de SIlo Paulo, "Cariocas inovam holografia", Sergio Adeodao, Silo Paulo, November 27,1988.

O Globo, "Primeira mostra de arte high tech", Frederico de Morais, Rio de Janeiro, April 6, 1986.

O Globo, "Holografla: o ator sai da tela e senta perto do espectador", Sheila Kaplan, Rio de Janeiro, December 13, 1984.

O Globo, "0 Sonho holografico de Alexander", Eduardo Kac, Rio de Janeiro, November 11, 1987.

O Globo, "Holofractal, a arte no futuro", Ligia Canongia,Rio de Janeiro, November 22, 1988.

Russian Titles

Bakhrakh, L.D. i S.D. Kremenetskii. Metody izmerenii parametrov iizluchaiushchikh sistem v blizhnei zone. Leningrad : Izd-vo "Nauka", Leningradskoe otdelenie, 1985.

Bakhrakh, L.D i V.A. Makeeva. Primenenie golografii v meditsine i biologii. Leningrad : Nauka, 1977.

Barachevskii, V.A.Svoistva svetochuvstvitel'nykh materialov i ikh primenenie v golografii: sbornik nauchnykh trudov. Otvetstvennyi redaktor Leningrad: Izd-vo "Nauka," Leningradskoe otd-nie, 1987.

____ . Neserebrianye i neobychnye sredy dlia golografii. Leningrad: "Nauka", Leningradskoe otd-ie, 1978.

Denisiuk, IU.N. Opticheskaia gologratiia : prakticheskie primeneniia . . Leningrad: Izd-vo "Nauka," Leningradskoe otd-nie, 1985.

____ .. Opticheskaia gologratiia: [Sb. statei]. AN SSSR, Fiz.-tekhn. in-t im. A.F. Ioffe, Nauch. sovet po probl. "Golografiia". Leningrad: Nauka, Leningr. otd-nie, 1979.

Derkach, M.F.Dinamicheskie spektry rechevykh signalov. L'vov: Izd-vo pri L'vovskom Gos. universitete Izdatel'skogo ob"edineniia "Vyshcha shkola", 1983.

Fizicheskie osnovy i prikladnye voprosy golografii : tematicheskii sbornik: [materialy XVI Vsesoiuznoi shkoly po fizicheskim 08novam golografiil redaktory G.V. Skrotskii, B.G. Turukhano, N. Turukhanol. Leningrad: Akademiia nauk SSSR, Fiziko-tekhn. ins-t im. A.F. Ioffe, 1984.

Gurevicha, S.B. i V.K. Sokol ova. Primenenie metodov opticheskoi obrabotki informatsii i golografii. Leningrad: LIIAF, 1980.

Gurevicha, S.B. Primenenie golografii v matsii, held in Riga, May 1980. Akademiia nauk SSSR. Ordena Lenina Fiziko-tekhnicheskii institut im. A.F. Ioffe.

Gurevicha, S.B., O.A. Potapova. "Gologratiia i opticheskaia obrabotka informatsii v geologii". Dokl. seminara. Leningrad: Akademiia Nauk SSSR, Fiziko-tekhnicheskii in-t im. A.F. Ioffe, 1980.

IAkovkin, I. B. Difraktsia sveta na akusticheskikh poverkhnostnykh volnakh. otv. redaktor S.V. Bogdanov. Novosibirsk: Izd-vo "Nauka", Sibirskoe otd ie, 1979.

IAroslavskii, L. P. and N.S. Merzliakov. TSifrovaia golograflia. Moskva: Izd-vo "Nauka", 1982.

International School on Coherent Optics and Holography,2nd: 1981: Varna, Bulgaria. Integral'naia optika, volokonnaia optika i golografiia: materialy vtoroi Mezhdunarodnoi shkoly po kogerentnoi optike i golografii--Varna

'81.28.09-03.10.1981. Varna. Bolgariia. Redakt-sionnaia kollegiia P. Simova •... Sofiia: Izd-vo Bolgarskoi akademii nauk. 1982.

Kirillov. N. 1.Vysokorazreshaiushchie fotoma-terialy dlia golografii i protsessy ikh obrabotki . Moskva: Nauka. 1979.

Klimenko. 1. S. Golografiia sfokusirovannykh izobrazhenii i spekl-interferometriia. Moskva: "Nauka," GJav. red. fiziko-matematicheskoi litri. 1985.

Klimkin. V. F. (Viktor Fedorovich) Opticheskie metody registratsii bystroprotekaiushchikh protsessov/ V.F.

Klimkin. A.N. Papyrin. R1. Soloukhin ; otvetst-vennyi redaktor N.G. Preobrazhenskii. Novosi-birsk : Izd-vo "Nauka," Sibirskoe otd-nie. 1980.

Kulakov. Sergei Viktorovich. Akustoopticheskie ustroistva spektral'nogo i korreliatsionnogo analiza signalov. Akademiia nauk SSSR. Nauchnyi sovet po probleme"Golografiia". Fiziko-tekhnicheskii institut imeni A.F. loffe. Leningrad:"Nauka". Leningradskoe otd-nie. 1978.

Petrashen. G. 1. Prodolzhenie volnovykh polei v zadachakh s seismorazvedki. Leningrad: "Nau-ka," Leningr. otd-nie. 1973.

Radiogolografiia i opticheskaia obrabotka in-formatsii v mikrovolnovoi tekhnike: [Sbornik statei]. Akademiia nauk SSSR. Otdelenie obsh-chei fiziki i astronomii. Nauchnyi sovet po pro-bIerne "Golografiia" Leningrad "Nauka," Lenin-gradskoe otd-nie. 1980.

Soboleva. G.A. Registriruiushchie sredy dlia izo-brazitelnoi golografii i kinogolografii: [Sb. statei]. AN SSSR. Otd-nie obshch. fiziki i astronomii. Nauch. sovet po probl. Golografiia. Leningrad: Nauka. Leningr. otd-nie. 1979_

Sokolov. A. V. i IA.A. Al'tmana. Primenenie metodov opticheskoi golografii dlia issledova-niia biologicheskikh mikroob"ektov. Leningrad: Nauka. Leningradskoe otd-nie.1978. Series title: Metody fiziologicheskikh issledovanii.

Voropaev. N. D. Anglo-russkii slovar' po kvanto-voi elektronike i golografii: Okolo 18000 termi-nov. Pod red. A. M. Leontovicha. Moskva: Rus. iaz .• 1977.

Vsesoiuznaia shkola po golografii. 6th: 1974: Yerevan. Armenian S.S.R. Materialy VI Vsesoi-uznoi shkoly po golografii:1l-17 fevralia 1974 g. redaktory. G.V. Skrotskii. B.G. Turukhano. N. Turukhano]. Leningrad: LIIAF. 1974.

Vsesoiuznaia shkola po golografii. 7th: 1975: ROBtoy. R.S.F.S.R Materialy VII Vsesoiuznoi shkoly po golografii: ianvar' 1975 g. [podgotov-leny k pechatki N. Turukhano]. Leningrad: Len-ingradskii in-t iadernoi fiziki. 1975.

Vsesoiuznaia shkola po fizicheskim osnovam go-lografii (14th:1982 : Dolgoprudnyi. R.S.F.S.R.). Prikladnye voprosy golografii: tematicheskii sbornik.[redaktory. G.V. Skrotskii. B.G. Turukh-ano. N. Turukhano]. Leningrad: LIIAF. 1982.

Zel'dovich. B. IA. N.F. Pilipetskii. & V.V.Shkunov. Obrashchenie volnovogo fronta. Moskva: "Nau-ka," GJav. red. fiziko-matematicheskoi lit-ry. 1985.

Swedish Titles

Holografi: det 3-dimensionella mediet. New York: Museum of Holography; Stockholm: distri-bution. AVC.1976.

Lasers And Holography

Advances in Laser Engineering: August 25-26. 1977. San Diego. California. Malcolm L. Stitch. Eric J . Woodbury. eds; presented by the Soci-ety of Photo-optical Instrumentation Engineers in conjunction with the IEEE Computer Society International Optical Computing Conference '77. Bellingham. WA: The Society. 1977.

Advances in nonlinear polymers and inorgan-ic crystals. liquid crystals. and laser media: 20-21 August 1987. San Diego. California. Solomon Musikant. chair/editor; sponsored by SPIE-The International Society for Optical Engineering. Bellingham. WA. USA: SPIE--the International Society for Optical Engi-neering.1988.

Beck. Rasmus. W. Englisch. and K. Gura. Table of Laser Lines in Gases and Vapors. 3d rev. and enl. ed. Berlin: Springer-Verlag. 1980. Series Title: Springer Series in Optical Sciences; v. 2.

Bennett. William Ralph Jr. Atomic gas laser transition data: a critical evaluation. New York: IFIlPlenum, 1979. Series: IFI data base library.

Chu. Benjamin. Laser Light Scattering. New York: Academic Press. 1974. Series: Quantum electronics--principles and applications.

Eichler. H. J .• P. Gunter. and D.W. Pohl. Laser-induced dynamic gratings. Berlin:Springer-Ver-lag. 1986. Series title: Springer Series in Optical Sciences; v. 50.

Electro-opticsllaser international '80 UK. Brighton. 25-27 March 1980: conference pro-ceedings. H.G. Jerrard. ed. Conference or-ganized by Kiver Communications Ltd. Gui-Jdford. Surrey. England: IPC Science and Technology Press. 1980.

Fundamentals of laser interactions II: proceed-ings of the fourth meeting on laser phenomena held at the Bundessportheim in Obergurgl. Aus-tria. 26 February-4 March 1989. F. Ehlotzky. ed. Berlin: Springer-Verlag. 1989. Series: Lecture notes in physics; 339.

ICALEO Technical Digest Eighty-Three; Vol. 36. Thledo. OH: Laser Institute.1983.

ICALEO Materials Processing Eighty-Three: Proceedings. Vol. 38. Thledo. OH: Laser Insti-tute. 1984.

ICALEO Holography & Information Processing Eighty-Three: Proceedings.Vol. 41. Toledo. OH: Laser Institute. 1984.

Industrial applications of laser technology: April 19-22.1983. Geneva Switzerland.William F. Fagan. chair.ed. Bellingham. WA: SPIE--the International Society for Optical Engineering. 1983. Series title: Proceedings ofSPIE--the In-ternational Society for Optical Engineering; v. 398.

The Industrial Laser Annual Handbook. David Belforte. Morris Levitt. editors; managing editor. Laureen Belleville. Tulsa. OK : Penn Well Books. 1987. Series: SPIE; v. 919.

Industrial Laser Interferometry: 14-16 January. 1987. Los Angeles. California. Ryszard J . Pry-putniewicz. chair/editor; sponsored by SPIE- the International Society for Optical Engineering. in cooperation with Center for Applied Optics. University of Alabama in Huntsville ... let al.]. Bellingham. WA .• USA : SPIE. 1987. Series: Proceedings of SPIE--the International Society for Optical Engineering.

Interferometric Metrology: 20-21 August 1987. San Diego. California. N.A. Massie. editor; sponsored by SPIE. the International Society for Optical Engineering; cooperating organizations. Applied Optics Laboratory/ New Mexico State University ... let al.]. Bell-ingham. WA .• USA: SPIE. 1988. Series: Crit-ical reviews of optical science and technology. SPIE; v. 816.

International Congress on Applications of La-sers and Electro-optics.1987: San Diego. Cali-fornia. Proceedings of the international confer-ence on optical methods in flow and particle diagnostics. Edited by Warren H. Stevenson; sponsored ... by the Laser Institute of America. Thledo. OH: Laser Institute of America. 1988. Series: LIA Contents: v. [1] Medicine & biology -- v. [2] Maters processing-- v. [3] Inspection. measurement & control. Laser diagnostics & photochemistry -- v. [4] Optical communications & information processing. Imaging & display technology.

International Symposium on Gas-Flow and Chemical Lasers (4th: 1982: Stresa. Italy). Gas flow and chemical lasers. Michele Onorato. ed. New York: Plenum Press.1984.

Kaminow. Ivan P. Laser devices and applica-tions. Ivan P. Kaminow and Anthony E. Sieg-man. eds. New York: IEEE Press.1973. Series title: IEEE Press selected reprint series.

Kock. Winston E. Engineering applications of la-sers and holography. New York: Plenum Press. 1975. Series title: Optical physics and engineer-ing.

____ . Lasers and holography; an introduction to coherent optics. 1st ed. Garden City. N.Y: Dou-bleday. 1969. Series title: Science study series ; [S62].

____ . Lasers & holography: an introduction to coherent optics. 2nd enl. ed. New York: Dover Publications. 1981.

Laser applications to optics and spectroscopy: based on lectures of the July 8-20.1973. Summer School. Crystal Mountain.Wash.

The Laser Marketplace in 1990: a seminar exam-ining recent trends and directions in the world-wide market for lasers. Morris R Levitt. Gary T. Forrest. editors; organized by Laser Focus World in cooperation with SPIE--the International So-ciety for Optical Engineering. Bellingham. WA .• USA: SPIE--the International Society for Optical Engineering. 1990.

Laser Measurements. Thledo. OH: Laser lnsti-tute. 1985.

Laser recording and information handling tech-nology. Proceedings of a seminar held August 21-22. 1974. San Diego. CA. Leo Beiser. ed. Palos Verdes Estates. CA: Society of Photo-Optical

Instnlmentation Engineers,1975_ Series title: Society of Photo-optical Instrumentation Engineers Seminar proceedings; 53_

Lasers and holographic data processing. N.G. Basov, ed_ Translated from the Russian by P.S. Ivanov. Moscow: Mir Publishers, 1984. Series title: Advances in science and technology in the USSR. Technology series.

Lasers and optical radiation. Geneva: World Health Organization; Albany, N.Y.: WHO Publications Centre USA [distributor],1982. Series: Environmental health criteria; 23.

Laser Window and Mirror Materials. Compiled by G. C. Battle, 'Ibm Connolly, and Anne M. Keesee; with a pref. by Charles S. Sahagian. New York: IFl!P1enwn,1977. Series: Solid State Physics Literature Guides; v. 9.

Lyons, Harold_ Lasers, quantum electronics, holography: part 1: Introduction to lasers: Engineering 823.1: a five-day short course, July 7- 11, 1975: lecture notes. Harold Lyons, coord. Los Angeles: University of California, University Extension, 1975.

_____ . Lasers, quantum electronics, holography: part 1, Introduction to lasers: Engineering 823.1, June 17-21, 1974 : lecture notes. Harold Lyons, coord. Los Angeles: University of California, University Extension, 1974.

The Marketplace for Industrial Lasers (no further bibliog information provided)

Menzel, R. Fingerprint Detection with Lasers. New York: Marcel Dekker, 1980.

National Conference on Measurements of Laser Emissions for Regulatory Purposes, Rockville, MD., 1974. National Conference on Measurements of Laser Emissions for Regulatory Purposes: proceedings of a conference held in Rockville, Maryland, June 4-7, 1974. Edited by Robert H. James; cosponsored by National Bureau of Standards and U.S. Dept. of Health, Education, and Welfare, Public Health Service, Food and Drug Administration, Bureau of Radiological Health. Rockville, MD.: The Bureau; Washington: for sale by the Supt. of Docs., U.S. Govt. Print. Off., 1976 i.e. 1977.

Nonlinear optical beam manipulation, beam combining, and atmospheric propagation: 11-14 January 1988, Los Angeles, California. Robert A. Fisher, chair/editor; sponsored by SPIE--the International Society for Optical Engineering. Bellingham, WA.: SPIE, 1988.

Tarasov, L. V. Laser age in optics. Translated from the Russian by V. Kisin. Moscow: Mir Publishers, 1981.

Trolinger, J . D. Laser applications in flow diagnostics. J .J . Ginoux, ed. Neuilly-sur-Seine, France: North Atlantic Treaty Organization, Advisory Group for Aerospace Research and Development,1988.

_____ Laser instrumentation for flow field diagnostics. S. M. Bogdonoff, ed. Neuilly-surSeine, France: North Atlantic Treaty Organization, Advisory Group for Aerospace Research and Development, 1974. Series title: AGARDograph; no. 186.

Ultrashort laser pulses and applications. Edited by W. Kaiser; with contributions by D.H.

Auston ... [et a1.]. Berlin: Springer-Verlag, [n.d.]. Series: 'Ibpics in applied physics; v. 18. Lasers--disserta tions

Cuendet, George Joseph. Optical path differences induced by absorption of laser energy in beam tubes wi th an axial flow, due to diffusive and acoustic effects. 1983.

Frehlich, Rodney George. Laser propagation in random media. 1982.

Kachen, George Ivan Jr.. Minimization of unwanted light in laser scattering experiments. 1965_

Sandstrom, Richard Lynne. An experimental study of optical phase fluctuations in a random refractive field ,1979.

This bibliography was compiled by Elizabeth Yerkes of Lyme, New Hampshire, who achieved order from chaos armed only with a Macintosh and The Chicago Manual of Style.

15

Glossary

Absorption Hologram: A hologram formed in a material which acquires a certain density in response to exposure. When the hologram is illuminated, part of the light which is not absorbed is diffracted into forming the image.

Achromatic: Free of color. Black and white. In optical systems, the term is used to describe lenses which correct for chromatic aberration.

Acoustical Holography: The making of holograms by using sound waves.

Additive Color Mixing: Means by which two or more frequencies are combined by superimposition to create more colors.

Ambient Light: Light present in the immediate environment. In holographic display, often used to describe background light that is not part of the hologram illumination and may interfere with the viewing of the image.

Amplitude: The maximum value of the displacement of a point on a wave front from its mean value. Graphically, the height or depth of the crest or trough of a wave from its zero point.

Amplitude Hologram: A hologram by which information is stored as variations in transmittance. Also called absorption hologram

Antihalation Backing (AH): A dark material placed on the back surface of a plate or film to prevent unwanted light from striking the emulsion. Useful to prevent the formation of "Newton Rings" in the hologram. Only to be used with transmission holograms.

Argon Laser: A laser which operates when argon gas is ionized and controlled by a magnetic field. Produces several blue and green frequencies.

Artistic Holography: Holograms created for the purpose of being seen and whose value derives at least in part from the image presented.

Astigmatism: An aberration caused by the horizontal and vertical aspects of an image forming in different planes.

Bandwidth: The range of frequencies over which a given instrument will operate.

Beamsplitter: An optical component which divides a beam into two or more separate beams. A 50:50 beamsplitter produces two beams of approximately equal intensities. A 90:10 beamsplitter transmits approximately 90% of the incident beam and reflects 10% into the second beam.

Benton Hologram: Another term for rainbow hologram. Named for its inventor, Steve Benton. A hologram produced by reducing vertical information in order to correct for image dispersion.

Biconcave: A lens which has both faces curving inward. A type of negative lens.

Biconvex: A lens which has both faces curving outward. A type of positive lens

Bleach: In holographic processing, a chemical used to change an absorption hologram into a phase hologram in order to improve efficiency (brightness).

Bragg Diffraction (Bragg's Law): Diffraction which is reinforced by reflection by a series of regularly spaced planes which correspond to a certain wavelength and angular orientation. The angle at which this reinforcement occurs is Bragg's angle.

BRH: Bureau of Radiological Health. U.S. government agency responsible for setting laser safety standards.

Brightness: A subjective term describing the amount of light perceived.

Cavity: Another name for optical cavity or laser cavity.

Chromatic Aberration: Lens or hologram irregularity due to the shifting of image position for each frequency. If severe enough, the image will appear to blur due to the lack of registration of the colors.

Coherence Length: The maximum path length difference (between the reference beam and the object beam) in a holographic set up that can be used and still obtain a clear, bright hologram. Coherence length depends on the type of laser used and how it is made. An etalon will increase the coherence length of a laser to about 30 feet.

Coherent Light: Light which is of the same frequency and vibrating in phase. The laser produces coherent light.

Collimated Light: Light which forms a parallel beam and neither converges nor diverges. Also referred to as collimated beam.

Collimator: A device used to produce collimated light by positioning a light source at the focal point of a lens or parabolic mirror. Such a device is called a collimating lens or collimating mirror, respectively.

Color Spread: The area over which a spectrum is dispersed.

Computer Generated Hologram: A synthetic hologram produced using a computer plotter. The binary structure is produced on a large scale and then photographically reduced into a given medium. The technique allows the production of impossible or nonexistent 3-dimensional forms.

Concave Lens: A lens with an inwardly curving surface which causes light to diverge. See also Negative lens.

Concave Mirror: A mirror with an inwardly curving surface which causes light to converge.

Constructive Interference: Coherent wave fronts of the same frequency are superimposed, and their instantaneous amplitudes add up to a greater amplitude than that of the component waves.

Continuous Wave Laser: A laser which emits a beam which does not vary over time.

Convergence: The optical bending of light rays toward each other, as by a convex lens or concave mirror.

Convex Lens: A lens with an outwardly curving surface which causes light to converge, usually to a focal point.

Copy Hologram: Another term for image plane hologram or any second generation hologram produced from a master hologram. A contact copy is produced by placing the plate in contact with the original.

Copy Plate: Another term for copy hologram. Usually refers to the plate before it is exposed.

Cross Hologram: Another name for the type of holographic stereogram which incorporates the advantages of rainbow holography. Named for Lloyd Cross.

Cross Talk: The phenomenon of spurious images formed by color holograms when an interference pattern formed by one color also reconstructs an image in another color.

Cylindrical Mirror/Lens: An optical component which causes light to focus as a slit or line by passing through or reflecting from a surface curved in one dimension.

Denisyuk Hologram: Another name for single beam reflection hologram. Named for its inventor, Y. N. Denisyuk.

Density: The amount of opacity or darkness of a medium.

Depth of Field: The area within which satisfactory resolution of

an image can be obtained. Also, in holography, used to describe the area within which any image can be formed, due to the constraints of coherence length.

Developer: A chemical solution which changes the latent image of a photographic image or holographic interference pattern (silver salts) into black metallic silver. The term development usually refers to the degree of effect of the developer or the cause of the amount of density.

Dichromated Gelatin (DCG): A light-sensitive gelatin made up of a solution of dichromate compound, usually ammonium dichromate, in the presence of a gelatin substrate. Exposure results in the cross-linking of gelatin molecules with each other.

Diffraction: The change in direction of a wave front at the edges of an aperture, caused by the wave nature of light. Diffraction is not the same process as reflection or refraction.

Diffraction Efficiency: In a hologram, the percentage of incident illumination light diffracted into forming the image. The greater the diffraction efficiency, the brighter the image will appear in a given light.

Diffraction Grating: A holographic diffraction grating is a hologram formed by the interference of two or more beams of pure, undiffused laser light.

Diffuse Reflector: An object that scatters illumination striking it. Most objects are diffuse reflectors.

Divergence: The bending of light rays away from each other, usually by concave lens or convex mirror, so that the light spreads out. Light will also diverge with a convex lens or concave mirror after it passes through the focal point.

Double Exposure: The formation of two holograms on the same recording medium. Used to cause either overlapping images or two discrete images to appear under different conditions.

Electromagnetic Radiation: Radiation emitted from vibrating, charged particles, all of which travels through space at the speed of light. Visible light is only a small part of the entire electromagnetic spectrum.

Embossed Hologram: A hologram copy made by pressing a metal surface relief master hologram into plastic film, or by using the master hologram in a mold.

Emulsion: A suspension of light sensitive silver salts (e.g. silver bromide) in gelatin, usually coated onto glass, polyester film, or by using the master hologram in a mold.

Exposure: The act or time of allowing light to impinge upon the emulsion.

Film Plane: The plane at which the recording material is located.

Fixer: A chemical solution which removes the unexposed silver salts from the emulsion to desensitize the emulsion after development.
f·number: The ratio of the focal length of a lens or curved mirror to its diameter.

Focal Length: The distance from the center plane of a lens or curved mirror to a position where a collimated beam is focused to a point. A lens with a negative focal length (a concave lens) appears to have a focus upstream from the lens, while a lens with a positive focal length (a convex lens) has a focus downstream from the lens.

Focused Image Hologram: Any hologram in which the image appears on the surface of the hologram or seems to intersect the surface of the hologram.

Fog: The darkening or exposing of film by inadvertently allowing ambient light to strike it. In holography, a fogged plate reduces fringe contrast, resulting in a less efficient image.

Fourier Transform Hologram: A hologram made using a reference beam diverging from a point at the same distance from the recording plate as the object. Also called Fraunhofer Hologram

Frequency: The number of crests of waves that pass a fixed point in a given unit oftime.

Fresnel Hologram: Another name for the common hologram. Defined as a hologram formed with an object located close to the recording medium.

Fringe: An individual interference band, made up of one cycle of constructive and destructive interference.

Front Surface Mirror: A mirror with the reflecting surface on the front. Conventional mirrors have their reflecting surfaces on the back of a piece of glass and are not useful for holography as the front surface produces a "ghost" reflection.

Gabor Hologram: An in-line hologram of the type invented by Dennis Gabor.

Gas Laser: A laser such as a Helium-Neon laser, in which the lasing medium is a gas.

Grating (also Diffraction Grating): A pattern of very fine lines of equal spacing, usually on the order of a few microns apart. A diffraction grating can redirect light and break white light into its component colors like a prism.

H-1: The first hologram made in the process of making a master hologram. An H-I has an image only viewable in laser light.

H-2: The second hologram made in the process of making master hologram. A master hologram is usually an H-2.

Helium-Neon (HeNe): The most common lasing material, which produces a continuous red beam at 632.8 nm.

HOE: Holographic Optical Element. A hologram which may be used to act as a lens, mirror, or a complex optical component.

Hologram: An interference pattern formed as a result of reference light encountering light scattered by an object and stored as such on a light sensitive emulsion.

Holographic Movie: The animation of a 3-dimensional holographic image by presentation of numerous holograms in rapid sequence in much the same way motion picture film operates. Unlike conventional cinema, it is only with extreme difficulty that the image can be projected, and true holographic movies are still very experimental. The term is often used, incorrectly, for holographic stereograms.

Holographic Stereogram: A hologram made by filming numerous angles of view of a scene and then storing the frames holographically. Each eye views a different frame, displaced so as to result in the illusion of a stereoscopic image. Also called a multiplex or integral hologram.

Image Plane Hologram: A second generation hologram formed by positioning a light sensitive plate in the plane of an image formed by a master hologram.

Incandescent Light: Light formed when an electric current passes through a resistant metal wire, usually situated in a vacuum bulb.

Incoherent Light: Light which is emitted with randomly varying phase and a mix of colors. Light from ordinary sources such as light bulbs is incoherent; laser light is coherent.

Index of Refraction: The ratio of the velocity of light in air to the velocity of light in a refractive material for a given wavelength.

Infrared: That part of the spectrum characterized by wavelengths somewhat longer than those of red light, which are not visible to the eye yet are often perceived as heat. Covers the spectrum from about 750 nm to 1000 micrometers.

In-Line Hologram: A Gabor hologram. Made by positioning the object and reference light along the same axis, resulting in a configuration practical only for making holograms of transparencies.

In Phase: The relationship of two waves of the same frequency when they travel through their maximum and minimum values simultaneously and are also polarized identically. Holograms must be made by waves which remain in phase during the course of an exposure.

Interference: The result of su perimposing two or more waves. The waves oscillate between negative and positive values, so when two waves are superimposed the positive values reinforce each other while negative values cancel out positive values.

Interference Pattern: A stationary pattern of interference that results when light waves are superimposed.

Interferometer: A device that utilizes interference of light to measure changes in systems with extreme accuracy. An interferometer can be used to test the stability of holographic systems.

Ion: An atom or molecule which has gained or lost an electron so that it acquires a negative or positive charge.

Ion Laser: A laser within which stimulated emission occurs as a result of energy changes between two levels of an ion. Argon and krypton are the two most common types of ion laser.

Krypton Laser: An ion laser that produces many frequencies which appear over a large part of the spectrum. The most common lines are blue, green, yellow, as well as a very strong red frequency.

LASER: From the acronym for "Light Amplification by Stimulated Emission of Radiation." A laser is usually in the form of a lightamplifying medium placed between two mirrors. Light not perfectly aligned with the mirrors escapes out the sides, but light perfectly aligned will be amplified. One mirror is made partially transparent. The result is an amplified beam of light that emerges through the partially transparent mirror.

Latent Image: The image or pattern stored in an emulsion before it is developed into a visible image.

Latent Image Decay: A condition that is common to fine grained silver emulsions, including the types used for holography. The decay occurs if the material is not processed soon after exposure, resulting in a lower density.

Light Meter: Any device used to sense and measure light. Usually used to sense intensity in order to determine exposure.

Line Spacing: The distance between individual interference fringes in a diffraction grating.

Lippman Hologram: Another name for reflection hologram.

Liquid Gate: A liquid-tight glass-sided plate holder in which the plate is exposed, processed and viewed with the cell fllied with liquid throughout.

Liquid Lens: A lens formed by filling a shaped glass or acrylic tank with liquid such as mineral oil. A liquid lens is often used in a system for making Cross holograms.

Master Hologram: The original H-2 hologram, from which copy holograms are made.

Mode: A laser can oscillate in a number of different modes. Spatial modes are the different repetitive patterns in which light can zigzag between the two laser mirrors. The optimum spatial mode for holography has no zigzag at all (called the TEMOO mode), resulting in a single emitted beam of light. Higher modes can result in a beam that gives a donu t-shaped spot, or a cloverleaf pattern. Longitudinal modes are the different wavelengths that are simultaneously emitted by a laser. A single-mode laser produces a single wavelength of light, and can only have a single spatial mode and single longitudinal mode.

Moire Pattern: A highly visible type of interference pattern formed when gratings, screens, or regularly spaced patterns are superimposed upon each other.

Monochromatic: Light or other radiation with one single frequency or wavelength. Since no light is perfectly monochromatic, the term is used loosely to describe any light of a single color over a very narrow band of wavelengths.

Motion: The effects of an object or holographic system not remaining rigidly fixed during exposure.

Multichannel Hologram: A hologram formed with two or more separate reference beams or angles.

Multiple Exposure: More than one exposure occurring on the same plate or film.

Multiplex: Another name for holographic stereogram.

NAH: A holographic plate without an antihalation backing. Also called an unbacked plate.

Negative Lens: A lens characterized by a concave surface which causes light to diverge. A negative lens has a negative focal length.

Newton's Rings: The series of rings or bands which appear due to interference between two nearly parallel surfaces. These rings often form as a result of light interacting between the front and back surfaces of a holographic plate.

Node: The part of a vibrating wave that is not moving - zero point. An antinode is a point on the wave of maximum displacement from the zero point.

Noise: Any unwanted light scattering by components in a holographic set-up or by particles in a holographic recording medium.

Object Beam: The light beam in a holographic set-up which illuminates the object. Also, the light reflected from or transmitted by the object which is recorded in the hologram.

Off-Axis: The type of hologram invented by Emmet Leith and Juris Upatnieks whereby object and reference beams approach the holographic plate at different angles.

On-Axis: Hologram formed with object and reference beams originating along the same axis. Also called an in-line or Gabor hologram.

Open Aperture: A transmission image plane hologram viewable in white light and characterized by both vertical and horizontal parallax and usually a brilliant white image.

Optical Cavity: The space between the two mirrors in a laser. The tube is located within the optic,?,l cavity.

Optical Component: An optical device consisting of the optics (lens, mirror, etc.) and a mount used to affix it to a vibration isolation table.

Optics: Those devices which change or manipulate light, including lenses, mirrors, beamsplitters, filters, etc. Also the science of electromagnetic radiation, its effects, and the phenomenon of vision.

Orthoscopic: Having the "right" appearance. Orthoscopic image has the correct appearance, whereas a "pseudoscopic" image appears to have its depth inverted.

Oscillator: Any device that converts energy into an alternating electromagnetic field, usually of constant period.

Overexposure: Improper exposure resulting from too much light or light reaching the plate or film for too long.

Parabolic Mirror: A mirror with a surface curved in the shape of a parabola. Used as a telescope mirror in astronomy or as a collimating mirror in holography.

Parallax: The difference between two different views of an object, obtained by changing viewing position.

Period: The time required for a wave to go through one complete cycle. The period of a typical light wave is about one trillionth of a second.

Phase: A wave oscillates from a positive value to a negative value and back to positive. The phase of a wave relates to where it is in its oscillation cycle at a particular moment. Usually only the phase difference between two waves is important.

Phase Hologram: A hologram which diffracts light by delaying the phase of certain portions of the lightwave, rather than absorbing certain portions. Bleached silver halide holograms, nCG holograms, and surface relief holograms are phase holograms, while unbleached silver halide holograms are amplitude holograms.

Phase Shift: The amount by which the phase of one light beam is delayed or advanced relative to another light beam.

Photochemistry: The branch of chemistry dealing with the effects of light on chemical reactions.

Photon: The smallest unit or quantum of electromagnetic energy known today.

Photopolymer: A material which "polymerizes" where it is exposed to light. Photo polymers are usually partially solidified plastics which finish solidifying when exposed.

Photoresist: A chemical substance made insoluable by exposure to light (usually ultraviolet). Although most often used to manufacture microcircuits, photoresist can be used to make holograms.

Pinhole: The small hole used to pass focused light from the objective in a spatial filter.

Plane Hologram: A hologram for which fringes are large with respect to the thickness of the emulsion, so that interference is mostly stored on the surface of the hologram.

Plano-Concave: A lens which has one concave surface and one flat surface.

Plano-Convex: A lens which has one convex surface and one flat surface.

Plateholder or Platen: Any device which holds a holographic plate or film in place during the exposure.

Polarization: The restriction of light or other radiation to vibration in only one plane.

Population Inversion: A condition whereby more atoms are in the excited state than in the ground state, resulting in the predominance of stimulated emission.

Positive Lens: A lens with an outwardly curving surface which causes light to converge. Also known as convex lens.

Processing: The entire chemical sequence, from development to final drying of the hologram.

Pseudocolor: The production of colors in a hologram which are not related to the true colors of the original objects. Usually used in connection with multicolor holograms.

Pseudoscopic: The opposite of orthoscopic. An image whose parallax is reversed.

Pulsed Hologram: A hologram produced with the short burst of light from the pulsed laser. May be used to make holograms of live subjects.

Pulsed Laser: A laser which emits radiation in a wave of short bursts and is inactive between bursts.

Quantum: The smallest amount that the energy of a wave may be divided into.

Rainbow Hologram: A white light viewable hologram with colors which shift through the spectrum as the hologram or the viewing angle is tilted. A rainbow hologram has no vertical parallax.

Real Image: An image that is formed in such a way that it actually comes to a focus. A real image is one that forms downstream from a lens; a virtual image appears to form upstream from a lens.

Real Time Holography: A technique whereby a holographic

image is superimposed over a real object in order to observe interference fringes generated by minute changes between the two.

Reconstruction Beam: Light directed at the finished hologram from which the object wave front will be recreated.

Recording Material: Any substance which may be used to record the interference pattern of the hologram.

Reference Angle: The angle at which the reference beam strikes the plate, usually measured in degrees from the plate surface.

Reference Beam: The unmodulated, pure laser light directed at the plate to interfere with the object light.

Reflection Hologram: A hologram made by allowing reference and object light to impinge on opposite sides of the plate. The finished hologram is viewed by allowing light to reflect from it to the observer.

Refraction: The bending of light which occurs when it passes from a medium of one refractive index to that of another. In a phase hologram, refraction causes a "phase delay" which corresponds to the original phase difference between the two stored wave fronts.

Refractive Index: Same as index of refraction.

Resolution: The ability of a film or an optical system to distinguish between two closely spaced points. Film resolution is usually expressed in terms of how many closely spaced lines per millimeter the film can record. Holographic films must be capable of high resolving capability since the interference fringes are often extremely small and closely spaced.

Resonance: A large amount of vibration in a system which is caused by a small stimulus with approximately the same period as the natural vibration period of the large system.

Resonant Cavity: Another name for optical cavity or laser cavity.

Scatter: Unwanted light which interferes with the making of a good quality hologram.

Settling Time: A period of time between the loading of the plate and the exposure in order to allow ambient vibrations time to dampen.

Set-up: The configuration of optical components used to produce a given hologram.

Shadowgram: A hologram made by deliberately moving an object during an exposure, or by using an inherently unstable object, in order to produce a 3-dimensional "holen or shadow where the object was once located.

Shutter: The device used to block the laser beam and then allow it to pass unobstructed for the desired exposure time.

Silver halide: The type of recording material which consists of light-sensitive silver particles suspended in gelatin.

Single beam hologram: A hologram made with one beam which acts as both reference and object illumination beam.

Slab Table: An optics table which uses a concrete slab as part of its inertial mass.

Slit Optics: Any optical device which causes light to be propagated into a line. Usually formed by light interacting with a cylindrical surface.

Solid State Laser: A laser which uses a solid material, such as ruby, as its lasing medium.

Space: The area between objects. In holography, the area between and including objects.

Spatial Filtering: The act of "cleaning up" the light of the laser beam by causing it to focus through a tiny aperture. Only the pure light can focus at the desired point, eliminating the effects of dust, optical surface scratches, etc.

Spatial Frequency: Often used with regard to line spacing in diffraction gratings. The spatial frequency is the reciprocal of line spacing, generally expressed in cycles per millimeter. See also resolution.

Speckle: The grainy appearance of an object, or a holographic image, viewed under laser light. It is caused by light reflecting from minute areas of the object and interfering with itself.

Spectral Reflection: Any reflection from a smooth, polished surface, such as a mirror. Also called specular reflection.

Splitbeam: The act of separating a beam of laser light into two components to separately control the action of reference and object illumination.

Squeegee: A device or action used to remove excess water from the emulsion to facilitate drying.

Stability: The requirement for holographic optical systems to remain motionless during an exposure.

Standing Wave: The result of superimposing two or more waves moving in different directions but having the same wavelength. An interference pattern is a slice through a standing wave pattern; and a hologram is a photograph of an interference pattern.

Stereogram: An image which creates a 3-dimensional illusion by presenting a different view of an object to each eye.

Stimulated Emission: Radiation produced by incoming radiation of the same phase, amplitude, and frequency.

Stop Bath: The chemical bath immediately following the developer which causes the developer to cease action.

Tem00: The lowest mode of a laser, characterized by a beam which is spatially coherent across the diameter of the beam.

Temporal Coherence: Coherence over time. The degree to which waves will remain coherent over time and distance.

Test Strip: A means of visually determining the correct exposure by making a series of individual exposures of varying times on the same plate. The proper time is determined by selecting the strip which yields the brightest or cleanest image.

Thermoplastic Film: A recording material which works due to the effects of electrostatic forces and heat to produce a deformation corresponding to the interference pattern exposed.

360-Degree Hologram: A hologram made by exposing recording material which completely surrounds an object.

Transfer Mirror: A mirror which redirects light from the laser toward the desired working area on the optics table.

Transmission Hologram: Any hologram viewed by passing light through it, toward the viewing side. Transmission holograms are made by allowing both object and reference light to impinge on the same side of the plate.

Transmittance: The proportion of light transmitted by a medium to that which is incident upon it.

Triethanolomine: A chemical used to change the thickness of the emulsion to produce different color playback, usually with reflection holograms.

Ultraviolet: An invisible part of the spectrum characterized by wavelengths somewhat shorter than violet (approx. 100-400 nm.).

Unbacked Plate (NAH): A holographic plate without an antihalation backing. Essential for reflection holograms.

Variable Beamsplitter (VBS): A beamsplitter whereby the ratio of transmitted to reflected beam changes as the beam intercepts the component at different points.

Vibration Isolation: The practice of removing a system from the effects of ambient vibrations which may induce changes, particularly in optical systems. Vibration isolation must be used in making a hologram to prevent the movement of interference fringes during an exposure.

Virtual Image: An image which appears "upstream" from a lens or hologram. A real image becomes visible when a piece of paper is placed in its location, but a virtual image is not accessible.

Volume Hologram: A hologram in which the thickness of the recording material is large compared to the spacing of the interference fringes. neG and photopolymer holograms are volume holograms; embossed holograms are "thin" or "surface" holograms.

Wave Form: The characteristic shape taken on by a wave front.

Wave Front: The surface of a propagating wave, where the phase of the wave is the same everywhere on the surface. A point source produces spherical wavefronts, a collimated beam consists of plane wavefronts, and light reflected from a complicated object has a wavefront with a very complicated shape.

Wavelength: The physical distance over which the complete cycle of one wave occurs. Wavelength is inversely proportional to frequency.

White Light Transmission Hologram: Any transmission hologram which can be displayed using ordinary white light.

YAG Laser: A solid state laser using Yttrium Aluminum Garnet as the lasing material.

Zone Plate: A pattern consisting of a central spot surrounded by concentric zones, alternating opaque and transparent, the total area of each zone being equal. A zone plate is equivalent to the hologram of a point object.

INDEX

Numerics

2D holograms 17
2-D model 17
2D3D 67, 68
360-degree Cross holographic stereogram 20
3D 55, 67
3-D glasses 8
3-D model 17, 64
3D object 58, 67
3M 75

A

A. H. Prismatic 83
absorption 36
achromatic 58
advertising 68, 79
Aerodynamics 50
Aerotech 40
agency 80
agents 79
Agfa-Gevaert 22, 26, 29, 75
agitation 29
alcove hologram 19, 21, 56, 58, 60
aluminized backing 73
Aluminum 66
ambient lighting 67
American Bank Note 64, 82, 84, 84
amplitude 36
amplitude hologram 29
anode 38, 71
applications 8
applicator machines 67
Applied Holographics 75
Argon 39
argon laser 38, 70, 84
Art, Science and Technology Institute 62
artistic 10, 63, 76
artistic displays 60
artistic holograms 19, 58
artist 79
artwork 17, 67, 80
astatine 26
atom 35, 36, 38
Australia 7, 64
automobile 46, 80

B

bank card 32, 76
banks 80
Bartolini 70

Bath PH 71
beam 8
beam splitters 8, 43, 44
Benton 13, 21, 56, 58, 68
Benton Hologram 13
Benton stereogram 19
Binar Detour Phase Hologram 56, 57, 60
binder 31
bindery 68, 73, 74
bleach 28, 29
bombs 35
books 32, 63, 74
Bose 74
Bragg 60
Bragg lane 10, 29
brochure 80
bromine 26
Burns 70, 71

C

California 82
Cambridge 30
camera 21, 39
camera-ready art 17
carrier sheet 73
CAT scans 60
cathode 38, 71
CD player 36
Central Michigan University 62
cereal boxes 63
CGH 55, 56, 60
chain stores 78
chamber 52
charge card 10
Charged-couple device 56
chemical dye lasers 36
chemicals 9, 28, 80
Chemistry 26
chlorine 26
Chroma Concepts 86
circuit boards 32
classification 10
Clay 70
closed circuit television camera 50
clothing 68
coated paper 67
Coating 70
coherence 39
coherence length 10, 39, 40
colleges 61, 62
collimated beams 43, 44
collimation 32

color stereogram 22
Colorization 17
combiners 43
commercial art 76
commercial jobs 13
commercial manufacture 26, 33
commercial market 63, 64
Commonwealth Scientific 82
compact discs 35
companies 75
composite structures 49
computer 32, 50, 56, 58
computer generated holograms 42, 55, 56
Computer monitor 56
computer generated holograms 21
computer generated images 21
Computer generated interferogram 56, 57, 60
concave lens 41
construction 35
Continuous Wave 16
convex 41
convex mirror 41
copyright 76, 78
corporate logo 17
counterfeiting 7, 64, 68, 80
courses 61
credit cards 7, 63, 64, 80
crests 39
Cross 19, 22
Corwn Roll Leaf 75
crystal growth 26
currency 7, 64, 82, 83
curved embossed holographic stereogram 21
custom 22
CW laser 16, 17
cyan filter 57
cylinder 20

D

daguerreotypes 16
Dai Nippon 75
daylight viewable 10, 14
DC current 71
DC rectifier 71
DCG 30, 33
dealer 79
decorations 68
defense 46
deformations 51

Dennison 82
Denysiuk 84
depolarized 50
depth 10, 32
Detour Phase Hologram 57
developer 26, 28, 29, 33, 70
dichromated gelatin 16, 25, 29
diffraction 9
diffraction efficiency 29
diffraction gratings 43, 46
digitized object 55
dip tanks, dipping 70
direct mail 66
discounts 78
distribution 75, 76, 78, 79, 80
doping agents 26
dose 33
double exposure 50, 51
Drivers License 82
DuPont 22, 30, 31, 75, 80

E

Ealing 40, 75
earthquake analysis 50
education 61, 62
Efficiency value 32
Einstein 35, 36
electrodes 38
Electroforming 68
electroforming bath 71
electroless nickel 70, 71
electromagnetic waves 36
electron beam 56
electron density 52
electorplating 70, 71
embossed hologram 7, 10, 14, 16, 17, 19, 20, 22, 33, 63, 64, 66, 68, 70, 80
Embossed stereogram 19
Embossing 58
embossing machines 73
emulsions 8, 10, 25, 39, 60
energy level 35, 36, 38
etalon 70
exposure 9, 16, 20, 22, 28, 29, 33, 39, 70
exposure photochemistry 26

F

fading 29
ferric nitrate 29
film 19, 22, 25, 26, 31, 32, 33, 39, 56

Filtration pump 71
fine art 79
fixer 28
flashlamp 36
Flat bed embossing 72
flat holographic stereogram 21
flat platen 68
flat stereograms 17
fluorescent light 14, 30, 33
fluorine 26
focal length 41, 42, 44
foil 83
food containers 32
Fourier Transform 56, 57
Fraunhofer 56
Fresnel 56
fringe patterns 8,28, 29
full 22

G
Gabor Zone 43
galleries 8, 14,61, 78, 79
gas lasers 36
gelatin 25, 26, 28, 30
ginshop8 78
gin wrap paper 82
glass 25, 31, 32
glass tube 38
grain 28
grain size 26, 28
granular 26
graphic art 17,68
graphics 31
gratings 43, 46, 55
gray scale 57
greeting cards 68
ground state 36

H
H-1 12,13,14,33
H-2 12, 32, 70
H-312
halides 26
heat transfer 50
heat-resistant optics 66
HeCd lasers 70
helium 38
Helium-Neon laser 36,38,39, 40,43,57,84
hi-fi speaker 52
hite light holograms 14
HOE 41, 42, 44, 55, 56, 63
Hoechst Celanese 75
Holodisk 19
Holographic and Media Institute
of Quebec. 62
holographic collimators 44
holographic lens 42, 43
holographic mirrors 44
Holographic Non Destructive Testing 49

Holographic Optical Elements 41
holographic stereogram 17
holographic stereograms 67
Holographic Studios 62
Holographix 84
Holography Institute 62
Holometry 49
Holos 83
horizontal parallax 13
hot foil blocking 72, 73
Hot foil stamping 73
hot plas tic 32
hot stamped hologram 64
hot stamping 67, 72,73,74,83
hot stamping foil 32, 64, 67
hot stamping machine 74
HUD46
Hughes 40

I
ICI75
idealized waveforms 56
Ilford 26, 28, 75
image area 66
image distortion 28
image plane 12, 32
image projection 10
imaginary object 55
immersion tank 33
import 78
industrial 46, 76
Integram 19
interference fringes 50, 55, 60
interference pattern 9, 16,39, 49
Interferometry 49
International Banknote Co 75,83
inventory 78
isolation table 17
isomer 33

J
jewelers 78
jig 71
Jodon 40
Joffa Institute 84
Jurewicz 61

K
Kinoform 56, 57, 60
Kluge 74
Kodachrome 57
Kodak 26
Krypton las.er 38, 39

L
label applicators 74
labeling machines 74
labels 7, 21, 64, 74

lacquer 73
Lake Forest College 61, 62
laminate 31
large format holograms 72
laser 8,16, 26,33,35, 38,50, 52
laser Iigh t 14
laser power 32
laser printers 84
laser transmission holograms 12
Laser-viewable 10
latent image 26, 28, 29
Law 61
lecture 61
Leith 36
Leningrad 84
lenses 8, 43, 55
Less-Than-360-degreeCurved-Cross-Holographic Stereogram 20

light bulb 9, 14, 16,20
Light Impressions 22, 83,84
light sources 14
light waves 9
Lighting 9, 14
line art 17, 68.
live subjects 21
Los Angeles School of Holography
62
Loughborough University 75

M
machine-applied 32
magazines 7, 32, 63, 64, 66
magnetic resonance imaging 60
Maiman 36
mandrel 71
manufacturer 40,76
ma nufacturer's ti ps 26
markets 80
maser 36
mask 20

mass manufacturing 22, 26, 30
Massachussetts Institute of Technology(MIT) 21, 56, 60, 62, 75
mass-produce 19
master 12, 13, 19,26,30,31, 33,42,57,64,66,68,72,73,76
Mastercard 64
McCormack 23
medical 35, 50
Melles 40
merchandise 63
metal master 71
metal mold 66
metal mother die 66

metallizing 66, 82
metastable 38
Metrologic 40
microcomputer 70
microdensitometer 55, 58
microwaves 36, 56
Military 46
Mirage hologram 30, 31, 32, 64
mirrored backing 14, 66
mirrors 8, 9, 38, 43, 44, 55 56,60,75
mixing vessels 28
mock 68
~eI26,31,33,64,66,80
Moire pattern 49
money 7,19
monomers 30, 31
motion picture footage 17, 22
movie camera 20
movie film 20
MRI60
multi-channel 22
mul ti-color rainbow hologram 14
Multiplex 19,60
multiplexed holograms 58
Museum of Holography 61, 62
Mylar 31

N
N233
nanosecond 17
National Geographic 64
near-field diffraction 56
neon 36, 38
New York School of Holography62
Newcastle Upon Tyne Polytechnic 61,62
Newport 75
nickel 66
Nickel anodes 71
nickel sulfamate 71
noise 29
noise-free 58
non-destructive testing 76
Note Printing Australia 82

O
object 8, 9, 39, 40, 41, 50, 51, 52,55,58,68
object beam 8, 9, 10,41,46
OE 80
OmniDex 30, SO
Ontario Science Centre 61, 62
opaque mask 13
optical elements 42, 80, 84
optical lens 42
optical system 42
optics 12
original masters 25

orthoscopic 16
output hardware 56
oven 32, 33
overlays 68

P

packaging 68
paint 17
palladium chloride 70
parallax 12, 17, 56
Paaadena City College 62
Patent Litigation 84
patents 84
peak sensitivity 29
Index
Perfect Shuffle opt system 46
phaae 39, 40, 57
phase hologram 29
photochemical reaction 26
Photochemistry 30
photograph 7, 17, 35, 67
photographic plate 12
photographs 68
photography 7, 30
photon 35, 36, 38
Photon League of Holographers 62
photopolymer 16,22, 25, 30, 31, 33,34,80
photopolymerization 31
photoreduction 57
photoresist 16, 25, 32-34, 66, 68, 70,72
photosensitizer 30
photosensitizing dye 31
photothermoplastic 84
Pilkington 75
plasma diagnostics 50, 52
plasticizing agents 31
plate 8, 26, 32, 39, 40, 41 , 84
plate holder 8
platens 72
plating 71
pIa ting bath 71
plotter 56
point 79,80
polarizer 50
Polaroid 30, 31,32,50, 64, 75
polyester 66, 72
polymer 33
population inversion 36, 38, 39
portraits 17
potassium 26
Potassium bromide 29
potassium dichromate 72
potassium iodide 29
potassium nitrate 26
power supply 39
precipitates 28
pre-coated plates 32
pre-swelled 22
Prince tennis racquets 64

printers 73, 74
printing press 33
printout 29
prism 43
processing 28, 29
processing solutions 28
production run 72
progrsmming 58
projected images 12, 14
pseudoscopic 16
public domain 76
pulsed embossing 67
pulsed emission 39
pulsed holograms 67
pulsed laser 16, 17, 50

Q

Q-switch 39
quantum level 35
quantum theory 35

R

R 60
radiation 36
rainbow hologram 13,14
rainbow transmission hologram 20, 66
Ralston 63
Raman-Nath diffraction 60
Real Time 50
real time NDT 50
Reconstruction 32
recording materials 16,22,26, 32,80
recording plate 8, 9, 12
red filter 57
reference 8
reference angle 22
reference beam 8, 9, 31, 41, 43,46,58,68
reference waves 51
reflection hologram 9, 10,21, 29
refractive index 30, 31
registration 22
reguJarly-scheduled programs 62
research 75
Reserve Bank of Australia 64, 82
Resist 32
resist 33
Resist Maaters 32, 70
retail 76, 78
Retailer 78
ripening 26
ROACH 56, 57
Roll embossing 72
roller 68
Roscoe 67
Royal College of Art 62

ruby 36
ruby laser 36

S

safeligh t 67
Saint Mary's College 62
sales representatives 78
sandwich hologram 30,51,52, 72
Saxby 70
scanners 76
scatter 28
SchoolofHologrsphy/Chicago 62
School ofthe Art Institute of Chicago 62
schools 61, 62
secondary images 13
security 63, 64, 73, 79, 80, 82, 83
sensitivity 28, 33
separations 68
set-up 8, 42
shelflife 33
shim 66, 71, 72
Shipley 32, 33, 70
shops 78
silver 66, 84
silver bromide 26, 28
silver chloride 26
silver halide 13, 16,25,26,28, 33,34,46
silver iodide 26
silver ions 28
Silver Spray 70
Singapore 64, 82
single beam 9
single filament 16
Solid mechanics 50
solid state lasers 36
solutions 28
solvent 33
spatial coherence 39, 40
spatial filter 43
speaker membrane 52
Spectrll Physics 40
spectral sensitivity 33
speed of light 39
SPIE 31
spin coating 33,70
spontaneous emission 35, 36
spotlights 14, 16
spray 33
stability 25
stage 20
stamping 72
stamping die 32, 66
stamps 7
stannous chloride 70
State Education Department of New York 61
State University of New York

61
stereo view 20
stereogram 14, 19,56, 58
stereosco pic 58
stickers 32, 33, 66, 67, 73
stimula ted emission 35, 36, 38,39
Stock images 33
Stop Bath 29
storefront 78
stress 50
stress analysis 50
stress loads 51
strobe light 52
structural weakness 49
students 61
studio 66
subjects 16, 17, 20,21
substrate 67
sulfuric acid solution 71
sunlight-viewable, 13,14
SUNY Buffalo 62
Superbowl 64
supermarket 7
suppliers 35
swelling agent ratios 22
system aberrations 42

T

table 8, 9, 39, 46
Tandem Optics 46
Taung child 64
Technoexan 84
television monitor 50
TEMoo4O
temperatures 29
temporal coherence 39
tensile stress 71
Texas State Technical Institute 62
textiles 68
thermoplastic 50, 72
thermoplastic roll embossing 73
Thick hologram 10
Thin hologrsm 10,17,55,60
Third Dimension 83
Three Dimensional CGH 57, 60
three-dimensional image 8, 58
tickets 64
time smear 20
time-averaged holographic interferometry 52
Towne 32
Towne Labs 70
Townes 36
trade shows 79
Transfer 12
Transfer copies 12, 13
transfer hologrsm 68
transition energy 35, 36

transmission hologram 9, 10, 13,14,41,43,49
troughs 39
tubes 39
turntable 22
two-channel 22
two-color hologram 22

U
U. S. Banknote 83
Uddelhom Steel 83
unit cost 42, 76
unit price 26
universities 61, 75
Upatienks 36
USSR 83

V
vacuoles 26
Vacuum-Deposited Silver 70
vertical parallax 13, 14
vibration 52
vibration isolation table 16
video 50
video display 21
Viewing 9
Vila Holographies 61, 62
visible light 36
volume hologram 58

W
warping 74
wash water 28
water 50
wave crest 36
wave patterns 55, 56
wave trough 36
wavelength 8, 29, 36
web coater 31
wet processing 50
white-light viewable 10, 12, 58
wholesaler 78, 79
Wolverhampton Polytechnic 62
workshops 61

Z
z axis 58
Zebra Books 64

DID YOU BORROW THIS COPY?

If so, now is the time to order your own personal copy of the Holography Market Place. This international directory for the holography industry is the first and only resource of its kind. You will refer to the HMP day after day, so don't you want one of your own?

Standing Order Plan:

If you would like to receive the *Holography Market Place* each year, ask to be put on our Standing Order Plan, where we will ship you one of the first copies each year.

COST:

For USA delivery:
US$38.00 (Includes UPS shipment)

For delivery outside USA:
We will accept your check drawn on any bank in the world for any one of the following amounts:
US$51.00 (Includes Airmail Shipment) or
£26.00 (Includes Airmail Shipment) or
DM76.00 (Includes Airmail Shipment)

TO ORDER:

Credit card customers can order by phone. Call: (800) 367 0930 or (415) 841 2474
OR FAX your order to: (415) 841 2695.

For orders by mail, fill out the information below and enclose check or credit card information.

NAME: _____

TITLE:_____

COMPANY:_____

ADDRESS: _____

CITY/STATE (PROVINCE)/: _____

COUNTRY/ZIP CODE: _____

PHONE/FAX:_____

FOR CREDIT CARD ORDER:

CHECK ONE: ☐ MASTER CARD - or ☐ VISA - or ☐ AMERICAN EXPRESS

CREDIT CARD NUMBER: _____

CREDIT CARD EXPIRATION DATE: _____

SINGATURE: _____

ADDRESS ALL CORRESPONDENCE TO:

HOLOGRAPHY MARKET PLACE
ROSS BOOKS
P.O. BOX 4340
BERKELEY, CA LIFORNIA 94704
USA

www.ingramcontent.com/pod-product-compliance
Lightning Source LLC
Chambersburg PA
CBHW051346200326
41521CB00014B/2490

* 9 7 8 0 8 9 4 9 6 0 9 6 3 *